# Elementary Semiconductor Device Physics

This book by two leading experts on integrated circuit design adopts an untraditional approach to introducing semiconductor devices to beginners. The authors use circuit theory to provide a digestible explanation of energy band theory and understanding of energy band diagrams.

After briefly summarizing the basics of semiconductors, the authors describe semiconductor devices from a circuit theoretic point of view, making the book especially suitable for circuit design students and engineers. Further to the emphasis on the circuit perspective, the book then uses circuit theory to introduce readers to the famously indigestible "energy bands" of crystalline solids. Additionally, the book explains how to read physics from "energy band diagrams" of semiconductor devices in great detail. The key to appreciating the real power of energy band diagrams is shown to lie in the understanding of the concept of the "quasi-Fermi levels," introduced in 1949 by William Shockley but remaining elusive to date and therefore often omitted from energy band diagrams. To rectify this, some of the energy band diagrams presented in this book, complete with quasi-Fermi levels, were drawn using a device simulator (a.k.a. technology computer-aided design; TCAD), offering quantitative information about device physics. The book could, therefore, also serve as a hands-on course text in TCAD-drawn band diagram reading.

Because no prior exposure to quantum mechanics is required and the book does not attempt to teach it, this book is ideal for students in various disciplines who may or may not be specializing in semiconductor devices. The numerous practical examples of reading TCAD-based energy-band diagrams are also invaluable to practicing semiconductor device engineers.

**Kazuya Masu** is President of Tokyo Institute of Technology (Tokyo Tech), a position he has held since 2018. He earned his bachelor's, master's, and doctoral degrees in engineering from Tokyo Tech in 1977, 1979, and 1982 respectively. He is a member of IEEE.

**Shuhei Amakawa** is Professor at Hiroshima University. He received his B.Eng., M.Eng., and Ph.D. degrees in engineering from the University of Tokyo, Tokyo in 1995, 1997, and 2001, respectively, and his M.Phil. degree in physics from the University of Cambridge in 2000. He served/serves as a Committee Member for the International Solid-State Circuits Conference and the International Microwave Symposium and is a member of IEEE.

"This book helps the reader resolve questions and gain fundamental understanding of the subject by carefully explaining it along with the historical background and limitations of the theory. The book also makes a unique attempt to explain band theory not by directly using quantum mechanics but by analogy with electric circuits, appealing to electrical engineers' intuition."

**Akira Matsuzawa,** *Professor Emeritus, Tokyo Institute of Technology*

"This book reframes the electronic properties of semiconductors from the perspective of circuit modeling and provides an intuitive understanding of the fundamentals of MOS devices, making the book easy to understand for beginners."

**Shinichi Takagi,** *Professor, The University of Tokyo*

# Elementary Semiconductor Device Physics

## Understanding Energy Band Formation Using Circuit Theory

Kazuya Masu and Shuhei Amakawa

CRC Press
Taylor & Francis Group
Boca Raton London New York

CRC Press is an imprint of the
Taylor & Francis Group, an **informa** business

Designed cover image: Energy band diagram of a reverse-biased p-n junction, showing lifetime-dependent quasi-Fermi level openings. Courtesy of Kosuke Otsuru.

First edition published 2025
by CRC Press
2385 NW Executive Center Drive, Suite 320, Boca Raton FL 33431

and by CRC Press
4 Park Square, Milton Park, Abingdon, Oxon, OX14 4RN

*CRC Press is an imprint of Taylor Francis Group, LLC*



Funding: Tokyo Institute of Technology



ISBN: 9781032574479 (hbk)
ISBN: 9781032574486 (pbk)
ISBN: 9781003439417 (ebk)

DOI: 10.1201/ 9781003439417

*Publisher's note:* This book has been prepared from camera-ready copy provided by the authors.

Typeset in Sabon
by Deanta Global Publishing Services, Chennai, India

*To the pioneers of semiconductor devices,*
*who also kindly left us with some gaps to fill.*

# Contents

# Preface

This book offers an easy-to-understand introduction to the electrical properties of semiconductors and electronic components fabricated from semiconductors, commonly known as *semiconductor devices* or simply *devices*. Our focus in this book is primarily on the latter. Given the abundance of excellent books on the subject, we will try to shed fresh light on some aspects of semiconductor devices that are not covered by ordinary books or that we struggled to understand. We assume that readers have a basic knowledge of mathematics, physics, and electric circuits at the level of a first- or second-year college student. Readers may or may not intend to specialize in semiconductor electronics in the future. We, therefore, have incorporated some cross-disciplinary viewpoints.

Acknowledging the ever-evolving nature of semiconductor technology, we have tried to focus on fundamentals that are not likely to become obsolete anytime soon, rather than trying to include the "latest topics" that are guaranteed to become outdated very quickly. We wrote this book hoping that readers will be able to achieve the following:

- To understand the functions of semiconductor devices as components of circuits.
- To recognize that wave phenomena in periodic structures underlie the electrical properties of crystals.
- To acquire the skill to interpret energy band diagrams of semiconductor devices.
- To understand the physics of the basic operation of p-n junction diodes and MOS transistors (MOSFETs).

To view semiconductor devices as components that provide certain functions is to view them from a circuit designer's perspective. Integrated circuit design is another subdiscipline of semiconductor electronics that we would like readers to be aware of. Designers are the users of semiconductor devices. It is always good to know about your users or customers.

Given the fact that we do not assume that readers have a knowledge of quantum mechanics, we will be looking at "wave phenomena in periodic structures" using periodic circuits. Although periodic circuits and solid crystals are governed by different laws of physics (circuit theory vs. quantum mechanics), there are striking similarities between them for a good reason. We take advantage of this fact.

So-called *energy band diagrams* are presented in almost all books on semiconductor devices. However, there are often no in-depth explanations as to why such diagrams are drawn and why they are so important. Actually, not all energy band diagrams are as informative and useful as they can be. We came to realize that *quasi-Fermi levels* (also known as *imrefs*) must be drawn in energy band diagrams for the latter to be able to display useful information about device physics. Quasi-Fermi levels are somewhat underappreciated and are often not covered by introductory books. Even in advanced books, it is difficult to find an explanation that really makes good sense. In fact, the inventor of quasi-Fermi levels, William Shockley, himself wrote that imref concepts were "hard to comprehend" and "sometimes hard to teach" [28]. However, a good grasp of imrefs is crucially important for an intuitive understanding of device operation and a mastery of reading energy band diagrams. Naturally, we will be discussing imrefs in great, possibly unprecedented, detail.

The *p-n junction* can be a device (i.e., diode) by itself, but it is also a fundamentally important structure contained in nearly all semiconductor devices. The *MOS transistor* is the most commonly used type of transistor today. It also contains another very important structure, the metal-oxide-semiconductor (MOS) structure. We will use these devices and structures as the platform for applying the basic principles that readers will learn.

Following "Introduction: Semiconductor Basics" as Chapter 1, Chapter 2, titled "Semiconductor Devices from a Circuit Theoretic Standpoint," describes the circuit-operational functions and other aspects of semiconductor devices from the standpoint of a circuit engineer. Note that it is possible to separate the function of a device in a circuit from how the function is actually realized using certain materials, structures, and fabrication techniques. As readers may be aware, device functions that were once realized with vacuum tubes are now achieved with semiconductor devices. While almost all electron devices are now implemented with semiconductors, there is still plenty of room for the development of new materials and device structures. When looking into possible new devices and/or materials, it might be a good practice to ask if there are benefits that circuit engineers will appreciate.

Chapter 3, titled "Waves in Periodic Structures," attempts to explain the formation of energy bands without assuming knowledge of quantum mechanics. The essentials of band theory are discussed using circuit theory, making it easily understandable to readers without a background in solid-state physics. In his famous book [27], Shockley briefly discussed the connection between solid crystals and periodic circuits. Our approach was motivated by it. The essence of the emergence of energy band structures lies in the periodic structure, and therefore band structures are not unique to solid crystals. If you are already familiar with energy bands, you can skip this chapter.

Chapter 4, titled "Physics of Semiconductors in Equilibrium," is where this book becomes somewhat ordinary compared with previous chapters. It looks at the carrier statistics of semiconductors. The chemical potential is explained and the physical meaning of the *Fermi level* is described. The connection of these concepts with the energy band diagram is emphasized throughout. It is important to understand that the Fermi level and the equilibrium carrier density are two sides of the same coin. The Fermi level can be drawn in energy band diagrams, so such diagrams provide you with information about the other side of the coin.

In Chapter 5, titled "Carrier Dynamics in Semiconductors," we move on to nonequilibrium conditions where steady current flows. The presentation goes along the "ordinary" lines of Chapter 4, but the extensive discussion of imrefs, including their crucial roles in energy band diagrams, might make this chapter somewhat out of the ordinary. As a steady current flows, the Fermi level, which is an equilibrium quantity, cannot be defined, necessitating the introduction of imrefs for electrons and holes separately. Just like the Fermi level, quasi-Fermi levels can be drawn in energy band diagrams. Such diagrams offer information about electron and hole densities and much more. Quasi-Fermi levels and energy band diagrams are key to the intuitive understanding of the operation of semiconductor devices.

Chapter 6, titled "p-n Junctions," focuses on the rectification characteristics of p-n junction diodes and their physics. It involves interpreting energy band diagrams, in which imrefs are drawn. Some of the energy band diagrams shown in this chapter are *quantitative* because they were drawn using semiconductor device simulation (also known as *technology computer-aided design* or *TCAD*). Surprisingly, this seems to be a new approach to semiconductor device teaching. TCAD is widely used today for analyzing various semiconductor devices. However, most TCAD users focus only on the current-voltage characteristics of their devices and do not use TCAD to draw energy band diagrams, although they might plot electrostatic potential and carrier densities on separate graphs. This is a pity because all these and more can be read from a single energy band diagram. Calculations without comprehending the true nature of device physics might not contribute to new knowledge. To design new device structures and improve performance, it is essential to recognize the physics behind current-voltage characteristics. In this chapter, we will also see that a standard assumption made in developing the theory of p-n junctions might not be physically correct.

In Chapter 7, titled "MOS Transistors," the basics of MOS capacitors and MOSFETs are described, and the characteristics of MOSFETs as four-terminal devices are derived. MOSFETs are the workhorses of today's digital world and are continuously evolving in structure and performance. But we will focus on planar, long-channel MOSFETs for reasons mentioned earlier. The presentation of this chapter is modeled on Tsividis' highly acclaimed book [33] and, importantly, is also consistent with our decision to make extensive use of quasi-Fermi levels. This approach starts from the analysis of the MOS capacitor and leads naturally to the proper treatment of the semiconductor *substrate*, or equivalently, the *back gate*—the fourth terminal, if any, of the MOSFET. Recent advanced MOSFETs might not have a back gate terminal that is accessible by the circuit designer, but that does not mean a region corresponding to the substrate does not exist within the device. The derivation of equations is quite long, and novice readers might find it quite difficult. However, we believe going through it is necessary for a solid understanding of the MOSFET operation. Based on our findings in Chapter 6, we also make minor reservations about the accepted theory for describing MOSFETs in the hope that some readers will look at it in the future.

The book is interspersed with a number of boxed columns. Some of these address questions that may naturally arise during the study of semiconductors but are hardly discussed. While this book may not serve as a replacement for conventional books due to its unconventional approaches, we hope that it will fill some gaps still left unattended. For those aspiring to specialize in semiconductor devices, we recommend that, after finishing reading this book, they graduate to more comprehensive, standard books on the subject. For readers already acquainted with semiconductor devices, this book's untraditional approach might offer moments of enjoyment. Specifically, we hope that those engineers utilizing TCAD for device analysis and development will discover the joy of reading energy band diagrams with imrefs.

When reading a book like this, readers are required to jump between pages to refer to equations and figures scattered across the book. To make it easier to do so, page numbers are given together with equation or figure numbers in cross references. The Symbol Index serves a similar purpose, where readers can look up mathematical symbols and find their definitions.

This book was derived from our book titled *Denshi bussei to debaisu (Elementary Solid-State Device Physics)*, written in Japanese and published by Corona Publishing Co. Ltd. in 2020 [20]. The feedback we received from its readers was essential in producing this book. Akira Matsuzawa and Shinichi Takagi were among them and also kindly provided helpful feedback on a draft of this book. We thank Kosuke Otsuru for drawing the book cover illustration, which shows an energy band diagram of a p-n junction. This work was supported in part by the MEXT Initiative for Establishing Next-Generation Novel Integrated Circuits Centers (X-NICS), Grant Number JPJ011438. We are indebted to Atsushi Hori, Kazuaki Sawada, Akinobu Teramoto, Hitoshi Wakabayashi, and Junichiro Yoshikawa for their support.

# Symbol Index

| Symbol | Description | Unit | Page |
|---|---|---|---|
| $A$, $a$ | This typeface is for scalars | – | |
| **A**, **a** | This typeface is for vectors | – | |
| **A**, **a** | This typeface is for matrices | – | |
| $\angle a$ | Argument of complex number $a$ | – | |
| $\dot{A}$ | Time derivative of $A$, namely $\mathrm{d}A/\mathrm{d}t$ or $\partial A/\partial t$ | – | |
| $\langle A \rangle$ | Statistical mean of $A$ | – | |
| $\Delta A$ | Increment or decrement of $A$ | – | |
| $\alpha$ | Current gain of CCCS | – | 41 |
| $\beta$ | Phase constant of transmission line | rad/m | 63 |
| $\beta_{\mathrm{A}}$ | Phase constant of transmission line A | rad/m | 68 |
| $\beta_{\mathrm{B}}$ | Phase constant of transmission line B | rad/m | 68 |
| $\gamma$ | Body-effect coefficient | $\mathrm{V}^{1/2}$ | 235 |
| $\Gamma(x)$ | Gamma function | – | 113 |
| $\epsilon$ | Permittivity | F/m | 153 |
| $\epsilon_0$ | Permittivity of free space ($\simeq 8.85 \times 10^{-12}$ F/m) | F/m | 165 |
| $\epsilon_{\mathrm{s}}$ | Permittivity of semiconductor | F/m | 155 |
| $\epsilon_{\mathrm{Si}}$ | Permittivity of silicon | F/m | 156 |
| $\zeta$ | Fermi level; (electro)chemical potential | J, eV | 89 |
| $\zeta_{1,\mathrm{final}}$ | Final value of Fermi level of system 1 | J, eV | 111 |
| $\zeta_{2,\mathrm{final}}$ | Final value of Fermi level of system 2 | J, eV | 111 |
| $\zeta_{1,\mathrm{initial}}$ | Initial value of Fermi level of system 1 | J, eV | 111 |
| $\zeta_{2,\mathrm{initial}}$ | Initial value of Fermi level of system 2 | J, eV | 111 |
| $\zeta_{\mathrm{A}}$ | Fermi level of A | J, eV | 169 |

| | | | |
|---|---|---|---|
| $\zeta_B$ | Fermi level of B | J, eV | 169 |
| $\zeta_{b1}$ | Chemical potential of black particles in system 1 | J | 108 |
| $\zeta_{b2}$ | Chemical potential of black particles in system 2 | J | 108 |
| $\zeta_{ext}$ | External chemical potential | J | 109 |
| $\zeta_G$ | Fermi level of gate metal | J, eV | 226 |
| $\zeta_{int}$ | Internal chemical potential | J | 109 |
| $\zeta'_{int}$ | Internal quasi-chemical potential | J | 118 |
| $\zeta_n$ | Quasi-Fermi level for electrons | J, eV | 125 |
| $\zeta_N$ | Fermi level in n-type semiconductor | J, eV | 174 |
| $\zeta_{n,ext}$ | External component of $\zeta_n$ | J, eV | 164 |
| $\zeta_{n,int}$ | Internal component of $\zeta_n$ | J, eV | 164 |
| $\zeta_{nN}$ | Electron quasi-Fermi level in n-type semiconductor | J, eV | 179 |
| $\zeta_{nP}$ | Electron quasi-Fermi level in p-type semiconductor | J, eV | 179 |
| $\zeta_p$ | Quasi-Fermi level for holes | J, eV | 126 |
| $\zeta_P$ | Fermi level in p-type semiconductor | J, eV | 174 |
| $\zeta_{pN}$ | Hole quasi-Fermi level in n-type semiconductor | J, eV | 179 |
| $\zeta_{pP}$ | Hole quasi-Fermi level in p-type semiconductor | J, eV | 179 |
| $\zeta_{tot}$ | Total chemical potential | J | 111 |
| $\zeta'_{tot}$ | Total quasi-chemical potential | J | 118 |
| $\zeta_{w1}$ | Chemical potential of white particles in system 1 | J | 108 |
| $\zeta_{w2}$ | Chemical potential of white particles in system 2 | J | 108 |
| $\kappa$ | Phase rotation | rad | 77 |
| $\kappa$ | Scaling factor ($\kappa > 1$) | – | 258 |
| $\lambda$ | Wavelength | m | 63 |
| $\mu$ | Voltage gain of VCVS | – | 41 |
| $\mu_1$ | Mobility of layer 1 | $m^2/(V\cdot s)$, $cm^2/(V\cdot s)$ | 165 |
| $\mu_{12}$ | Apparent mobility of two layers | $m^2/(V\cdot s)$, $cm^2/(V\cdot s)$ | 165 |
| $\mu_2$ | Mobility of layer 2 | $m^2/(V\cdot s)$, $cm^2/(V\cdot s)$ | 165 |

| | | | |
|---|---|---|---|
| $\mu_n$ | Electron mobility | $m^2/(V{\cdot}s)$, $cm^2/(V{\cdot}s)$ | 135 |
| $\mu_p$ | Hole mobility | $m^2/(V{\cdot}s)$, $cm^2/(V{\cdot}s)$ | 136 |
| $\nu$ | Frequency | Hz | 60 |
| $\xi$ | Sum of diagonal elements of 2× 2 matrix divided by 2 | – | 272 |
| $\pi$ | Circular constant ($\simeq 3.14159\cdots$) | – | 59 |
| $\rho$ | Resistivity | $\Omega{\cdot}m$, $\Omega{\cdot}cm$ | 139 |
| $\rho_n$ | Resistivity associated with electron conduction | $\Omega{\cdot}m$, $\Omega{\cdot}cm$ | 138 |
| $\rho_p$ | Resistivity associated with hole conduction | $\Omega{\cdot}m$, $\Omega{\cdot}cm$ | 138 |
| $\rho_t$ | Trap charge density | $C{\cdot}m^{-3}$, $C{\cdot}cm^{-3}$ | 153 |
| $\sigma$ | Conductivity | S/m, S/cm | 139 |
| $\sigma_1$ | Conductivity of layer 1 | S/m, S/cm | 165 |
| $\sigma_{12}$ | Apparent conductivity of two layers | S/m, S/cm | 165 |
| $\sigma_2$ | Conductivity of layer 2 | S/m, S/cm | 165 |
| $\sigma_n$ | Conductivity associated with electron conduction | S/m, S/cm | 138 |
| $\sigma_p$ | Conductivity associated with hole conduction | S/m, S/cm | 139 |
| $\tau$ | Time | s | 30 |
| $\tau$ | Delay | s | 258 |
| $\tau_{drn}$ | Dielectric relaxation time in n-type | s | 155 |
| $\tau_{drp}$ | Dielectric relaxation time in p-type | s | 121 |
| $\tau_e$ | Mean free time for electrons | s | 135 |
| $\tau_g$ | Mean free time for gas molecules | s | 142 |
| $\tau_h$ | Mean free time for holes | s | 136 |
| $\tau_n$ | Lifetime of electrons (as minority carriers) | s | 150 |
| $\tau_p$ | Lifetime of holes (as minority carriers) | s | 151 |
| $\Phi$ | Flux linkage | Wb | 32 |
| $\phi(x)$ | Wave function | – | 70 |
| $\Phi_0$ | Initial value of flux linkage | Wb | 32 |

| | | | |
|---|---|---|---|
| $\varphi_{AB}$ | Contact potential between A and B measured from B | V | 169 |
| $\varphi_{B}$ | Bulk potential | V | 171 |
| $\varphi_{BA}$ | Contact potential between A and B measured from A | V | 169 |
| $\varphi_{bi}$ | Built-in potential | V | 177 |
| $\varphi_{B,N}$ | Bulk potential of n-type semiconductor | V | 172 |
| $\varphi_{B,P}$ | Bulk potential of n-type semiconductor | V | 172 |
| $\varphi_{i,j}$ | Contact potential between materials *i* and *j* | V | 173 |
| $\Phi_{n}$ | Electron flux density | $m^{-2}s^{-1}$ | 130 |
| $\varphi_{o}$ | Potential difference across gate oxide due to $Q_o$ | V | 224 |
| $\varphi_{ox}$ | Electrostatic potential difference across gate oxide | V | 234 |
| $\Phi_{p}$ | Hole flux density | $m^{-2}s^{-1}$ | 130 |
| $\varphi_{s}^{(1)}$ | Band bending in MOS capacitor (depletion) | V | 294 |
| $\varphi_{s}^{(2)}$ | Band bending in MOS capacitor (weak inversion) | V | 295 |
| $\varphi_{s}^{(3)}$ | Band bending in MOS capacitor (strong inversion onset) | V | 295 |
| $\varphi_{s}^{(4)}$ | Band bending in MOS capacitor (strong inversion) | V | 296 |
| $\varphi_{SM}$ | Contact potential between Si substrate and gate metal | V | 223 |
| $\varphi_{th}$ | Thermal voltage | V | 236 |
| $\varphi_{W,A}$ | Work function of A (in volts) | V, eV | 170 |
| $\varphi_{W,B}$ | Work function of B (in volts) | V, eV | 170 |
| $\chi$ | Electron affinity (in volts) | V, eV | 171 |
| $\psi$ | Electrostatic potential | V | 97 |
| $\psi_{A}$ | Electrostatic potential of A | V | 169 |
| $\psi_{B}$ | Electrostatic potential of B | V | 169 |
| $\psi_{F}$ | Fermi potential | V | 97 |
| $\psi_{n}$ | Quasi-Fermi potential for electrons | V | 123 |

| | | | |
|---|---|---|---|
| $\psi_p$ | Quasi-Fermi potential for holes | V | 123 |
| $\psi_s$ | Surface potential | V | 225 |
| $\psi_{sT}$ | Approximate surface potential under strong inversion | V | 236 |
| $\omega$ | Angular frequency | rad/s | 31 |
| $\omega(\beta)$ | Dispersion relation of transmission line | rad/s | 75 |
| $\omega_c$ | Cut-off angular frequency | rad/s | 58 |
| $\omega_c$ | Carrier angular frequency | rad/s | 75 |
| $\omega_s$ | Signal angular frequency | rad/s | 75 |
| $a$ | Lattice constant | m, Å | 80 |
| $a$ | Acceleration | N/kg | 138 |
| $A$ | (1, 1) element of ABCD matrix | – | 269 |
| Å | angstrom. 1 Å equals $1\times10^{-10}$ m | – | 26 |
| $a_i$ | Traveling-wave phasor incident on port $i$ | $V\cdot\Omega^{-1/2}$ | 271 |
| $B$ | (1, 2) element of ABCD matrix | Ω | 269 |
| $B$ | Magnitude of **B** | $Wb/m^2$ | 159 |
| **B** | Magnetic flux density vector | $Wb/m^2$ | 159 |
| $b_i$ | Outgoing traveling-wave phasor at port $i$ | $V\cdot\Omega^{-1/2}$ | 271 |
| $C$ | Capacitance | F | 30 |
| $C$ | Per-unit-length capacitance of transmission line | F/m | 51 |
| $C$ | (2, 1) element of ABCD matrix | S | 270 |
| $C_{ch}$ | Chord capacitance | F | 37 |
| $C_d$ | Per-unit-area depletion capacitance | $F/m^2$ | 188 |
| $C_{gate}$ | Gate capacitance | F | 258 |
| $C_{inc}$ | Incremental (or small-signal) capacitance | F | 37 |
| $C_{ox}$ | Per-unit-area gate oxide capacitance | $F/m^2$ | 220 |
| d | Differential symbol, used in the form d$x$ | – | 30 |
| $d$ | Distance between conductors | m | 61 |
| $D$ | (2, 2) element of ABCD matrix | S | 270 |
| $\partial$ | Partial differential symbol | – | 44 |

| | | | |
|---|---|---|---|
| $D_A^+$ | Coefficient in wave function (region A) | – | 71 |
| $D_A^-$ | Coefficient in wave function (region A) | – | 71 |
| $d_b$ | Depletion layer thickness in MOS structure | m | 229 |
| $D_B^+$ | Coefficient in wave function (region B) | – | 71 |
| $D_B^-$ | Coefficient in wave function (region B) | – | 71 |
| $d_{dep}$ | Depletion layer thickness | m | 186 |
| $D_g$ | Diffusion coefficient for gas | $m^2/s$ | 143 |
| $D_n$ | Diffusion coefficient for electrons | $m^2/s$ | 140 |
| $D_p$ | Diffusion coefficient for holes | $m^2/s$ | 141 |
| e | $= \exp(1) \simeq 2.718\cdots$ | – | 63 |
| $E$ | Electron energy | J, eV | 11 |
| $\mathbf{E}$ | Electric field vector | V/m | 159 |
| $\mathcal{E}$ | Electrostatic field; electric field | V/m | 14 |
| $E(k)$ | $E$-$k$ dispersion relation of solid | J, eV | 79 |
| $E_A$ | Acceptor energy level | J, eV | 22 |
| $E_c$ | Conduction band bottom energy | J, eV | 12 |
| $E_D$ | Donor energy level | J, eV | 20 |
| $E_F$ | Fermi energy | J, eV | 91 |
| $E_g$ | Energy gap | J, eV | 12 |
| $E_i$ | Intrinsic Fermi level | J, eV | 96 |
| $\mathcal{E}_m$ | Magnitude of maximum electric field | V/m | 184 |
| $\mathcal{E}_s$ | Normal electric field at the substrate surface | V/m | 238 |
| $E_v$ | Valence band top energy | J, eV | 12 |
| $\mathcal{E}_x$ | $x$ component of electric field | V/m | 249 |
| $\mathcal{E}_y$ | $y$ component of electric field | V/m | 249 |
| $f$ | Frequency | Hz | 60 |
| $F$ | Force | N | 118 |
| $\mathbf{F}$ | ABCD matrix (or F matrix) | See p. 270 | 269 |
| $f(E)$ | Distribution function for electrons | – | 89 |

| | | | |
|---|---|---|---|
| $f_c$ | Cut-off frequency | Hz | 69 |
| $f_c(E)$ | Distribution function for conduction band states | – | 125 |
| $f_{clk}$ | Clock frequency | Hz | 258 |
| $f_h(E)$ | Distribution function for holes | – | 93 |
| $\mathbf{F}_L$ | Lorentz force vector | N | 159 |
| $f_v(E)$ | Distribution function for conduction band states | – | 125 |
| $g$ | Acceleration of gravity | $m/s^2$ | 111 |
| $G$ | Conductance | S | 29 |
| $G_{ch}$ | Chord conductance | S | 34 |
| $g_{ds}$ | Drain conductance | S | 44 |
| $G_{inc}$ | Incremental (or small-signal) conductance | S | 34 |
| $g_m$ | Transconductance | S | 42 |
| $g_n$ | Electron generation rate | $s^{-1}m^{-3}$, $s^{-1}cm^{-3}$ | 148 |
| $g_{n0}$ | Electron generation rate in equilibrium | $s^{-1}m^{-3}$, $s^{-1}cm^{-3}$ | 150 |
| $g_p$ | Electron generation rate | $s^{-1}m^{-3}$, $s^{-1}cm^{-3}$ | 148 |
| $g_{p0}$ | Hole generation rate in equilibrium | $s^{-1}m^{-3}$, $s^{-1}cm^{-3}$ | 151 |
| $h$ | Planck constant ($\equiv 6.62607015\times 10^{-34}$ J·s) | J·s | 59 |
| $h$ | Height | m | 111 |
| $\hbar$ | $= h/2\pi$. Reduced Planck constant | J·s/rad | 59 |
| i | Imaginary unit, $i^2 = -1$ | – | 73 |
| $i$ | Current | A | 62 |
| $i$ | Positive integer; Port number | – | 271 |
| $I$ | Current | A | 29 |
| $\Im(z)$ | Imaginary part, $y$, of complex number $z = x + jy$ | – | 55 |
| $I^+$ | Forward current-traveling-wave phasor | A | 64 |
| $I^-$ | Backward current-traveling-wave phasor | A | 64 |
| $I_1$ | Current flowing into port 1 | A | 42 |
| $I_2$ | Current flowing into port 2 | A | 42 |
| $I_B$ | Current flowing into back gate | A | 249 |

| | | | |
|---|---|---|---|
| $I_{\mathrm{D}}$ | Current flowing into drain | A | 249 |
| $I_{\mathrm{DS}}$ | Drain-to-source current | A | 220 |
| $I_{\mathrm{DS0}}$ | Drain-to-source current at threshold | A | 260 |
| $I_{\mathrm{DSsat}}$ | Saturated drain-to-source current | A | 220 |
| $I_{\mathrm{G}}$ | Current flowing into gate | A | 249 |
| $I_{\mathrm{S}}$ | Current flowing into source | A | 249 |
| j | Imaginary unit, $\mathrm{j}^2 = -1$ | – | 72 |
| $j$ | Positive integer | – | xxix |
| $J$ | Magnitude of **J** | A/m$^2$ | 161 |
| **J** | Current density vector | A/m$^2$ | 159 |
| $J_{\mathrm{drift}}$ | Drift current density | A/m$^2$ | 139 |
| $J_{\mathrm{n}}$ | Electron current density | A/m$^2$ | 130 |
| $J_{\mathrm{n,drift}}$ | Electron drift current density | A/m$^2$ | 138 |
| $J_{\mathrm{p}}$ | Hole current density | A/m$^2$ | 130 |
| $J_{\mathrm{p,drift}}$ | Hole drift current density | A/m$^2$ | 138 |
| $J_{\mathrm{s}}$ | Reverse saturation current density | A/m$^2$ | 191 |
| $k$ | Boltzmann constant ($k \simeq 1.38 \times 10^{-23}$ J/K) | J/K | 26 |
| $k$ | Spring constant | N/m | 51 |
| $k$ | Wave number | rad/m | 63 |
| **k** | Wave number vector | rad/m | 63 |
| $k_{\mathrm{A}}$ | Wave number in region A | rad/m | 71 |
| $k_{\mathrm{B}}$ | Wave number in region B | rad/m | 71 |
| $k_x$ | $x$-component of **k** | rad/m | 63 |
| $k_y$ | $y$-component of **k** | rad/m | 63 |
| $k_z$ | $z$-component of **k** | rad/m | 63 |
| $\ell$ | Length of transmission line | m | 68 |
| $\ell_{\mathrm{A}}$ | Length of transmission line A | m | 68 |
| $\ell_{\mathrm{B}}$ | Length of transmission line B | m | 68 |
| $L$ | Inductance | H | 32 |
| $L$ | Per-unit-length inductance of transmission line | H/m | 61 |
| $L$ | Channel (or gate) length of MOSFET | m, nm, µm | 20 |
| $L'$ | Effective channel (or gate) length of MOSFET | m, nm, µm | 253 |

| | | | |
|---|---|---|---|
| $L_{ch}$ | Chord inductance | H | 39 |
| $L_D$ | Debye length | m | 158 |
| $l_e$ | Electron mean free path | m | 135 |
| $L_{inc}$ | Incremental (or small-signal) inductance | H | 39 |
| $L_n$ | Electron diffusion length (as minority carriers) | m, μm | 194 |
| $L_p$ | Hole diffusion length (as minority carriers) | m, μm | 194 |
| $m$ | Mass | kg | 111 |
| $m_0$ | Rest mass of electron ($\simeq 9.1 \times 10^{-31}$ kg) | kg | 27 |
| $m_c$ | Density-of-states effective mass of electron | kg | 88 |
| $m_e$ | Effective mass of electron | kg | 80 |
| $m_h$ | Effective mass of hole | kg | 136 |
| $m_v$ | Density-of-states effective mass of hole | kg | 88 |
| $n$ | Conduction electron density | $m^{-3}$, $cm^{-3}$ | 8 |
| $n$ | Positive integer | – | 76 |
| $n$ | Gate-control efficiency factor | – | 267 |
| $N$ | Number of repetitions | – | 66 |
| $N$ | Number of electrons | – | 134 |
| $\Delta n$ | Difference between electron and hole densities | $m^{-3}$, $cm^{-3}$ | 99 |
| $\Delta n$ | Excess electron density. $\Delta n \gtrless 0$ | $m^{-3}$, $cm^{-3}$ | 149 |
| $n_1$ | Carrier density of layer 1 | $m^{-3}$, $cm^{-3}$ | 165 |
| $n_{12}$ | Apparent carrier density of two layers | $m^{-3}$, $cm^{-3}$ | 165 |
| $n_2$ | Carrier density of layer 2 | $m^{-3}$, $cm^{-3}$ | 165 |
| $N(E)$ | Density-of-states function | $J^{-1}m^{-3}$, $eV^{-1}cm^{-3}$ | 89 |
| $N_A$ | Acceptor density | $m^{-3}$, $cm^{-3}$ | 99 |
| $N_A^-$ | Acceptor ion density | $m^{-3}$, $cm^{-3}$ | 99 |
| $N_A^{-\prime}$ | Effective acceptor ion density | $m^{-3}$, $cm^{-3}$ | 99 |
| $N_c$ | Effective density of states of conduction band | $m^{-3}$, $cm^{-3}$ | 93 |
| $N_D$ | Donor density | $m^{-3}$, $cm^{-3}$ | 99 |
| $N_D^+$ | Donor ion density | $m^{-3}$, $cm^{-3}$ | 99 |

| | | | |
|---|---|---|---|
| $N_D^{-\prime}$ | Effective donor ion density | $m^{-3}$, $cm^{-3}$ | 99 |
| $n_g$ | Density of gas | $m^{-3}$ | 142 |
| $N_g$ | Reference density of gas | $m^{-3}$ | 142 |
| $n_i$ | Intrinsic carrier density | $m^{-3}$, $cm^{-3}$ | 8 |
| $n_i'$ | Effective intrinsic carrier density | $m^{-3}$, $cm^{-3}$ | 124 |
| $n_N$ | Electron density of n-type semiconductor | $m^{-3}$, $cm^{-3}$ | 150 |
| $n_{N0}$ | Equilibrium electron density of uniform n-type | $m^{-3}$, $cm^{-3}$ | 151 |
| $n_P$ | Electron density of p-type semiconductor | $m^{-3}$, $cm^{-3}$ | 149 |
| $n_{P0}$ | Equilibrium electron density of uniform p-type | $m^{-3}$, $cm^{-3}$ | 149 |
| $n_s$ | Surface electron density | $m^{-3}$, $cm^{-3}$ | 228 |
| $n_v$ | Electron density of valence band | $m^{-3}$, $cm^{-3}$ | 288 |
| $N_v$ | Effective density of states of valence band | $m^{-3}$, $cm^{-3}$ | 94 |
| $p$ | Momentum | kg·m/s | 51 |
| $p$ | Hole density | $m^{-3}$, $cm^{-3}$ | 7 |
| $\Delta p$ | Excess hole density. $\Delta p \gtrless 0$ | $m^{-3}$, $cm^{-3}$ | 149 |
| $p_N$ | Hole density of n-type semiconductor | $m^{-3}$, $cm^{-3}$ | 150 |
| $p_{N0}$ | Equilibrium hole density of uniform n-type | $m^{-3}$, $cm^{-3}$ | 150 |
| $p_P$ | Hole density of p-type semiconductor | $m^{-3}$, $cm^{-3}$ | 149 |
| $p_{P0}$ | Equilibrium hole density of uniform p-type | $m^{-3}$, $cm^{-3}$ | 149 |
| $p_s$ | Surface hole density | $m^{-3}$, $cm^{-3}$ | 228 |
| $q$ | Elementary charge ($\simeq$ 1.6 $\times$ $10^{-12}$ C). Electron charge is $-q$ | C | 12 |
| $Q$ | Charge stored in capacitor | C | 29 |
| $Q_0$ | Initial charge stored in capacitor | C | 30 |
| $Q_b$ | Per-unit-area depletion charge of MOS capacitor | $C/m^2$ | 235 |
| $Q_d$ | Per-unit-area fixed charge (unsigned) in depletion layer | $C/m^2$ | 184 |
| $Q_G$ | Per-unit-area gate charge | $C/m^2$ | 224 |

| | | | |
|---|---|---|---|
| $\Delta Q_G$ | Change in $Q_G$ | $C/m^2$ | 225 |
| $Q_{gi}$ | Per-unit-area induced charge in the substrate of a MOS capacitor | $C/m^2$ | 224 |
| $\Delta Q_{gi}$ | Change in $Q_{gi}$ | $C/m^2$ | 225 |
| $Q_{inv}$ | Per-unit-area inversion charge of MOS capacitor | $C/m^2$ | 235 |
| $Q_o$ | Per-unit-area interface charge between oxide and substrate | $C/m^2$ | 224 |
| $r$ | Radius | m | 276 |
| $R$ | Resistance | Ω | 29 |
| $\Re(z)$ | Real part, $x$, of complex number $z = x + jy$ | – | 55 |
| $r_B$ | Bohr radius ($\simeq 5.3 \times 10^{-11}$ m) | m, Å | 276 |
| $R_{ch}$ | Chord resistance | Ω | 33 |
| $r_H$ | Hall factor | – | 161 |
| $R_H$ | Hall coefficient | $m^3/C$ | 161 |
| $R_{H12}$ | Apparent Hall coefficient of two layers | $m^3/C$ | 165 |
| $R_{H2}$ | Hall coefficient of layer 2 | $m^3/C$ | 166 |
| $R_{inc}$ | Incremental resistance | Ω | 33 |
| $r_m$ | Transresistance | Ω | 42 |
| $r_n$ | Electron recombination rate | $s^{-1}m^{-3}$, $s^{-1}cm^{-3}$ | 148 |
| $r_{n0}$ | Electron recombination rate in equilibrium | $s^{-1}m^{-3}$, $s^{-1}cm^{-3}$ | 150 |
| $R_{on}$ | On resistance of MOSFET | Ω | 258 |
| $r_p$ | Hole recombination rate | $s^{-1}m^{-3}$, $s^{-1}cm^{-3}$ | 148 |
| $r_{p0}$ | Hole recombination rate in equilibrium | $s^{-1}m^{-3}$, $s^{-1}cm^{-3}$ | 151 |
| $R_{ref}$ | Reference resistance | Ω | 272 |
| $\mathbf{S}$ | S-matrix | – | 272 |
| $S$ | Subthreshold swing | mV/dec | 267 |
| $S_{ji}$ | $(j, i)$-element of S-matrix | – | 271 |
| $t$ | Time | s | 29 |
| $T$ | Absolute temperature | K | 26 |
| $t_0$ | Initial time | s | 30 |
| $t_1$ | Thickness of layer 1 | m | 165 |
| $t_{12}$ | Thickness, $t_{12} = t_1 + t_2$ | m | 165 |

| | | | |
|---|---|---|---|
| $t_2$ | Thickness of layer 2 | m | 165 |
| $t_{ox}$ | Thickness of gate oxide | m, nm | 258 |
| $U$ | Scaling factor ($U > 1$) | – | 258 |
| $U(x)$ | Potential energy function | J, eV | 70 |
| $U_A$ | Potential energy in region A | J, eV | 70 |
| $U_B$ | Potential energy in region B | J, eV | 70 |
| $U_n$ | Electron generation-recombination rate | $s^{-1}m^{-3}$, $s^{-1}cm^{-3}$ | 148 |
| $U_N(x)$ | Chebyshev polynomial of the second kind | – | 274 |
| $U_p$ | Hole generation-recombination rate | $s^{-1}m^{-3}$, $s^{-1}cm^{-3}$ | 148 |
| $v$ | Voltage | V | 61 |
| $v$ | Magnitude of **v** | m/s | 160 |
| $\mathbf{v}$ | Velocity vector | m/s | 160 |
| $V$ | Voltage | V | 29 |
| $V'$ | Voltage | V | 37 |
| $\Delta V$ | Voltage difference | V | 111 |
| $V^+$ | Voltage-traveling-wave phasor traveling in a positive direction | V | 63 |
| $V^-$ | Voltage-traveling-wave phasor traveling in a negative direction | V | 63 |
| $V_0$ | Voltage amplitude | V | 31 |
| $V_1$ | Voltage of port 1 | V | 41 |
| $V_2$ | Voltage of port 2 | V | 41 |
| $V_A^+$ | Forward voltage-traveling-wave phasor (line A) | V | 71 |
| $V_A^-$ | Backward voltage-traveling-wave phasor (line A) | V | 71 |
| $V_B^+$ | Forward voltage-traveling-wave phasor (line B) | V | 71 |
| $V_B^-$ | Backward voltage-traveling-wave phasor (line B) | V | 71 |
| $V_{BS}$ | Back gate-source voltage | V | 256 |
| $V_{CB}$ | Channel potential (back gate-referenced) | V | 242 |
| $V_{DB}$ | Drain-back gate voltage | V | 248 |

| | | | |
|---|---|---|---|
| $V_{dd}$ | Supply voltage for MOSFET | V | 257 |
| $v_{drift}$ | Drift velocity | m/s | 121 |
| $V_{DS}$ | Drain-source voltage | V | 220 |
| $V_{fb}$ | Flat-band voltage | V | 224 |
| $v_g$ | Group velocity | m/s | 75 |
| $V_{GB}$ | Gate-back gate voltage | V | 225 |
| $V_{GB}^{(1)}$ | Gate-back gate voltage (depletion) | V | 294 |
| $V_{GB}^{(2)}$ | Gate-back gate voltage (weak inversion) | V | 295 |
| $V_{GB}^{(3)}$ | Gate-back gate voltage (onset of strong inversion) | V | 295 |
| $V_{GB}^{(4)}$ | Gate-back gate voltage (strong inversion) | V | 295 |
| $V_{GB}^{(a)}$ | Gate-back gate voltage (accumulation) | V | 294 |
| $V_{GS}$ | Gate-source voltage | V | 220 |
| $V_H$ | Hall voltage | V | 161 |
| $v_n$ | Electron velocity | m/s | 130 |
| $\mathbf{v}_n$ | Electron velocity vector | m/s | 160 |
| $v_{n,drift}$ | Electron drift velocity | m/s | 135 |
| $v_p$ | Hole velocity | m/s | 130 |
| $V_P$ | Pinch-off voltage | V | 245 |
| $\mathbf{v}_p$ | Hole velocity vector | m/s | 160 |
| $v_{p,drift}$ | Hole drift velocity | m/s | 136 |
| $v_{ph}$ | Phase velocity | m/s | 75 |
| $V_{SB}$ | Source-back gate voltage | V | 248 |
| $V_T$ | Threshold voltage of MOSFET | V | 219 |
| $V_{T0}$ | Threshold voltage of MOS capacitor | V | 237 |
| $V_{TB}$ | Threshold voltage of 3-terminal MOS (back gate-referenced) | V | 244 |
| $V_{TC}$ | Threshold voltage of 3-terminal MOS (channel-referenced) | V | 247 |
| $v_{th}$ | Thermal velocity | m/s | 134 |
| $W$ | Width | m | 160 |
| $W$ | Channel or gate width of MOSFET | m, µm | 220 |

| | | | |
|---|---|---|---|
| $W_m$ | Magnetic energy | J | 39 |
| $W_s$ | Electrostatic energy | J | 39 |
| $x$ | Mixture ratio ($0 < x < 1$) | – | 4 |
| $x$ | $x$ coordinate | m | 14 |
| $x$ | Real number; Real part of complex number | – | 78 |
| $\Delta x$ | Infinitesimal section in $x$ direction | m | 61 |
| $x_N$ | Depletion layer thickness in n-type region | m | 183 |
| $x_P$ | Depletion layer thickness in p-type region | m | 183 |
| $y$ | $y$ coordinate | m | 153 |
| $y$ | Real number; Imaginary part of complex number | – | 78 |
| $Y$ | Admittance | S | 56 |
| $y_C$ | Channel depth or thickness of inversion layer | m, nm | 251 |
| $z$ | $z$ coordinate | m | 153 |
| $Z$ | Impedance | Ω | 56 |
| $Z_0$ | Characteristic impedance of transmission line | Ω | 64 |
| $Z_{0A}$ | Characteristic impedance of transmission line A | Ω | 67 |
| $Z_{0B}$ | Characteristic impedance of transmission line B | Ω | 67 |
| $Z_{in}$ | Input impedance | Ω | 56 |
| $Z'_{in}$ | Input impedance | Ω | 58 |

CHAPTER 1

# Introduction

## *Semiconductor Basics*

What is the most abundant artifact (artificial object) on Earth? In this chapter, we will go over the basics of semiconductors and consider this question.

## 1.1 WHAT ARE SEMICONDUCTOR DEVICES?

The word "device" has various meanings. In this book, *devices* are considered *circuit elements* that constitute an electronic circuit. There are various types of circuit elements, but the ones discussed in this book are *semiconductor devices* or *electron devices*. These two terms are often used interchangeably, although only the latter includes vacuum tubes.

Semiconductor devices are devices made of materials classified as *semiconductors*. Examples include two-terminal *diodes* with a rectifying action and three-terminal *transistors* with an amplifying or switching action. Semiconductor devices also include light-emitting devices that emit light and light-receiving devices that sense light. Solar cells and image sensors are also light-receiving devices.

Then, what are semiconductors?

## 1.2 CLASSIFICATION OF SOLIDS

Before we get into semiconductors, let's start with a more general discussion of solids. There are many possible ways to classify solids. They could be classified, for example, according to:

- Resistivity or conductivity
- Arrangement of atoms (crystalline or noncrystalline)

DOI: 10.1201/ 9781003439417-1

TABLE 1.1 Classification of Solids According to Resistivity (After [22])

| Name | Resistivity | Electron density | Examples |
|---|---|---|---|
| Conductor | $10^{-6} \sim 10^{-5} \Omega \cdot \text{cm}$ | $10^{22} \sim 10^{23}\ \text{cm}^{-3}$ | copper, aluminum, gold |
| Semiconductor | $10^{-2} \sim 10^{9} \Omega \cdot \text{cm}$ | $10^{6} \sim 10^{17}\ \text{cm}^{-3}$ | silicon, germanium, gallium arsenide |
| Insulator | $10^{14} \sim 10^{22} \Omega \cdot \text{cm}$ | $1 \sim 10\,\text{cm}^{-3}$ | diamond, glass, rubber |

- Purity (pure material or mixed with impurities)

These are not independent of one another; both the atomic arrangement and purity affect electrical properties.

Table 1.1 shows a rough classification of solids based on resistivity. Resistivity is the reciprocal of conductivity. Since conductivity is related to the conduction electron density, it is also given in Table 1.1. There are also materials with intermediate resistivities not listed in Table 1.1 and materials that become superconductors at cryogenic temperatures.

Most *conductors*, which conduct electricity very well, are metals. In silicon integrated circuits, where many devices are integrated on a silicon chip and interconnected, copper and aluminum are mainly used as materials for interconnects. Gold is often used as an interconnect material in compound semiconductor (p. 4) integrated circuits.

*Insulators* do not conduct electricity. There are various types of insulators. Diamonds are crystals, so their atoms are arranged regularly, while glass is amorphous, so its atoms are not arranged regularly. Rubber is an organic polymer. The name "insulator" comes from the fact that current hardly flows through it. An insulator is also called a *dielectric* when the focus is on its capacity to accommodate electric lines of force. The *dielectric constant* or *permittivity* is the material parameter related to it. It is an important technical issue in integrated circuits to reduce the parasitic capacitance between interconnects by minimizing the dielectric constants of the dielectrics that fill the space between the interconnects. Conversely, dielectrics with high permittivity are required for the insulator (gate dielectric film) used in the MOS transistor, which is described in Chapter 7.

*Semiconductors* are literally materials whose conductivities are somewhere between those of conductors and insulators. However, simply stating that a material having an electrical conductivity somewhere between that of a conductor and an insulator fails to capture the remarkable characteristics of semiconductors. Semiconductors have the following features:

- By adding appropriate impurities, the polarity of the charged particles responsible for electrical conduction can be selected, and the conductivity can be changed dramatically—by orders of magnitude.
- By combining the addition of impurities and appropriate structures, it is possible to select the polarity of the charged particles and achieve a significant change in conductivity by electrical means (e.g., by applying a voltage).

The second point can be considered a description of semiconductor devices. The ability to change conductivity by many orders of magnitude suggests that something like a switch could be made. Furthermore, by making good use of the existence of positive and negative mobile charges, a variety of semiconductor devices have been invented and utilized.

Changing conductivity by many orders of magnitude is usually not possible with metals or insulators. It is, therefore, not possible to use metals or insulators to make devices that operate by manipulating the conductivity of the material itself. As is well known, switches made of metals and insulators (including air) change the resistance of the path through which a current flows (or tries to flow but cannot) by mechanical operation. The same is true for relays, which combine electromagnets with mechanical switches to enable electrical on/off.

Vacuum tubes, which function similarly to transistors, are made of metals and insulators but use electric discharges in a vacuum to control the current. However, it is difficult to create a good off state with negligible leakage current.

It should be clear from the above that semiconductors are of overwhelming importance when discussing the properties of solids in relation to device applications. Some typical semiconductor materials are listed in Table 1.2.

*Silicon* is the most widely used semiconductor material. We will also use the chemical formula, Si, to refer to silicon. Note that *silicone* is a word similar to silicon, but it refers to a completely different

TABLE 1.2 Typical Semiconductor Materials

| Chemical formula | Name | Description or application |
|---|---|---|
| Si | Silicon | Most widely used semiconductor; VLSI |
| Ge | Germanium | First transistor was made of germanium |
| SiC | Silicon carbide | High-voltage devices |
| GaAs | Gallium arsenide | Optical devices, low-noise transistors |
| GaN | Gallium nitride | Optical devices, high-power transistors |
| InP | Indium phosphide | Optical devices, high-frequency transistors |
| $Si_{1-x}Ge_x$ | Silicon germanium | Semiconductor alloy; bipolar transistors |

substance group (see Problem 1.1 on p. 26). Silicon is a key semiconductor material for integrated circuits and power electronics. It is also used for solar cells and displays that require a large area because of its low cost. Table 1.3 shows some material parameters of silicon for reference. Since the values vary somewhat from literature to literature, they are rounded to two significant digits.[1] The meaning of each parameter will be explained later.

Like silicon, *germanium* is an *elemental semiconductor*—a semiconductor composed of a single element. It was a material extensively studied in the early days of semiconductor devices, and the first transistor was made of germanium. Although it is more difficult to handle than silicon as a material, it has some superior electrical properties.

*Compound semiconductors* are composed of multiple elements and there are a variety of materials. Gallium arsenide is more expensive than silicon, but it has excellent light-receiving and light-emitting capabilities and low noise. It is used for optical devices and low-noise transistors. Some material parameters of gallium arsenide are listed in Table 1.3. The chemical formula of gallium arsenide is GaAs.

*Silicon germanium* in Table 1.2 is special in that, unlike SiC, GaAs, GaN, and InP in the same table, the ratio of silicon to germanium does not need to be a fixed integer. Therefore, it is considered to be a homogeneous mixture of two solids, mixed at the atomic level, rather than a compound semiconductor. This type of semiconductor is called

[1] This does not mean that the values vary from the third digit onward.

TABLE 1.3 Material Parameters of Silicon and Gallium Arsenide at Room Temperature (After [30])

| | Si | GaAs |
|---|---|---|
| Relative permittivity | 12 | 13 |
| Energy gap $E_g$ (eV) | 1.1 | 1.4 |
| Electron affinity $\chi$ (eV) | 4.0 | 4.0 |
| Lattice constant $a$ (Å) | 5.4 | 5.6 |
| Atom density ($cm^{-3}$) | $5.0 \times 10^{22}$ | $4.4 \times 10^{22}$ |
| Intrinsic carrier density $n_i$ ($cm^{-3}$) | $1.0 \times 10^{10}$ | $2.1 \times 10^{6}$ |
| Conduction band effective density of states $N_c$ ($cm^{-3}$) | $2.8 \times 10^{19}$ | $4.7 \times 10^{17}$ |
| Valence band effective density of states $N_v$ ($cm^{-3}$) | $2.6 \times 10^{19}$ | $7.0 \times 10^{18}$ |
| Electron effective mass* $m_e/m_0$ | 0.98<br>0.19 | 0.063 |
| Hole effective mass** $m_h/m_0$ | 0.16<br>0.49 | 0.076<br>0.5 |
| Electron mobility $\mu_n$ ($cm \cdot V^{-1}s^{-1}$) | $1.5 \times 10^{3}$ | $8.0 \times 10^{3}$ |
| Hole mobility $\mu_p$ ($cm \cdot V^{-1}s^{-1}$) | $5.0 \times 10^{2}$ | $4.0 \times 10^{2}$ |

* *There are two effective masses depending on the direction of movement in a crystal.*
** *There actually are two types holes: heavy holes and light holes.*

a *semiconductor alloy*. Since the properties of semiconductor alloys depend on the mixing ratio, the ratio is chosen to achieve the desired properties. Silicon germanium is often abbreviated as "SiGe," but this is not its chemical formula. Another example of a semiconductor alloy is aluminum gallium arsenide, $Al_xGa_{1-x}As$, which is an alloy between GaAs and aluminum arsenide (AlAs).

When the semiconductor materials in Table 1.2 are compared with the periodic table of elements, we notice that all the elements belong to one of the few groups of elements. Table 1.4 shows the relevant part of the periodic table. First, the elemental semiconductors Si and Ge are both group 14 elements. SiC and SiGe are also composed only of elements of group 14. The other materials in Table 1.2 are all compounds composed of equal numbers of elements from group 13 and group 15. The above suggests that the properties of semiconductors are related to valence electrons. We will come back to this point in §1.3. In the field of semiconductor devices, groups of elements are often designated by Roman numerals (see Table 1.4). Following this convention, Si and Ge are *group IV semiconductors* and GaAs is a *III–V semiconductor*.

TABLE 1.4 Partial Periodic Table Related to Semiconductors

| Group name | 13 | 14 | 15 |
|---|---|---|---|
| Old group name | III | IV | V |
| | B | C | N |
| | Al | Si | P |
| | Ga | Ge | As |
| | In | Sn | Sb |

## 1.3 PROPERTIES OF SEMICONDUCTORS

The goal of this section is to provide a quick overview of the properties of semiconductors without using too many mathematical formulas. More in-depth explanations will be provided in Chapter 4 onward.

### 1.3.1 Arrangement of Atoms

Most semiconductors used in devices are *single crystals* grown by sophisticated growth technology. When we simply say "crystals," we usually mean single crystals. Crystalline semiconductors have excellent electrical properties. Since defects in crystals adversely affect electrical properties, it is important to grow crystals with as few crystalline defects as possible.

Polycrystalline silicon (poly-Si) is often used for solar cells and display devices that require large areas. *Polycrystal* is an agglomeration of many fine crystal grains with different orientations. Polycrystalline semiconductors are inferior to crystalline semiconductors in terms of electrical performance, but they can be fabricated into large-area devices more easily and inexpensively than single crystals. Poly-Si is often used as a material for the gate electrode of MOS transistors (Chapter 7).

Unlike crystals, some semiconductors do not have a regular arrangement of atoms and are called *amorphous semiconductors*. Although their performance is even lower than that of polycrystalline semiconductors, they are easier to produce and are used in situations where high performance is not required.

Hereafter, semiconductors are assumed to be crystals unless otherwise specified.

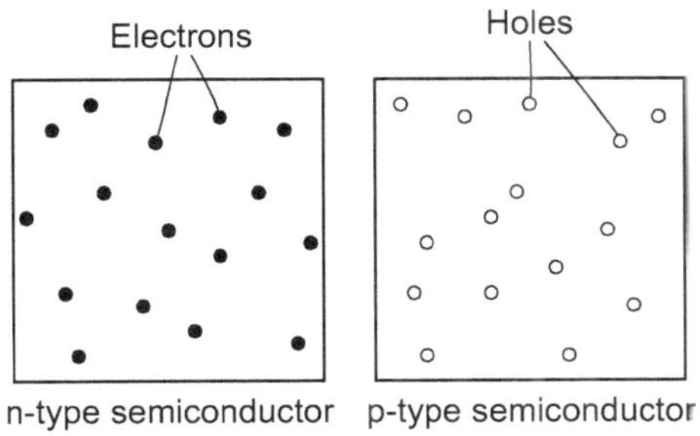


FIGURE 1.1 n-Type and p-type semiconductors. Black-filled circles (•) represent electrons. Open circles (○) represent holes.

### 1.3.2 Intrinsic and Doped Semiconductors

As mentioned in §1.2, semiconductors used in devices usually have their electrical characteristics manipulated by adding a small amount of certain carefully chosen impurities (not just any impurities). In contrast, pure semiconductors with no intentionally added impurities are called *intrinsic semiconductors*. Impure semiconductors doped with appropriate impurities are called *extrinsic semiconductors* or *doped semiconductors*.

Doped semiconductors are divided into two types: n-type and p-type semiconductors. In n-type semiconductors, negatively charged *electrons* are responsible for electrical conduction (Fig. 1.1). In contrast, positively charged particles called *positive holes* or simply *holes* are responsible for electrical conduction in p-type semiconductors. Electrons and holes are collectively called *mobile charge carriers* or *carriers*.

### 1.3.3 Carriers in Intrinsic Semiconductors

An intrinsic semiconductor is a pure semiconductor that has no impurities added to make it n- or p-type. An intrinsic semiconductor contains equal numbers of electrons and holes. When discussing solid materials including semiconductors, it is more convenient to consider the number of carriers per unit volume, i.e., the density or concentration of carriers, instead of discussing the absolute number of carriers. So let $n$ and $p$, respectively, denote the electron and hole densities contributing to electrical conduction. Then, the electron and hole

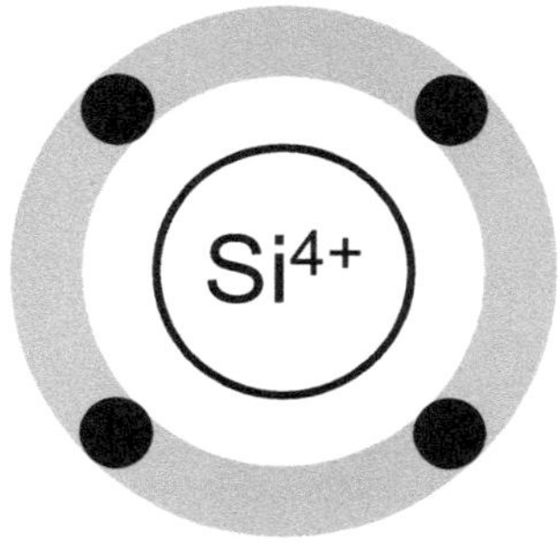


FIGURE 1.2 The silicon atom has four valence electrons.

densities of intrinsic semiconductors can be written as

$$n = p = n_{\mathrm{i}}, \qquad \text{(Carrier densities of intrinsic semiconductor)} \quad (1.1)$$

where $n_{\mathrm{i}}$ is called the *intrinsic carrier density*. Note, however, that (1.1) does not hold at all times in all intrinsic semiconductors. Specifically, (1.1) may not hold in those parts of an intrinsic semiconductor that have "bent energy bands" (Chapter 6). Numerical examples of $n_{\mathrm{i}}$ are given in Table 1.3 (p. 5). Intrinsic semiconductors are semiconductors, but their conductivities at room temperature are not very high. They are more like insulators and are sometimes described as being *semi-insulating*. The three symbols $n$, $p$, and $n_{\mathrm{i}}$ in (1.1) are universally used in the semiconductor literature, not only in this book.

In order to make the following discussion more concrete, we will consider intrinsic silicon as an example of an intrinsic semiconductor. Since silicon is an element of group 14 (or group IV), a silicon atom has four valence electrons. The atomic number of silicon is 14, so the total number of electrons is 14. But we can think of it as having four electrons around a tetravalent cation. The cartoon in Fig. 1.2 shows this situation.

Silicon atoms share valence electrons to form a covalent crystal. The crystal structure is the so-called *diamond structure* (see Problem 1.2 on p. 26). A pair of valence electrons can be thought of as being shared by two atoms, as shown in the cartoon in Fig. 1.3.

### 1.3.4 Energy Band Formation

In the cartoon of Fig. 1.3, it seems as if two valence electrons are trapped or localized between two neighboring atoms, but the

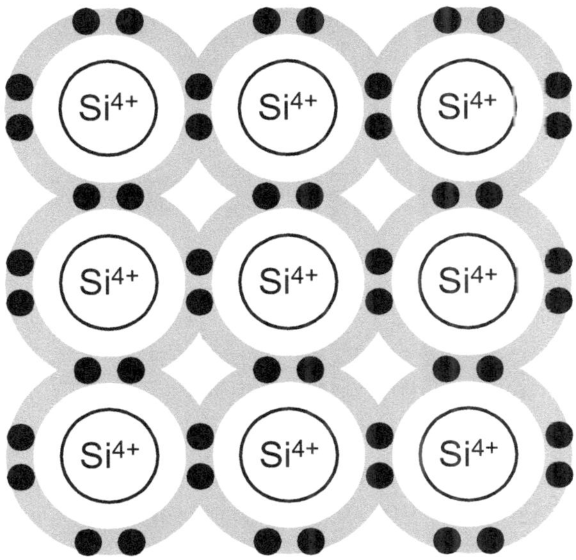


FIGURE 1.3 A model of silicon crystal.

actual situation is somewhat different. To understand this, let us first consider how valence electrons reside in a single silicon atom.

Considering by analogy with Bohr's hydrogen-like atom model, an atom has a number of orbitals that accommodate electrons. Each electron in an orbital then has certain energy associated with the orbital. Electrons in orbitals near the nucleus have lower energy, whereas valence electrons that are far from the nucleus and are involved in the formation of chemical bonds and chemical reactions have higher energy. These energies associated with orbitals are called *energy levels*. Based on the above discussion, Fig. 1.4 shows a very rough conceptual drawing, with the curves representing an electrostatic potential due to the positively charged nucleus. Rigorous treatment of atoms requires quantum chemistry, which is far beyond our scope, but an important point here is that the atomic energy levels are discrete and countable.

Next, suppose we have a very large number of silicon atoms. In order to direct our discussion toward crystals, let us assume that these atoms are arranged regularly with equal spacing. If the atoms were spaced more than a few centimeters apart, they could be considered independent of each other and the energy levels of the orbitals in which electrons reside would be the same as in Fig. 1.4.

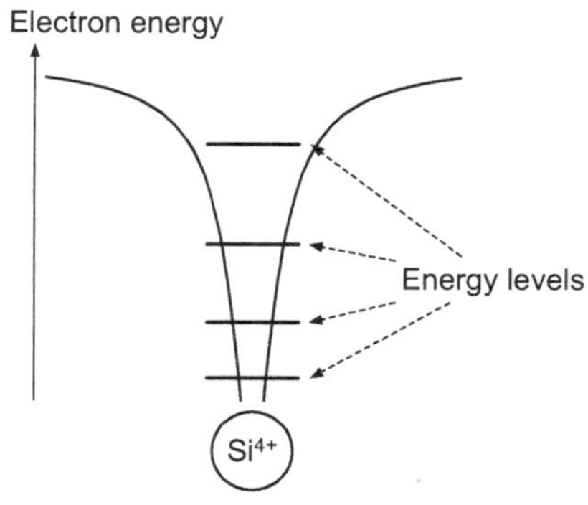


FIGURE 1.4 Energy levels of an isolated atom.

What would happen if we made the distance between atoms smaller and smaller? When the atomic spacing becomes very small, the presence of other atoms can no longer be ignored. What would happen then to the orbitals of each atom? The "shape" of each orbital would change, but the total number of orbitals contributed by participating atoms would be maintained. However, the orbitals from each atom cannot stay at the same energy levels. They would assume different energy levels that do not overlap with each other. Fig. 1.5 depicts this situation, with the horizontal axis representing the atomic spacing and the vertical axis representing the electron energy. As atomic spacing decreases, energy levels spread with very small energy spacing, forming *bands* of densely distributed energy levels. In fact, when silicon atoms are arranged regularly, a silicon crystal should form, and the atomic spacing cannot be changed arbitrarily. The known crystal lattice spacing (or lattice constant) of silicon is the atomic spacing that can actually be realized.

If the number of atoms is very large, there will be a virtually continuous distribution of energy levels on the energy axis. In this situation, individual electrons do not belong to any particular atom but to a mass of atoms or the crystal. That is, the electrons that were originally localized to atoms become delocalized throughout the crystal. A set of energy levels densely distributed on the energy axis is called an *energy band*. The region in Fig. 1.5 where there are no energy levels is called the *forbidden band*, *energy gap*, *energy band gap*, or simply *bandgap*. The energy bands are also called the *allowed bands* in contrast to the forbidden bands.

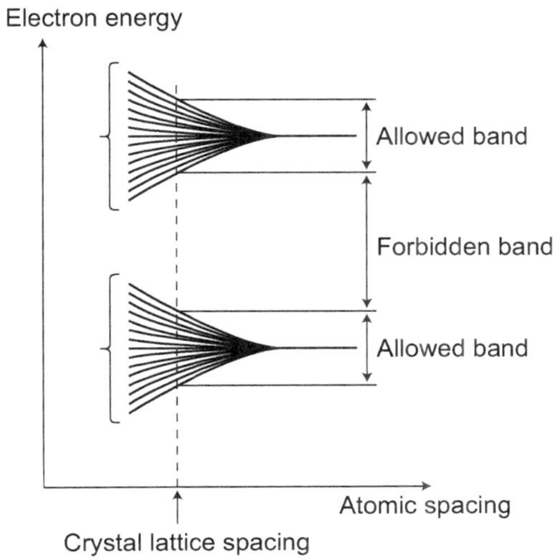


FIGURE 1.5 Relation between atomic spacing and energy levels.

In general, solids may have multiple allowed and forbidden bands, but it is usually sufficient to consider only two allowed bands and the forbidden band sandwiched between them when discussing semiconductors. The allowed band that lies above the forbidden band is called the *conduction band*, and the other allowed band that lies below the forbidden band is called the *valence band*.

The reason why energy band formation occurs as a result of the regular arrangement of indistinguishable identical objects (atoms in this case) has to do with the symmetry of the structure (regularity of atomic arrangement) and can be discussed mathematically. We will discuss energy band formation further in Chapter 3 using circuit theory.

### 1.3.5 Properties of Intrinsic Semiconductors

As mentioned in §1.3.4, there are two allowed bands in semiconductors. Orbitals in the valence band are almost completely filled with electrons at room temperature, but a small number of orbitals near the top of the band are vacant. Fig. 1.6 illustrates this situation. The conduction band is almost empty, but a small number of orbitals near the bottom of the band are filled with electrons. The vertical axis in Fig. 1.6 is the electron energy $E$. The shaded area indicates the presence of electrons. The horizontal axis is a spatial coordinate (e.g.,

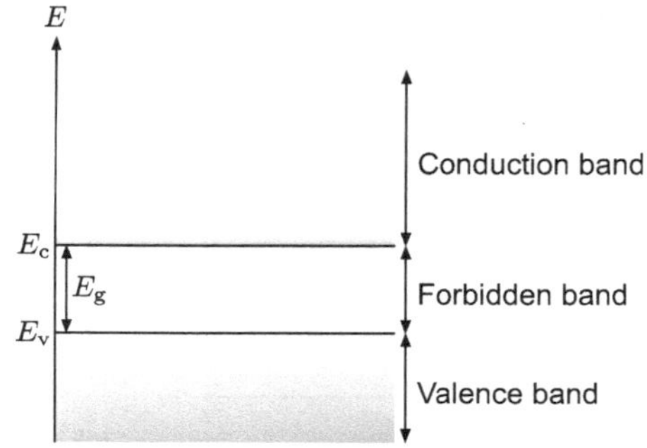


FIGURE 1.6 An energy band diagram of an intrinsic semiconductor (*E*-*x* diagram).

*x*-coordinate), but this is not specified at this time because a spatially uniform crystal is assumed. Such a diagram is called an *energy band diagram,* or *band diagram* for short.

As shown in Fig. 1.6, the electron energy at the top of the valence band is denoted by $E_v$, the bottom of the conduction band is denoted by $E_c$, and the magnitude of the energy gap (or bandgap energy) is denoted by $E_g$. These three symbols are widely used in the

The source of the small number of electrons near the bottom of the conduction band of an intrinsic semiconductor is the valence band. These electrons were originally occupying orbitals near the top of the valence band but were thermally excited and settled into orbitals near the bottom of the conduction band, as shown in Fig. 1.7. Therefore, empty orbitals are left near the top of the valence band. In fact, these "holes" near the top of the valence band behave as positively charged particles—known as *positive holes,* or simply *holes.* If the electron charge is $-q$, the hole charge is $+q$. From the above, (1.1) on p. 8 holds for the electron and hole densities in intrinsic semiconductors.

The "small number" of electrons and holes per unit volume is

$$n_i \simeq 1 \times 10^{10} \text{cm}^{-3} \qquad \text{(Intrinsic carrier density of silicon)} \qquad (1.2)$$

for intrinsic silicon at room temperature (see Table 1.3 on p. 5). It may not be so obvious whether the value in (1.2) can be called "a small number" or "low density," but the atomic density of crystalline silicon is about $5 \times 10^{22}$ cm$^{-3}$ (see Table 1.3). The number of carriers, therefore, is smaller than the number of atoms by ten orders of magnitude, and in this sense, the number of carriers can be regarded as small.

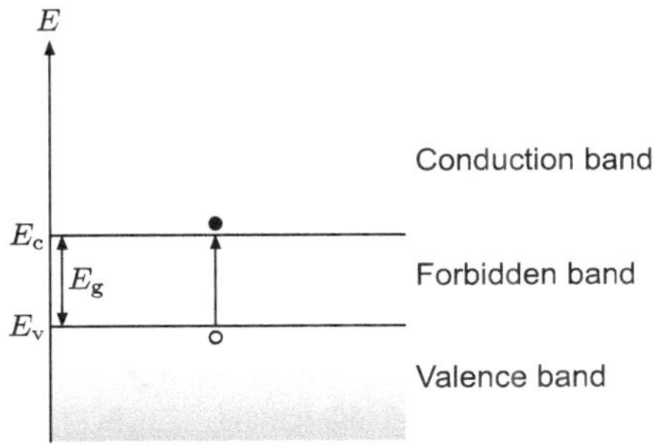


FIGURE 1.7 Electrons in a conduction band of an intrinsic semiconductor originate from the valence band.

In contrast, the conduction electron density of metals is comparable to the atomic density.

The magnitude of intrinsic carrier density $n_i$ is related to the ease with which thermal excitation of electrons from the valence band to the conduction band occurs. The larger the energy gap

$$E_g = E_c - E_v, \qquad \text{(Energy gap)} \tag{1.3}$$

the less likely thermal excitation is to occur, and hence the smaller intrinsic carrier density $n_i$ (§4.2.3). The value of $E_g$ depends on the material. According to Table 1.3, $E_g \simeq 1.1\,\text{eV}$ (electron volts) for silicon. Compare this with thermal energy $kT \simeq 26\,\text{meV}$ (millielectron volts) corresponding to room temperature, $T = 300\,\text{K}$ (see Problem 1.3 on p. 26). We see that $E_g \gg kT$, and therefore thermal excitation of valence band electrons into the conduction band does not occur easily at room temperature (§5.6.1). This is the reason why the carrier density of intrinsic silicon is so small.

### 1.3.6 Energy Band Diagrams

An energy band diagram is a diagram in which the vertical axis is the electron energy $E$ and the horizontal axis is a spatial coordinate, as shown in Figs. 1.6 and 1.7, where quantities like the conduction band bottom energy $E_c$ and the valence band top energy $E_v$ are plotted. Note that shading (as in Figs. 1.6 and 1.7) and drawing of particles (as in Fig. 1.7) are not always done. Electrons in the conduction band have higher energy than those in the valence band. As considered in

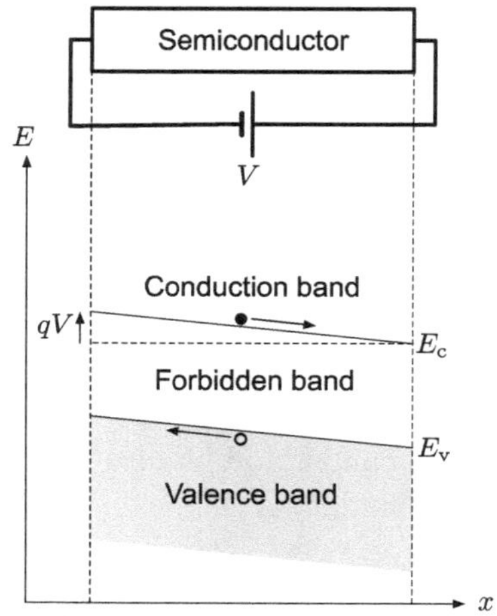


FIGURE 1.8 An energy band diagram of a semiconductor in an external electrostatic field (*E*-*x* diagram).

connection with Fig. 1.4 (p. 10), this energy is related to the electrostatic potential. Since holes are positively charged, the energy of a hole increases as it goes down the vertical axis.

Charged particles in an electrostatic field have potential energy that depends on their location. Therefore, if a semiconductor piece is placed in an electrostatic field, the energy of the carriers in the semiconductor will also depend on their positions in it. As shown in Fig. 1.8, when a DC voltage $V$ is applied to a semiconductor piece of length $L$ in the $x$-direction, the electrostatic field $\mathcal{E}$ in the semiconductor is considered to be given by

$$\mathcal{E} = \frac{V}{L}. \tag{1.4}$$

Note that we used different typefaces to distinguish between the electron energy $E$ and the electrostatic field $\mathcal{E}$. Therefore, there is a difference, $qV$, in the potential energy of an electron (or a hole) between the right and left ends of the semiconductor piece in Fig. 1.8. The further to the right in the energy band diagram, the smaller the potential energy of an electron due to the electrostatic field $\mathcal{E}$, so the band diagram is downward sloping toward the right as shown.

Since an electrostatic force $q\mathcal{E}$ acts on electrons in a rightward (→) direction, electrons in the conduction band can be interpreted as trying to roll down the slope of $E_c$ in the energy band diagram. Similarly, a leftward (←) force $q\mathcal{E}$ acts on holes. Considered upside down, holes in the valence band can be interpreted as trying to go down the slope of $E_v$. Thus, if we have an energy band diagram, we can read the directions of the electrostatic force acting on carriers from the slope of $E_c$ or $E_v$.

There is a reason we wrote "we can read the directions of the force acting on carriers" and not "we can see the directions in which carriers move." In some cases, *carriers may move in the opposite direction to the force acting on them.* You might recall a common situation in which a ball thrown upward temporarily moves upward against gravity due to inertia. However, in semiconductor devices, another mechanism allows carriers to flow *steadily* in the opposite direction to the force acting on them. We will come across such an example in 6. And the example is not a special, contrived one, but a p-n junction diode, one of the simplest semiconductor devices. Understanding the physics of carriers in connection with energy band diagrams is among the most important goals of this book.

Now, as you can infer from Fig. 1.8 (p. 14), $E_c$ and $E_v$ in the energy band diagram are related to the electrostatic potential $\psi(x)$ as follows:

$$E_c(x) = -q\psi(x) + \text{const.}, \tag{1.5}$$

$$E_v(x) = -q\psi(x) - E_g + \text{const.}, \tag{1.6}$$

where "const." in (1.5) and (1.6) represents a constant term. Also, (1.3) on p. 13 was assumed to hold regardless of the value of $x$. It does not hold if the material changes depending on $x$. The values of the constant terms in (1.5) and (1.6) are the same in both equations.

**TWO KINDS OF ENERGY BAND DIAGRAMS**

The term "energy band diagram" is used to refer to two different kinds of diagrams.

Fig. 1.6 (p. 12) and Fig. 1.8 (p. 14) correspond to what are known as the *E-x diagrams* [22]. The *E-x* diagram has a spatial coordinate (or *x*-coordinate) on the abscissa and is used when $E_c$ and $E_v$ vary spatially. Since $E_c$ and $E_v$ of semiconductor devices may vary spatially due to changes in materials, impurity doping (§1.3.7), and external fields, *E-x* diagrams are an indispensable tool for studying semiconductor devices. When semiconductor device engineers refer to "energy band diagrams," they usually mean $E$-$x$ diagrams.

The other kind of energy band diagram is the $E$-$k$ diagram [22], where $k$ is the *angular wave number* (or *wave number* for short) when electrons in a solid are treated as waves in the quantum mechanical sense. $E$-$k$ diagrams are also sometimes called dispersion curves (§3.3). $E$-$k$ diagrams are used to present the properties of solid materials, including semiconductors. When material scientists refer to "energy band diagrams," they usually mean $E$-$k$ diagrams.

Not surprisingly, $E$-$x$ and $E$-$k$ diagrams are related to each other. The $E$-$x$ diagram depicts the position dependence of a certain point (typically a maximum or minimum point) in the $E$-$k$ diagram (see Fig. 3.21 (p. 80)).

Energy band diagrams in this book are $E$-$x$ diagrams unless otherwise stated.

Energy band diagrams are often used to study semiconductor devices. From (1.5) and (1.6), we can also read the electrostatic potential (within a constant) from $E_C(x)$ or $E_V(x)$ in an energy band diagram. The quantities plotted on the energy band diagram are not limited to $E_C(x)$ and $E_V(x)$. If they were, we would not be able to tell the direction in which carriers are actually moving! At this point, it may not be so clear why we should want to draw such diagrams, but their importance will become clearer as we proceed to later chapters.

### 1.3.7 n-Type and p-Type Semiconductors

As already mentioned in §1.3.2, the electrical properties of semiconductors depend very sensitively on the content of certain impurities. When an intrinsic semiconductor is doped with a certain impurity, its properties change significantly (see Problem 1.4 on p. 26).

#### *1.3.7.1 n-Type Semiconductors*

Let us consider the addition of a very small amount of phosphorus, an element of group 15 (or group V), to intrinsic silicon (see Table 1.4 on p. 6). The phosphorus (P) atom has five valence electrons as shown in Fig. 1.9.

FIGURE 1.9 A model of a phosphorus atom.

When a phosphorus atom fits into a lattice point of a silicon crystal, as shown in Fig. 1.10, four valence electrons are used to form covalent bonds with surrounding silicon atoms, leaving one valence electron unused. This excess electron is freed as a conduction electron that can move freely in the crystal (see Problem 1.5 on p. 27 for a possible problem with the cartoon in Fig. 1.10). The phosphorus becomes positively charged as a univalent cation. Roughly speaking, the silicon crystal is provided with nearly as many conduction electrons as the number of phosphorus atoms added. As a result, the electron density $n$ exceeds the hole density $p$ by many orders of magnitude ($n \gg p$) and the silicon becomes n-type. The *majority carriers* in n-type semiconductors are electrons, and the *minority carriers* are holes. Fig. 1.11 depicts this situation on an energy band diagram.

Impurities of group 15 (group V) that ionize in semiconductors to provide electrons, such as phosphorus in this example (see Table 1.4 on p. 6), are called *donors*. Typical donors added to silicon are phosphorus (P) and arsenic (As). Be careful not to confuse the polarity of majority carriers with the polarity of impurity ions. Donor ions are positively charged cations. Note that donor ions are *fixed charges* and cannot move about because they are embedded in the crystal lattice. Since the number of conduction electrons supplied by donors is equal to the number of resulting donor ions, a uniform n-type semiconductor is, in general, electrically neutral.

As mentioned earlier, donor atoms in semiconductors fit into crystal lattice points and are part of the crystal. At temperatures much lower than room temperature, an electron is loosely bound to a donor nucleus, as shown in Fig. 1.12. This can be thought of as a hydrogen-like atom with very low binding energy. At room temperature, the

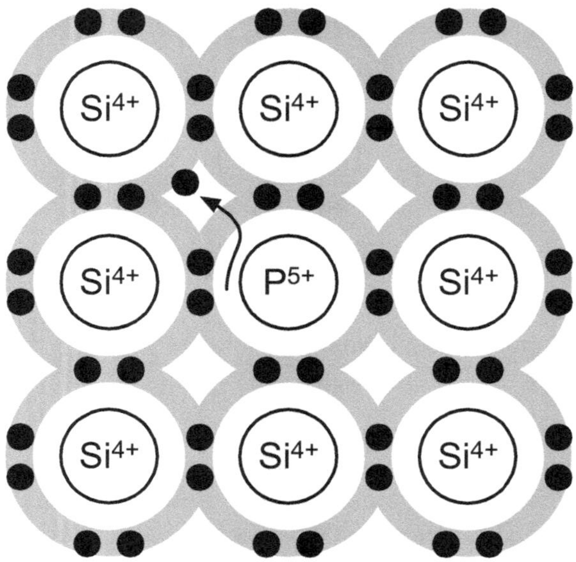


FIGURE 1.10 A model of n-type silicon doped with phosphorus.

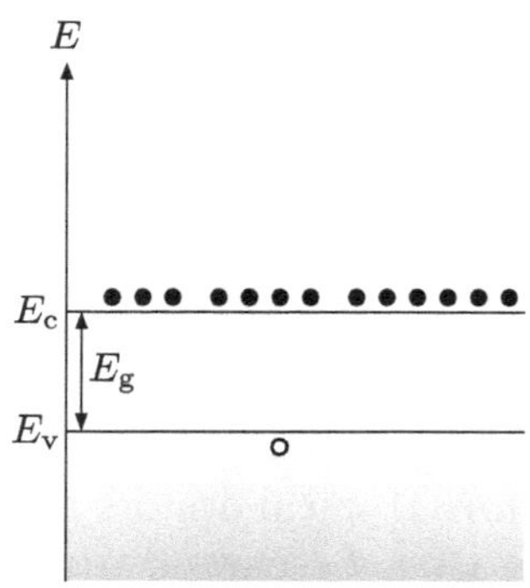


FIGURE 1.11 Carriers in conduction and valence bands of n-type semiconductor.

electron is free to move around in the crystal, shaking off the binding of the donor ion. This means that the ionization energy of the donor is at most comparable to the thermal energy of room temperature (26 meV at $T = 300\,\mathrm{K}$).

Based on this, the *donor level* $E_D$ can be written in an energy band diagram as shown in Fig. 1.13. Note that the orbital belonging to a donor is not delocalized throughout the crystal like those belonging

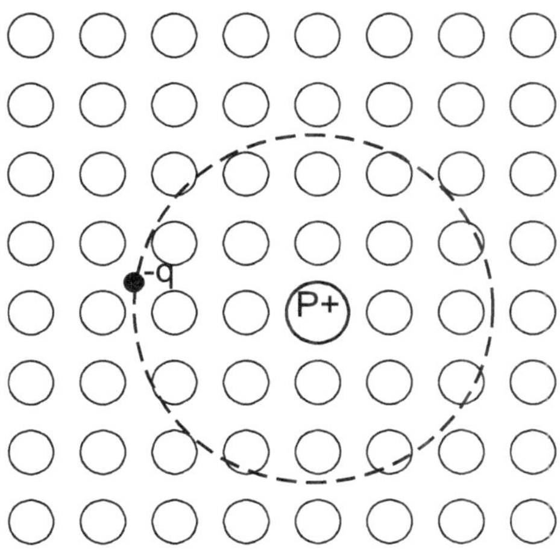


FIGURE 1.12 Phosphorus ion in silicon crystal.

to the conduction or valence band. Fig. 1.13 shows two donor levels at the same energy at two different locations. In Fig. 1.13, filled circles (•) represent electrons and open circles (◦) represent holes (p. 7 onward). The neutral donor atom has an electron and a hole in it and the donor ion has a hole. It may seem strange to have a hole in a neutral donor or a donor ion, but electrically, there is no problem with these representations of the two states of a donor. More importantly, it is also consistent with the more general entity known as the *donor-type trap* (p. 143), which not only emits/captures an electron to/from the conduction band but can also emit/capture a hole to/from the valence band. We will see in §5.6.2 that the donor is in fact a kind of donor-type trap. For silicon, $E_g \simeq 1.1\,\text{eV}$ (see Table 1.3 on p. 5), whereas $E_c - E_D$ is typically only several tens of millielectron volts, comparable to the thermal energy at room temperature (see Problem 1.5 on p. 26).

### 1.3.7.2 p-Type Semiconductors

Let us now consider the addition of a small amount of boron (B), an element of group 13 (group III), to intrinsic silicon. The boron atom has three valence electrons as shown in Fig. 1.14. When a boron atom fits into a lattice point of a silicon crystal as shown in Fig. 1.15, covalent bonds are formed with surrounding silicon atoms. Each covalent bond needs two electrons, so the formation of four covalent bonds

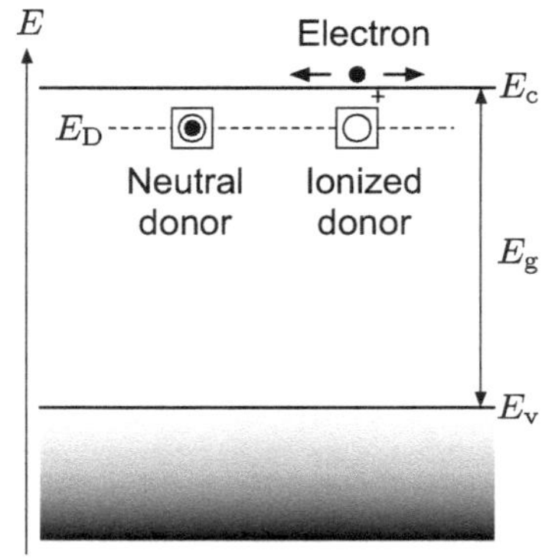


FIGURE 1.13 Donor levels in the forbidden band.

FIGURE 1.14 A model of a boron atom.

implies that the participating atoms are one electron short. The result is a positively charged *hole*, which can move freely through the crystal. Then, the boron becomes a monovalent anion. The silicon crystal is supplied with roughly as many holes as the number of boron atoms added. As a result, the hole density $p$ exceeds the electron density $n$ by many orders of magnitude ($p \gg n$), and the silicon becomes p-type. Therefore, the majority carriers in a p-type semiconductor are holes, and the minority carriers are electrons. Fig. 1.16 depicts this situation on an energy band diagram.

Group 13 (group III) impurities, such as boron, that ionize in semiconductors to provide holes, are called *acceptors* (see Table 1.4). Boron is a typical acceptor added to silicon. Acceptor ions in semiconductors are negatively charged anions. Acceptor ions are also embedded in crystals and are immobile, fixed charges. Since the number of

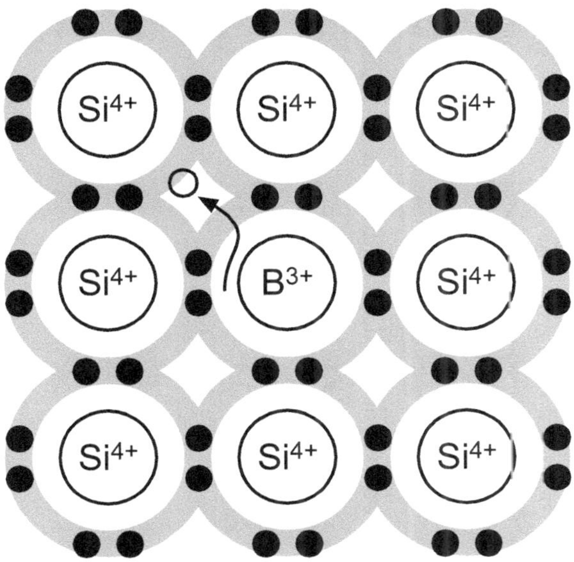


FIGURE 1.15 A model of p-type silicon doped with boron.

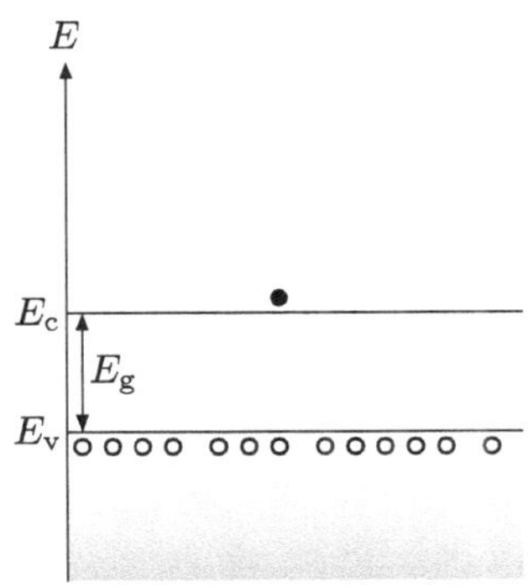


FIGURE 1.16 Carriers in conduction and valence bands of a p-type semiconductor.

holes supplied by acceptors is equal to the number of resulting acceptor ions, a uniform p-type semiconductor is, in general, electrically neutral.

An acceptor nucleus loosely binds a hole at much lower temperatures than room temperature, as shown in Fig. 1.17. At room temperature, holes are free to move through the crystal, free from the acceptor ions. The acceptor level $E_A$ is located slightly above $E_v$ in the

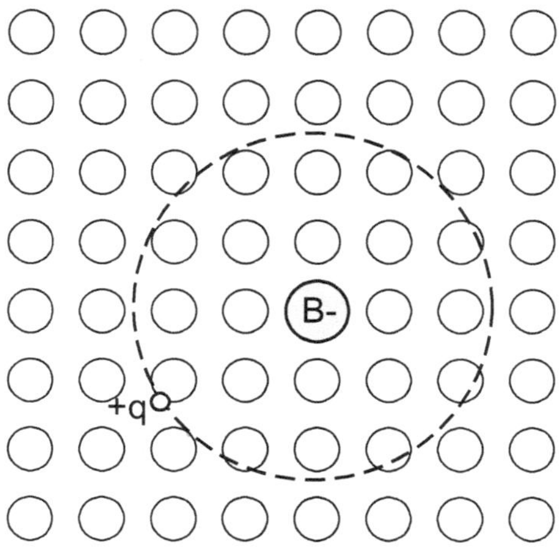


FIGURE 1.17 A boron ion in silicon crystal.

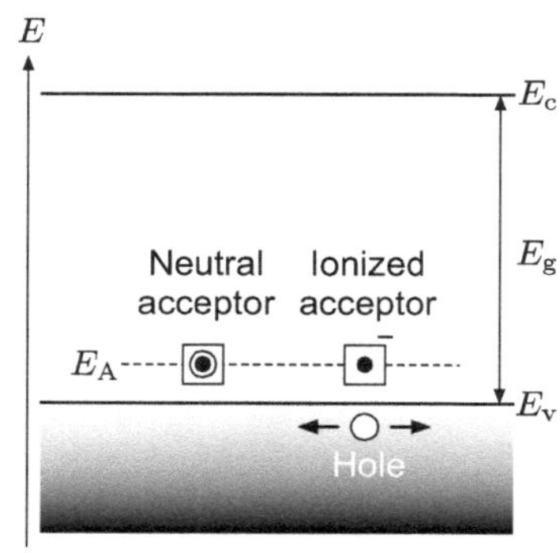


FIGURE 1.18 Acceptor levels in the forbidden band.

forbidden band, as shown in Fig. 1.18. Typically, $E_A - E_v$ is several tens of millielectron volts.

### *1.3.7.3 Impurities versus Dopants*

Here are some notes on terms related to impurity doping. First, donors and acceptors are also collectively called *dopants*. If we simply say "impurities," it is not clear whether they are intentionally added donors and/or acceptors or unintentional contaminants. But if we say "dopants," then we know that they are donors or acceptors that are intentionally "doped" [24]. This is also the reason we use the term "doped semiconductors" in this book rather than "impure semiconductors."

**IMPURITY DOPING IN SEMICONDUCTORS**

We discussed "adding" impurities or dopants to semiconductors, but semiconductors are basically solids. It should be quite different from dissolving salt in water or adding silicone as a food additive to cooking oil or coffee. How would we actually dope a semiconductor with donors or acceptors?

There are two basic approaches. One is to mix dopants when growing semiconductor crystals. There are various methods of crystal growth, and the liquid-phase growth method may be somewhat similar to dissolving sugar in water and then freezing it. This method does not allow fine control of the spatial distribution of dopant density.

The other is to add dopants to a semiconductor crystal after crystal growth while fabricating a device, typically by a method called *zion implantation*. In this method, dopant atoms are ionized and accelerated in a high electric field and shot into the semiconductor. The type and density of dopants can be changed locally. Impurity atoms cannot be just injected and left as they are, they must be *annealed* to fit into crystal lattice points so that they become part of the crystal and function as dopants.

## 1.4 WHAT IS THE MOST ABUNDANT ARTIFACT ON EARTH?

What is the most abundant artifact (or artificial object) on Earth? The fact that there are so many of them means that they cannot be large. They must be very small. And there should be more constituent parts that make something up than those that are made up of multiple parts.

Probably the most abundant artifact on Earth is the metal-oxide-semiconductor (MOS) transistor, a semiconductor device that makes up today's integrated circuits (ICs). How many MOS transistors are there on Earth? The number of shipments per year well exceeded the number of ants (see Table 1.5) before the end of the 1990s [18]. Shipments have continued to increase year after year. Of course, the cumulative number of transistors exceeds the number of shipments

TABLE 1.5 Big Numbers around Us

| | |
|---|---|
| Population of Earth | $8 \times 10^9$ |
| Number of ants on Earth | $10^{16} \sim 10^{17}$ |
| Age of the Universe | $5 \times 10^{17}$ s |
| 1 mole | $6 \times 10^{23}$ |
| Number of MOS transistors on Earth | ? |

per year. How many MOS transistors do you think exist on Earth when you read this book (Problem 1.7 on p. 27)?

Incidentally, why are there so many MOS transistors?

First, MOS transistors are tiny. To give an example, the length of one side of a MOS transistor may be about 0.1 μm = 0.0001 mm. The thickness of a hair is about 0.1 mm. So $10^6$ transistors can, in principle, be put on the cross-sectional area of a hair.

MOS transistors are inexpensive. The price of a MOS transistor can be incredibly low. The main material of MOS transistors is silicon (Si). Since silicon is a major constituent of rock and sand, it is virtually inexhaustible. Silicon is therefore very cheap as a material. The lower limit of cost reduction for a given product is determined by the price of raw materials. If the main material were gold (Au), for example, no matter how much you cut processing costs, transportation costs, labor costs, and various other expenses, the price of gold itself (expensive!) remains and cannot be reduced any further. But silicon is cheap. Therefore, MOS transistors can be made cheaper in terms of raw material prices, and in fact, they are inexpensive.

MOS transistors can be made tiny. Silicon *integrated circuits* consisting of numerous MOS transistors are actually not assembled from individually manufactured MOS transistors. A large number of transistors are manufactured into a single-crystal silicon substrate at once, and several layers of wiring that connect the transistors are built on top. Manufacturing techniques similar to printing technology are used. In the traditional way of making electronic circuits, circuits were assembled by soldering individual components on a printed circuit board (PCB). The new method puts as many different circuit blocks as possible on a single IC chip. Then, the finished integrated circuits and large circuit components that are difficult to integrate into an IC are arranged on a PCB.

Then, why did we miniaturize MOS transistors so much? As we will discuss in §7.5, this was because it was known that the smaller the MOS transistor, the higher the performance, especially its operation speed. Importantly, even if many smaller MOS transistors are packed into an IC, the power consumption of the entire circuit does not increase very much. In contrast, bipolar transistors (p. 45), which may also be made of silicon, outperform MOS transistors, but they consume more power and generate more heat, so the level of integration could not be increased beyond a certain level. By using many small, high-performance MOS transistors and increasing the level of integration, we can create higher-performance circuits, especially computers. These ICs are the building blocks of the hardware side of today's advanced information society.

The degree of integration of ICs has grown rapidly since its infancy. Gordon Moore, a co-founder of semiconductor manufacturer Intel, projected in 1965 that the number of components on an integrated circuit would double every year. Later, this was slightly modified, and the number of transistors on an integrated circuit was projected to double every two years (1975). Leaving aside specific ways of phrasing it, the key point is that the number of transistors and circuit performance increase exponentially over the years. This is known as *Moore's law*. The technical background of Moore's law is the scaling law for MOS transistors, discussed in §7.5. Moore's law was initially a future projection. However, as semiconductor technology progressed and the industry prospered, nearly according to the "law," it became the guiding principle for the semiconductor industry. In other words, Moore's law has come to serve as a self-fulfilling prophecy.

The social impact of Moore's law has been dramatic. The cost of information processing by computers has dropped exponentially. The amount of information that can be processed has increased exponentially. In tandem with these, the capacity of magnetic storage devices has increased exponentially, and the speed of communication has also increased exponentially. Semiconductor devices, the subject of this book, lie at the heart of the hardware technology that made this possible.

## 1.5 SUMMARY

In this chapter, we gave a quick overview of semiconductors.

- Solids can be classified into conductors, semiconductors, and insulators based on conductivity.

- The conductivity of a semiconductor can be changed by many orders of magnitude by appropriate impurity doping and/or electrical means.
- *Devices* in this book refer to circuit elements, and those made of semiconductors are called *semiconductor devices*.
- The majority of carriers in an n-type semiconductor, doped with donors, are electrons, whereas the majority of carriers in a p-type semiconductor, doped with acceptors, are holes.
- The rapid progress of semiconductor devices integrated circuits composed of semiconductor devices provided the hardware basis for the rise of information technology.

## 1.6 PROBLEMS

1.1 Find out what kind of substance *silicone* (not silicon) is. Use search engines as appropriate. Does it contain silicon?

1.2 The crystal structure of silicon is a diamond structure. Find out what kind of three-dimensional structure this is.

1.3 Show that the thermal energy of room temperature, $T = 300\,\text{K}$, is about 26 meV (millielectron volts). Let the Boltzmann constant be $k = 1.38\times10^{-23}$ J/K and the elementary charge be $q = 1.6\times10^{-19}$ C.

1.4 The amount of dopant added to doped semiconductors is very small. As an example, suppose that boron atoms are added at a ratio of one boron atom for every $10^5$ silicon atoms. Refer to Table 1.3 (p. 5) and find the acceptor (boron) density in this case. If the hole density of the resulting p-type silicon is equal to the acceptor density, how many times is the hole density greater than the intrinsic carrier density?

1.5 Let's model a donor after Bohr's hydrogen-like atom (see Fig. 1.12 on p. 19). Assume that there is a donor in silicon and that the space between the donor nucleus and an electron bound to it has the permittivity of silicon (see Table 1.3 on p. 5) First, calculate the ratio of the radius of the ground-state electron orbital to the Bohr radius $r_B \simeq 0.53\,\text{Å}$). Then, find the ionization energy from the ground state. The donor level is considered to be located this much below the conduction band bottom energy, $E_c$. How many times greater is this value than the ionization energy of the ground-state hydrogen atom, 13.6 eV? Let the permittivity of the vacuum be $\epsilon_0 = 8.85 \times 10^{-12}$ F/m, the effective mass of the electron in silicon (p. 80) be $m_e = 0.98m_0$, the mass of the electron be $m_0 = 9.1 \times 10^{-31}$ kg, the elementary charge be $q = 1.6 \times 10^{-19}$ C, and the Planck constant be $h = 6.6 \times 10^{-34}$ J·s.

1.6 In n-type semiconductors, $n > p$ holds for electron density $n$ and hole density $p$. It should also be obvious that $n > n_i$. Then, what about the relationship between $p$ and the intrinsic carrier density $n_i$? Make a guess based on what you have learned in this chapter.

1.7 Make a very rough estimate of the number of MOSFETs on Earth in Table 1.5 (p. 24).

CHAPTER 2

# Semiconductor Devices from a Circuit-Theoretic Standpoint

In this chapter, we classify circuit elements and consider what semiconductor devices are in terms of circuit theory. Here, we will start with a review of circuit elements in linear circuit theory. As we proceed, semiconductor devices are shown to be classified as nonlinear circuit elements. However, not all nonlinear elements are made of semiconductors. The reason for going through the trouble of explaining in this somewhat unusual manner is to separate the *functions* of semiconductor devices as circuit elements from the *impact resulting from the use of semiconductors* for implementing nonlinear circuit elements. The distinction between linear and nonlinear elements also turns out to be relevant when considering the physics of semiconductors.

## 2.1 LINEAR CIRCUIT ELEMENTS

"Circuit theory" taught in the first year of college engineering courses is basically *linear* circuit theory. Most of the circuit elements that appear there are linear two-terminal elements. A circuit consisting of linear circuit elements and voltage and/or current sources is a linear circuit. As will be explained later, sources are generally not linear circuit elements. The characteristics of a linear two-terminal device are mathematically represented by "a straight line passing through the origin" in a two-dimensional plane. The type of circuit element (resistor, capacitor, inductor, etc.) is determined by the pair of coordinate axes that span the plane.


DOI: 10.1201/ 9781003439417-2

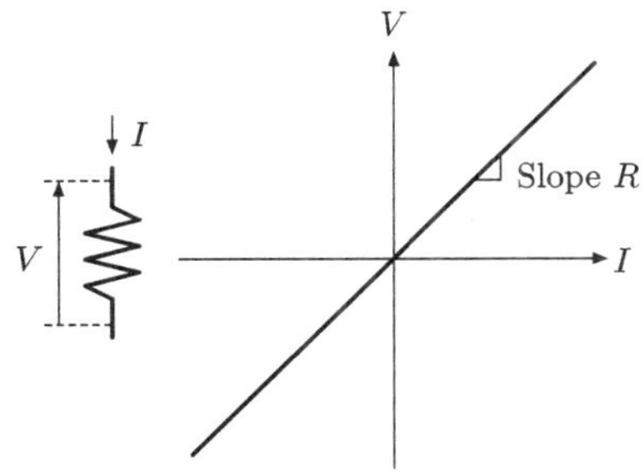


FIGURE 2.1 An element whose characteristics are given by a single straight line passing through the origin in the $I$-$V$ plane is a linear resistor.

### 2.1.1 Linear Resistors

At time $t$, the current $I(t)$ through a linear resistor and the voltage $V(t)$ across its terminals satisfy the following *constitutive relation*:

$$V(t) = RI(t). \qquad \text{(Constitutive relation for linear resistor)} \qquad (2.1)$$

This represents the so-called Ohm's law. If (2.1) is plotted on a plane with $I$ on the horizontal axis and $V$ on the vertical axis, it becomes a straight line passing through the origin, as shown in Fig. 2.1. The slope, $R$, of this line is the resistance in ohms ($\Omega$). Note that (2.1) holds for any voltage waveform $V(t)$ and current waveform $I(t)$, not just for DC. Conversely, a two-terminal element whose characteristics are given by "a straight line passing through the origin in the $I$-$V$ plane" may be defined as a linear resistor. In this book, we use the zigzag schematic symbol, shown in Fig. 2.1, for resistors.

If we swap the vertical and horizontal axes and consider the plane with $I$ on the vertical axis and $V$ on the horizontal axis, the slope of the line, $G = 1/R$, is the conductance, in units of siemens (S).

### 2.1.2 Linear Capacitors

At time $t$, the charge $Q(t)$ stored in a linear capacitor and the voltage $V(t)$ across its terminals satisfy the following constitutive relation:

$$Q(t) = CV(t). \qquad \text{(Constitutive relation for linear capacitor)} \qquad (2.2)$$

If (2.2) is plotted on a plane with $V$ on the horizontal axis and $Q$ on the vertical axis, it becomes a straight line passing through the origin,

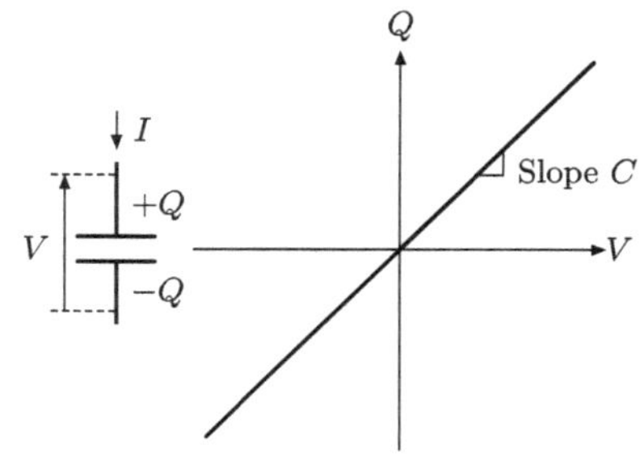


FIGURE 2.2 An element whose characteristics are given by a single straight line passing through the origin in the $V$-$Q$ plane is a linear capacitor.

as shown in Fig. 2.2. The slope, $C$, of this line is the capacitance in units of farads (F). Conversely, an element whose characteristics are given by "a straight line passing through the origin in the $V$-$Q$ plane" may be defined as a linear capacitor.

Let us also check the relationship between (2.2) and the current $I(t)$. Since the charge is the integral of the current over time, $Q(t)$ is given by

$$Q(t) = Q_0 + \int_{t_0}^{t} I(\tau)\,\mathrm{d}\tau, \tag{2.3}$$

where $Q_0$ is the charge stored at some time $t_0$. By differentiating (2.3) by $t$, we obtain the following familiar relationship between the current and the voltage:

$$I(t) = \frac{\mathrm{d}Q(t)}{\mathrm{d}t} = C\frac{\mathrm{d}V(t)}{\mathrm{d}t}. \tag{2.4}$$

However, if the point $(I(t), V(t))$ is plotted on an $I$-$V$ plane, the locus will be complicated depending on the current and voltage waveforms. This makes it practically impossible to define a linear capacitor using a locus in the $I$-$V$ plane. In contrast, the locus of a point $(V(t), Q(t))$ lies on the line given by (2.2), passing through the origin, shown in Fig. 2.2, regardless of the current and voltage waveforms. This indicates that the constitutive relation (2.2) is, indeed, a more fundamental relation than (2.4).

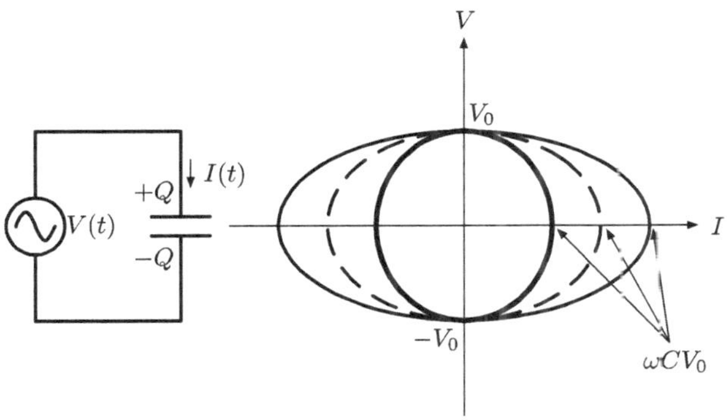


FIGURE 2.3 Loci on the *I-V* plane of a linear capacitor driven by a sinusoidal voltage source.

*Example: Locus of Linear Capacitor in the I-V Plane*

Assume that a linear capacitor of capacitance $C$ is connected to an AC voltage source of voltage $V(t) = V_0 \cos\omega t$. Consider the locus of the point $(I(t), V(t))$, where $\omega$ is the angular frequency. Using (2.4), the current is given by $I(t) = -\omega CV_0 \sin\omega t$. The point on the *I-V* plane, therefore, is given by $(I(t), V(t)) = (-\omega CV_0 \sin\omega t, V_0 \cos\omega t)$. Using the trigonometric identity $\sin^2\omega t + \cos^2\omega t = 1$, the equation for the locus is given by

$$\left[\frac{I(t)}{\omega CV_0}\right]^2 + \left[\frac{V(t)}{V_0}\right]^2 = 1. \tag{2.5}$$

Equation (2.5) represents an ellipse in the *I-V* plane. However, (2.5) includes parameters $V_0$ and $\omega$ that are not attributes of the linear capacitor. That is, the locus on the *I-V* plane depends on the waveform, as shown in Fig. 2.3.

In this example, the voltage waveform was a sine wave, so the locus on the *I-V* plane could be found analytically. However, in general, the equation of the locus cannot be found analytically. In contrast, the locus on the *V-Q* plane is always the straight line shown in Fig. 2.2, regardless of the waveform. The equation of the locus can be expressed using only the circuit element value, $C$, as in (2.2). ■

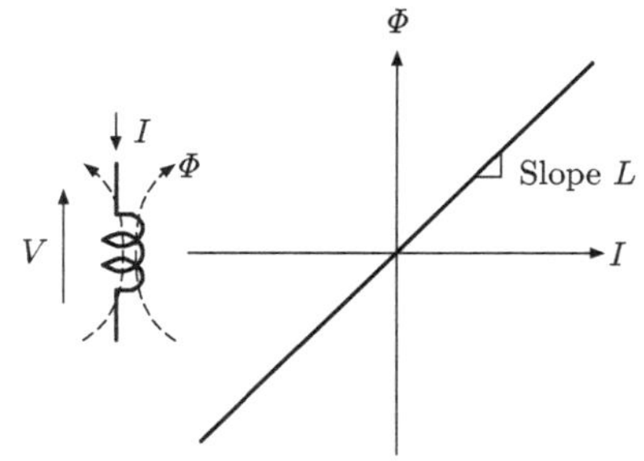


FIGURE 2.4 An element whose characteristics are given by a single straight line passing through the origin in the $I$-$\Phi$ plane is a linear inductor.

### 2.1.3 Linear Inductors

At time $t$, the magnetic flux $\Phi(t)$ penetrating a linear inductor and the current $I(t)$ through the inductor satisfy the following constitutive relation:

$$\Phi(t) = LI(t). \qquad \text{(Constitutive relation for linear inductor)} \quad (2.6)$$

If (2.6) is plotted on a plane with $I$ on the horizontal axis and $\Phi$ on the vertical axis, it becomes a straight line passing through the origin, as shown in Fig. 2.4. The slope, $L$, of the line is the inductance in units of henries (H). Conversely, an element whose characteristics are given by "a straight line passing through the origin in the $I$-$\Phi$ plane" may be defined as a linear inductor.

The magnetic flux $\Phi(t)$ and the voltage $V(t)$ satisfy a dual relationship to (2.3) as follows:

$$\Phi(t) = \Phi_0 + \int_{t_0}^{t} V(\tau)\,\mathrm{d}\tau, \quad (2.7)$$

where $\Phi_0$ is the magnetic flux at time $t_0$. By differentiating (2.7) with respect to $t$, we obtain the following relationship between the voltage and the current:

$$V(t) = \frac{\mathrm{d}\Phi(t)}{\mathrm{d}t} = L\frac{\mathrm{d}I(t)}{\mathrm{d}t}. \quad (2.8)$$

## 2.2 NONLINEAR CIRCUIT ELEMENTS

We saw in §2.1 that linear circuit elements are all represented by a single straight line passing through the origin in a plane spanned by certain variables. The characteristics of nonlinear circuit elements are similarly considered on an appropriate plane, but the locus representing the characteristics is not "a straight line passing through the origin."

### 2.2.1 Nonlinear Resistors

The characteristics of a two-terminal nonlinear resistor are represented by a single, usually curved, line on a plane with $I$ on the horizontal axis and $V$ on the vertical axis. This line does not need to pass through the origin. This line may also be a single straight line or consist of straight line segments.

#### *Example: A Nonlinear Resistor Represented by a Single-Valued Function*

Fig. 2.5 shows the current-voltage characteristics of a nonlinear resistor and its schematic symbol [8]. The schematic symbol allows one to distinguish the two terminals, which is necessary when the $I$-$V$ curve is not point-symmetric about the origin. The horizontal axis is $I$ and the vertical axis is $V$ in Fig. 2.5 as before, but it is more common to put $V$ on the horizontal axis and $I$ on the vertical axis. This is probably because the power supply used in experiments is usually more like a voltage source—it is more natural to consider $V$ as the independent variable. From here on, we will also use graphs with the vertical and horizontal axes interchanged. ■

If the voltage $V$ is given by a single-valued function $V(I)$ of the current $I$ as in Fig. 2.5, then

$$R_{\text{inc}}(I) \equiv \frac{\mathrm{d}V(I)}{\mathrm{d}I} \qquad \text{(Incremental resistance)} \tag{2.9}$$

is the slope of the tangent line to the $I$-$V$ curve and is called the *incremental resistance*, *small-signal resistance*, or *differential resistance*. On the other hand, the slope of the line connecting a point on the $I$-$V$ curve and the origin is given by

$$R_{\text{ch}}(I) \equiv \frac{V(I)}{I} \qquad \text{(Chord resistance)} \tag{2.10}$$

and is called the *chord resistance* [6, 35]. *Chord* is a mathematical term that refers to a straight line segment connecting two points on a

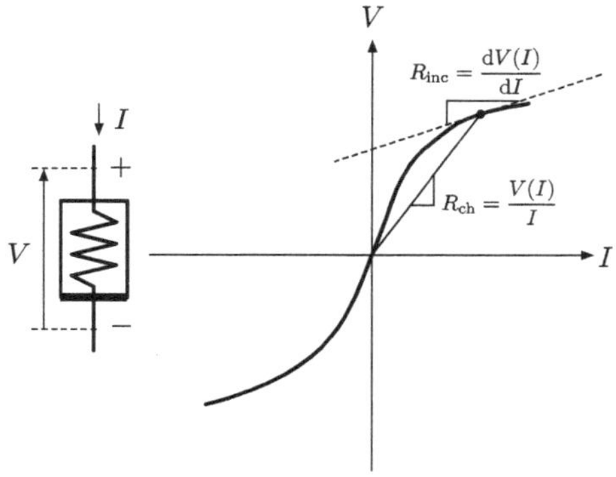


FIGURE 2.5 An element whose characteristics are given by a single (usually curved) line in the *I*-*V* plane is a nonlinear resistor.

curve. $R_{ch}$ ($I$) represents the resistance of a line segment strung from the origin like a string (or chord) of a stringed musical instrument. Note, however, that if the *I*-*V* curve does not pass through the origin, the line segment is not a chord in the mathematical sense. Although chord resistance is not a very widely used term, we use it in this book to clarify the distinction from incremental resistance. Note that in nonlinear resistors, the term "resistance" only has a qualitative meaning.

In Fig. 2.5, since the current $I$ is also a single-valued function of the voltage $V$, it is possible to write it as

$$I(V) = V^{-1}(V), \tag{2.11}$$

where $V^{-1}(V)$ is the inverse function of $V(I)$. The incremental conductance $G_{inc}$ ($V$) and the chord conductance $G_{ch}$ ($V$) are defined similarly to (2.9) and (2.10), respectively.

### *Example: Ideal Rectifier*

An *ideal rectifier* is a circuit element that conducts current in only one direction. It is also called the *ideal diode*. Its *I*-*V* characteristics are shown in Fig. 2.6. There is no voltage drop across an ideal rectifier when current flows through it. This defines the *forward direction* of the current. The voltage across it is zero ($V = 0$) when the nonzero current, $I > 0$, is flowing, regardless of the value of $I$. Conversely, when a voltage is applied in the *reverse direction* ($V < 0$), the current is constantly zero ($I = 0$).

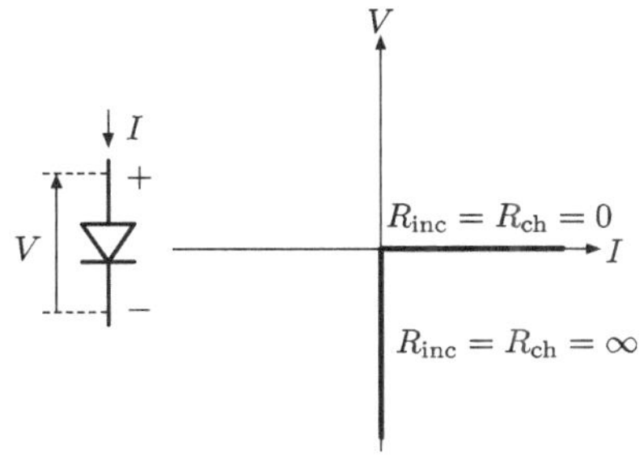


FIGURE 2.6 *I-V* plane characteristics of an ideal rectifier.

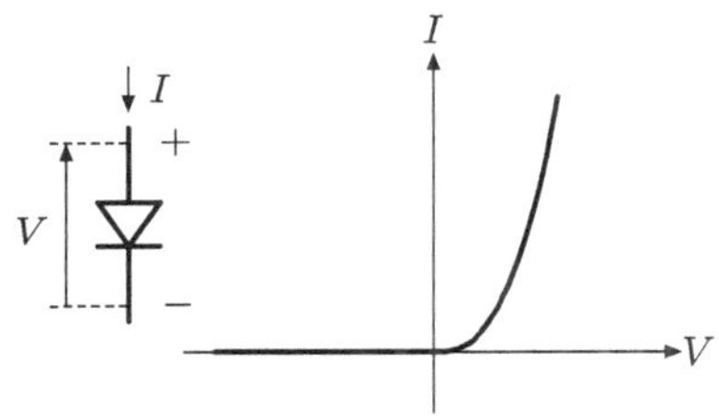


FIGURE 2.7 *I-V* plane characteristics of a p-n junction diode.

Fig. 2.6 shows a piecewise linear line (consisting of two straight line segments) on the *I-V* plane. It is not "a straight line that goes through the origin," so the ideal rectifier is a nonlinear resistor. The incremental and chord resistances of an ideal rectifier take the same value. ■

*Example: p-n Junction Diodes*

*p-n junction diodes* are two-terminal semiconductor devices with rectifying action, composed of p-type and n-type semiconductors. They are quite often called just "diodes." *I-V* characteristics of a p-n junction diode, shown in Fig. 2.7, indicate that it is a nonlinear resistor. We will discuss the physics of p-n junction diodes in Chapter 6. ■

*Example: DC Voltage Source*

As shown in Fig. 2.8, characteristics of a DC voltage source are represented by "a straight horizontal line on the *I-V* plane that does not pass through the origin" (it passes through the origin only when the

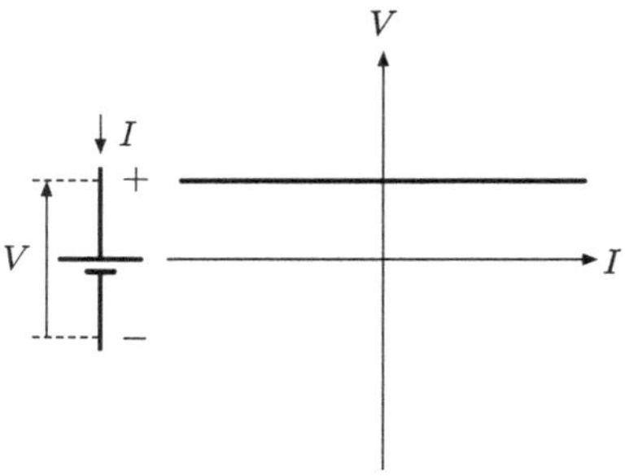


FIGURE 2.8 *I-V* plane characteristics of a DC voltage source.

output voltage is 0 V). The DC voltage source, therefore, is a nonlinear resistor. The incremental resistance of the DC voltage source is $R_{inc} = 0$ regardless of the value of $I$, but the value of the chord resistance depends on $I$. Usually, the *internal resistance* of the voltage source is said to be 0. Obviously, this resistance is not the chord resistance but the incremental resistance.

You might have thought that the current flowing through the voltage source in Fig. 2.8 is reversed, but in general, the current flowing through a circuit element is defined to have a positive value when the current flows into the "positive terminal," so Fig. 2.8 is correct. According to this definition, known as the *associated reference direction*, the value of the current flowing through a voltage source is usually negative ($I < 0$). ■

*Example: DC Current Source*

The DC current source is also a type of nonlinear resistor, as shown by the voltage-current characteristics in Fig. 2.9. The incremental conductance of the DC current source is given by $G_{inc} = 0$ regardless of the value of $V$, but the value of the chord conductance varies with $V$. According to the associated reference direction, the current and voltage are defined as in Fig. 2.9, so that $V < 0$. ■

### 2.2.2 Nonlinear Capacitors and Inductors

The definitions of a nonlinear capacitor and nonlinear inductor should be obvious by analogy to the nonlinear resistor.

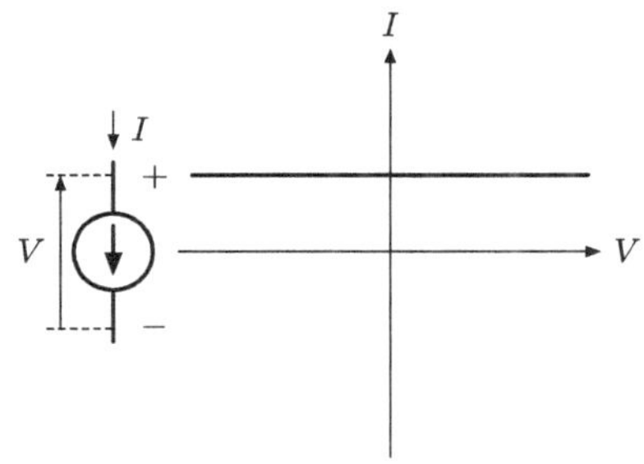


FIGURE 2.9 *V-I* plane characteristics of a DC current source.

A nonlinear capacitor is a circuit element whose characteristics are given by a single (typically curved) line in the *V-Q* plane. If the charge $Q$ is given by a single-valued function $Q(V)$ of voltage $V$, the incremental capacitance (aka small-signal capacitance or differential capacitance) can be written as

$$C_{\mathrm{inc}}(V) \equiv \frac{\mathrm{d}Q(V)}{\mathrm{d}V}. \qquad \text{(Incremental capacitance)} \qquad (2.12)$$

The chord capacitance is given by

$$C_{\mathrm{ch}}(V) \equiv \frac{Q(V)}{V}. \qquad \text{(Chord capacitance)} \qquad (2.13)$$

Equation (2.3) on p. 30 holds for the charge $Q$ of a nonlinear capacitor, too. Note, however, that (2.4), which involves a voltage-independent capacitance $C$, does not hold. Using the incremental capacitance $C_{\mathrm{inc}}(V)$, $Q(V)$ can be written as follows.

$$Q(V) = Q_0 + \int_0^V C_{\mathrm{inc}}(V')\,\mathrm{d}V', \qquad (2.14)$$

where $Q_0$ is the charge when $V = 0$. By differentiating (2.14) with respect to time $t$, we obtain

$$I(t) = \frac{\mathrm{d}Q(t)}{\mathrm{d}t} = \frac{\mathrm{d}Q(V)}{\mathrm{d}V}\frac{\mathrm{d}V(t)}{\mathrm{d}t} = C_{\mathrm{inc}}(V)\frac{\mathrm{d}V(t)}{\mathrm{d}t}. \qquad (2.15)$$

### THE FOURTH BASIC CIRCUIT ELEMENT

Dr. Masamitu Kawakami, who served as President of the Tokyo Institute of Technology and Nagaoka University of Technology, presented in his book on electric circuits [13] a chart that nicely summarized the relationship between physical quantities related to circuits and linear two-terminal circuit elements (Fig. 2.10). He called it the *OK chart* ("O" for *Omoto, Yoshikazu* and "K" for *Kawakami, Masamitu*).

Incidentally, you might have noticed that there is no circuit element that directly connects the charge $Q$ and the magnetic flux $\Phi$ at the bottom. Can we think of a circuit element defined by "a line on the $Q$-$\Phi$ plane"? For linear capacitors, the constitutive relation is (2.2) on p. 29, from which follows $C = Q/V$. Likewise, for linear inductors, $L = \Phi/I$ follows from (2.6) on p. 32. Now, assuming linear elements, the ratio of $Q$ to $\Phi$ is $Q/\Phi = LI/CV$. $L/C$ has the dimensions of the square of resistance (see (3.30) on p. 64), and $I/V$ has the dimensions of the reciprocal of resistance. So, the *linear* circuit element that connects $\Phi$ and $Q$ is the linear resistor.

However, if we extend the discussion to the nonlinear case, the conclusion changes. L. O. Chua noticed that there was no circuit element that connected $Q$ and $\Phi$, and by considering the nonlinear case as well, he discovered a new nonlinear circuit element—*memristor* [7]. A memristor is a circuit element whose characteristics are given by "a curve on the $Q$-$\Phi$ plane." A memristor is a nonlinear resistor with memory and is different from the usual nonlinear resistor defined by "a curve on the $I$-$V$ plane" (§2.2.1). Memristors did not win widespread acceptance for decades after their discovery, but they came into the limelight when it was recognized that memory devices now known as "resistance change memories" were memristors.

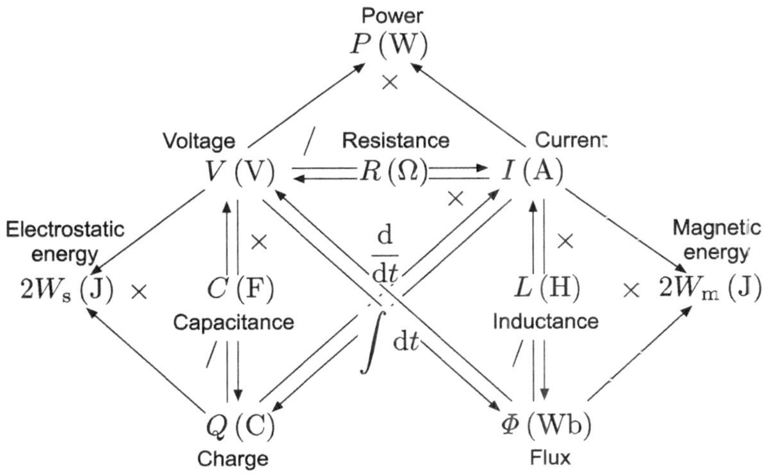


FIGURE 2.10 OK chart.

Today, it is known that there are, in theory, infinitely many possible nonlinear circuit element types other than the memristor. The *memcapacitor* and the *meminductor* are among those that have been named. As for linear elements, it is known that elements called *frequency-dependent negative resistors (FDNRs)* can be considered [5].

Equation (2.15) looks similar to (2.4), but the same form of equation does not hold if $C_{inc}(V)$ is replaced with $C_{ch}(V)$. The distinction between the incremental capacitance and the chord capacitance is critically important.

Similarly, a nonlinear inductor is a circuit element whose characteristics are given by a single (typically curved) line in the $I$-$\Phi$ plane. Equation (2.7) on p. 32 holds for the magnetic flux $\Phi$ of a nonlinear inuctor, too. If $\Phi$ is given by a single-valued function $\Phi(I)$ of current $I$, the incremental inductance (aka small-signal inductance or differential inductance) can be written as

$$L_{inc}(I) \equiv \frac{d\Phi(I)}{dI}. \tag{2.16}$$

The chord inductance is given by

$$L_{ch}(I) \equiv \frac{\Phi(I)}{I}. \tag{2.17}$$

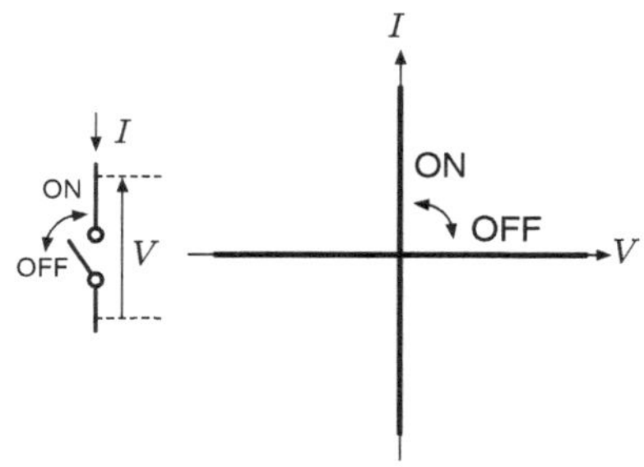


FIGURE 2.11 *I-V* plane characteristics of an ideal switch.

## 2.3 TIME-INVARIANT AND TIME-VARYING CIRCUIT ELEMENTS

The circuit elements we have considered so far are all *time-invariant* circuit elements whose characteristics do not change with time. In contrast, circuit elements whose characteristics change in a time-dependent manner are called *time-varying* circuit elements.

### *Example: Time-Invariant Linear Resistor*

The linear resistor defined by (2.1) on p. 40 is a *linear time-invariant* (LTI) circuit element because the resistance $R$ does not depend on time $t$. ■

### *Example: Ideal Switch*

An ideal switch is a circuit element whose characteristics can be switched between the on and off states at appropriate times. As is clear from Fig. 2.11 (p. 40), at a given instant, the characteristics of the switch are represented by "a straight line passing through the origin in the *I-V* plane," so the switch can be regarded as a *variable linear resistor* or a *linear time-varying* (LTV) resistor. ■

Although a linear time-varying element is a linear element, it can also play the role of a nonlinear element, depending on how its characteristics are varied over time. Such a circuit includes the *FET resistive mixer* [19], which is a kind of frequency mixing circuit. "FET" stands for field-effect transistor (p. 44).

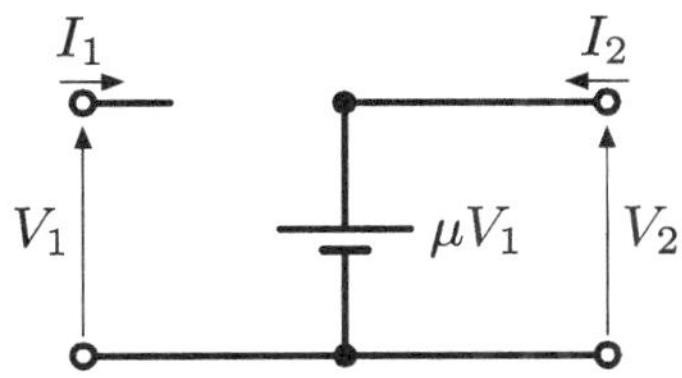


| | | |
|---|---|---|
| Input | $V_1$ | |
| Output | $V_2 = \mu V_1$ | ($\mu$: voltage gain) |
| Input resistance | $\infty$ | ($I_1 = 0$) |
| Output incremental resistance | 0 | |

FIGURE 2.12 Voltage-controlled voltage source.

## 2.4 MULTITERMINAL ELEMENTS AND CONTROLLED SOURCES

A circuit element with three or more terminals is called a *multiterminal circuit element*. Since the general theory of multiterminal elements is very difficult, we will only discuss a type of three-terminal element, known as *controlled sources* or *dependent sources*. Controlled sources are basically voltage or current sources, as the name suggests, but their output voltage or current depends on the input to the control terminal. Since sources are a kind of nonlinear resistor, as we saw in §2.2.1, controlled sources can be regarded as three-terminal variable nonlinear resistors. Controlled sources are often used to describe or *model* the characteristics of semiconductor devices, such as transistors. They are, therefore, important circuit elements when considering semiconductor devices.

*Example: Voltage-Controlled Voltage Source*

The voltage-controlled voltage source (VCVS), shown in Fig. 2.12, outputs a voltage $V_2$ proportional to the input voltage $V_1$. The proportionality coefficient $\mu = V_2/V_1$ is called the *voltage gain.* ■

*Example: Current-Controlled Current Source*

The current-controlled current source (CCCS), shown in Fig. 2.13, outputs a current $I_2$ proportional to the input current $I_1$. The proportionality coefficient $\alpha = I_2/I_1$ is called the *current gain.* ■

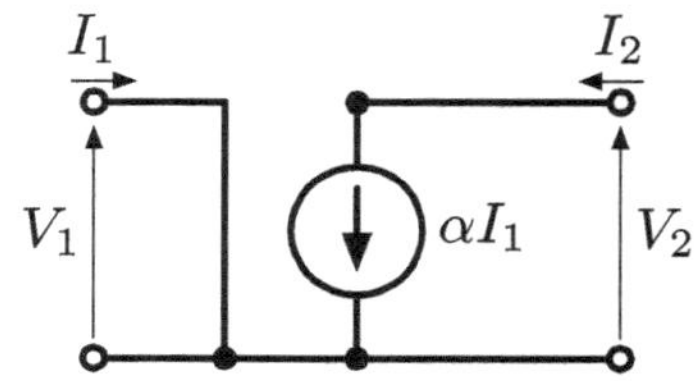


| | | |
|---|---|---|
| Input | $I_1$ | |
| Output | $I_2 = \alpha I_1$ | ($\alpha$: current gain) |
| Input resistance | 0 | ($V_1 = 0$) |
| Output incremental resistance | $\infty$ | |

FIGURE 2.13 Current-controlled current source.

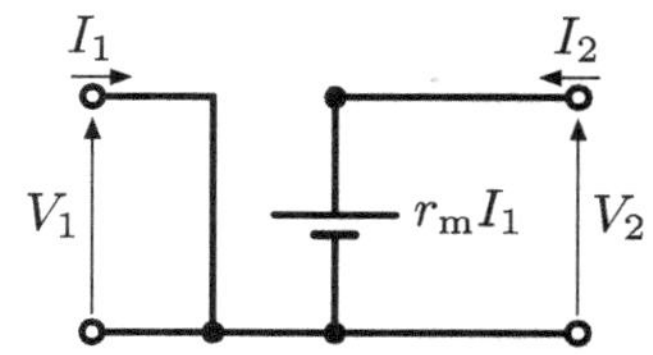


| | | |
|---|---|---|
| Input | $I_1$ | |
| Output | $V_2 = r_m I_1$ | ($r_m$: transresistance) |
| Input resistance | 0 | ($V_1 = 0$) |
| Output incremental resistance | 0 | |

FIGURE 2.14 Current-controlled voltage source.

*Example: Current-Controlled Voltage Source*

The current-controlled voltage source (CCVS), shown in Fig. 2.14, outputs a voltage $V_2$ proportional to the input current $I_1$. The proportionality coefficient $r_m = V_2/I_1$ has the dimensions of resistance and is called the *transresistance*. The subscript “m” comes from “mutual” as in the “mutual inductance” of a transformer. ■

*Example: Voltage-Controlled Current Source*

The voltage-controlled current source (VCCS), shown in Fig. 2.15, outputs a current $I_2$ proportional to the input voltage $V_1$. The proportionality coefficient $g_m = I_2/V_1$ has the dimensions of conductance and is called the *transconductance*.

The $V_2$-$I_2$ characteristics of the VCCS for several values of the control voltage $V_1$ are shown in Fig. 2.16. ■

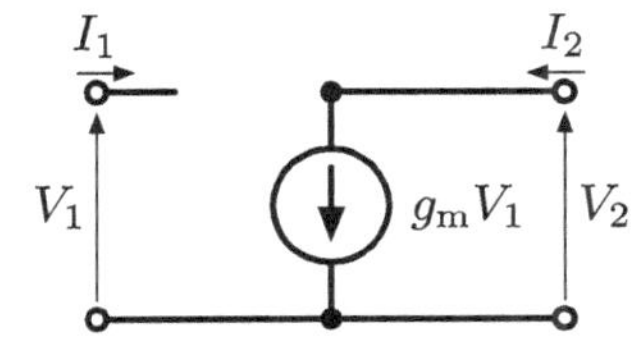


| | | |
|---|---|---|
| Input | $V_1$ | |
| Output | $I_2 = g_m V_1$ | ($g_m$: transconductance) |
| Input resistance | $\infty$ | ($V_1 = 0$) |
| Output incremental resistance | $\infty$ | |

FIGURE 2.15 Voltage-controlled current source.

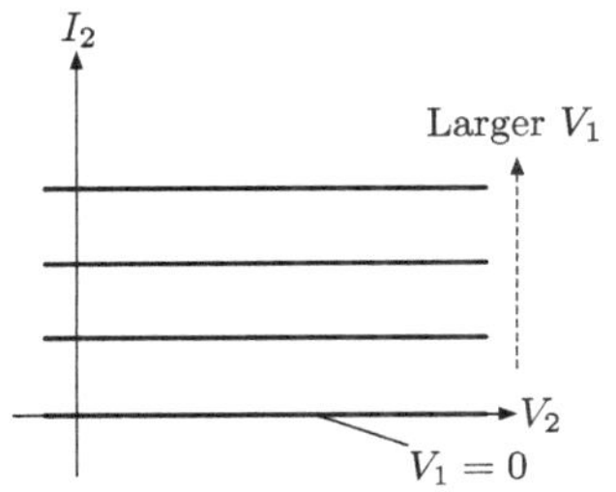


FIGURE 2.16 $V_2$-$I_2$ characteristics of a voltage-controlled current source.

If a DC voltage or current is given as an input to a controlled source, the output is just a DC voltage or current. In general, an input that varies with time is often given to a controlled source. In such a case, the controlled source can be regarded as a time-varying circuit element.

## 2.5 TRANSISTORS

We saw in §2.2.1 that the p-n junction diode, shown in Fig. 2.7 (p. 35), is a real element that approximates the ideal rectifier, shown in Fig. 2.6 (p. 35). Similarly, the controlled sources discussed in §2.4 are ideal elements. The corresponding real elements are three-terminal variable nonlinear resistors whose characteristics can be varied according to the input applied to the control terminal. Of such variable resistors, those made of semiconductors are called *transistors*. This is the circuit-theoretic definition of the transistor.

The word "transistor" is said to be derived from "transresistance," a name that suggests a connection with the CCVS (Fig. 2.14 on p. 42). However, their actual characteristics are more like those of the VCCS

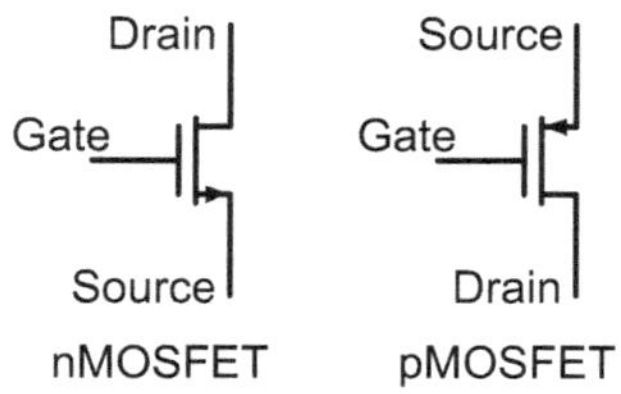


FIGURE 2.17 Schematic symbols of an nMOSFET and a pMOSFET. Other symbols are also in use.

(Fig. 2.15 on p. 43), which suggests that they could have been called "transductors." But history did not unfold that way.

*Example: MOSFETs*

MOS transistors or *MOSFETs* (metal-oxide-semiconductor field-effect transistors) are the most widely used transistors today. Metal-oxide-semiconductor (MOS) comes from their structure (§7.1). MOSFETs are classified as a type of transistor called the *field-effect transistor* (FET). As shown in Fig. 2.17, there are two types of MOSFETs—nMOSFET and pMOSFET. The arrows in the schematic symbols represent the direction of current flow. In nMOSFETs, electrons carry the current. In pMOSFETs, holes carry the current. nMOSFETs and pMOSFETs are often used in a *CMOS* (complementary MOS) configuration that combines both types of MOSFETs.

Between the *drain* and *source* terminals of a FET is a variable nonlinear resistor. The *gate* is the control terminal. Approximate current-voltage characteristics of nMOSFET are shown in Fig. 2.18. $I_{\mathrm{DS}}$ is the current that flows from the drain to the source (or drain-source current). $V_{\mathrm{GS}}$ is the voltage between the gate and the source (or gate-source voltage), and $V_{\mathrm{DS}}$ is the voltage between the drain and the source (or drain-source voltage). Shown in Fig. 2.18 are the transconductance

$$g_{\mathrm{m}} = \frac{\partial I_{\mathrm{DS}}}{\partial V_{\mathrm{GS}}} \qquad \text{(Transconductance)} \qquad (2.18)$$

and the drain-source conductance

$$g_{\mathrm{ds}} = \frac{\partial I_{\mathrm{DS}}}{\partial V_{\mathrm{DS}}}. \qquad \text{(Drain-source conductance)} \qquad (2.19)$$

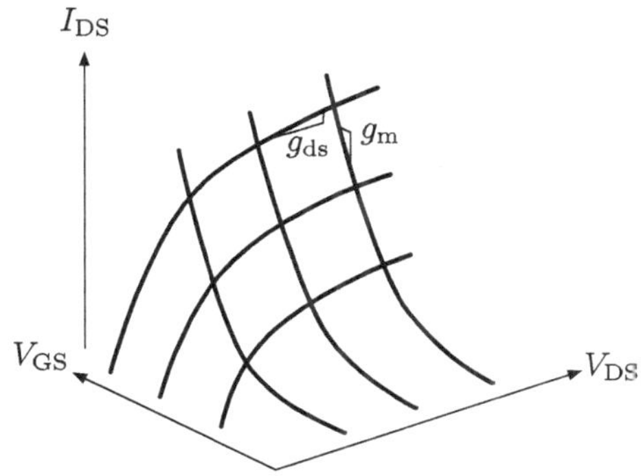


FIGURE 2.18 Characteristics of an nMOSFET.

Since $I_{DS}$ depends both on $V_{GS}$ and $V_{DS}$ as is clear from Fig. 2.18, (2.18) and (2.19) are partial derivatives. In the region where $g_{ds}$ is close to 0, the characteristics of the nMOSFET are similar to those of the VCCS (Fig. 2.16 on p. 43). ■

*Example: Bipolar Transistors*

Bipolar transistors had been the mainstay of transistors before MOSFETs took over. In the past, the word "transistor" referred to a bipolar transistor. Low power consumption and large-scale integration are more difficult to achieve with bipolar transistors than with MOSFETs, so they are no longer used as much as the latter in integrated circuits. However, they are superior to MOSFETs in terms of gain, noise, and high-frequency performance, and are used in situations where these are important. Between the *collector* and *emitter* electrodes is the variable nonlinear resistor of a bipolar transistor. The *base* is the control terminal. The word "bipolar" refers to the fact that both electrons and holes are involved in the operation of these transistors. As shown in Fig. 2.19, there are two types of bipolar transistors, npn and pnp, which are made by combining n-type and p-type semiconductors. The arrows in the schematic symbols represent the direction of current flow. Current-voltage characteristics of bipolar transistors are similar to those of MOSFETs. ■

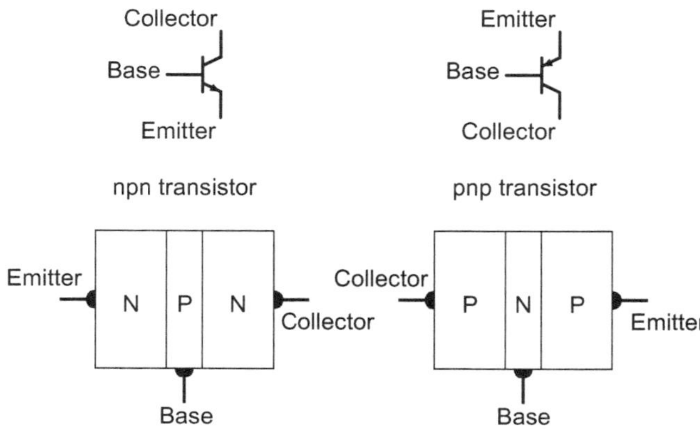


FIGURE 2.19 Schematic symbols and structures of npn and pnp bipolar transistors.

## 2.6 CIRCUIT-THEORETIC POSITIONING OF SEMICONDUCTOR DEVICES

The semiconductor devices mentioned so far are limited to p-n junction diodes, MOSFETs, and bipolar transistors. These and all other semiconductor devices are nonlinear circuit elements. In other words, semiconductor devices are nonlinear circuit elements made of semiconductors.

Not all nonlinear elements need to be made from semiconductors, as far as their functions in a circuit are concerned. In fact, devices with characteristics similar to those of semiconductor devices can be made as vacuum tubes and the like. However, the use of semiconductors as materials for nonlinear circuit elements has resulted in tremendous advances in both the devices themselves and the resulting circuits (§1.4).

Compared with vacuum tubes made of metal and glass (and vacuum), semiconductor devices have the following remarkable characteristics:

- Sizes of devices and circuits: Semiconductor devices can be made very small and overall circuit dimensions can also be very small.
- Ease of manufacturing: Printing-based manufacturing technology makes it easy to manufacture ICs consisting of a large number of devices.
- Operating voltage: Vacuum tubes require tens of volts, but semiconductor devices can operate at one to several volts.

- Performance: Small means that it takes less time for a signal to pass through a device or a circuit, so it can operate faster and perform better.
- Power consumption: Small size and low supply voltage mean lower current flow and lower power consumption.
- Price: For very small transistors, the price per transistor is very low, which could also make ICs low cost. Note, however, that some optical devices for optical communications and large devices for power electronics are quite expensive.

If some novel material is discovered, it may be worthwhile to consider whether it has the above advantages.

**LINEAR CIRCUIT ELEMENTS MADE OF SEMICONDUCTORS**

We explained that a semiconductor device is a "nonlinear circuit element made of semiconductor." Then, are there any linear circuit elements made of semiconductors?

In silicon ICs, use is often made of resistors made of polycrystalline silicon (§1.3.1). These are meant as linear circuit elements made of semiconductors, but not many people bother to call them semiconductor devices. However, if we want to investigate their characteristics in detail, we would have to treat them as semiconductor devices. As a matter of fact, resistors of this kind show some nonlinearity. So strictly, they too are nonlinear circuit elements after all.

## 2.7 SUMMARY

In this chapter, we considered what semiconductor devices are from a circuit-theoretic viewpoint.

- Introductory courses in "circuit theory" mainly cover linear circuit theory.
- "Incremental resistance" and "chord resistance" must be considered for the quantitative discussion of nonlinear resistors.
- Voltage and current sources are nonlinear circuit elements.
- Controlled sources are nonlinear, nonreciprocal circuit elements.

- Diodes are nonlinear resistors with rectifying action.
- Transistors are three-terminal variable nonlinear resistors made of semiconductors.
- Semiconductor devices are nonlinear circuit elements made of semiconductors.

**CLASSIFICATION OF CIRCUIT ELEMENTS**

In the above, we classified circuit elements from the following perspectives:

- Linear versus nonlinear
- Time-invariant versus time-varying
- Two-terminal versus multiterminal

Other perspectives include:

- Passive versus active

A *passive* circuit element is a circuit element that may consume part or all of the power it receives from a circuit. If elements that consume no power at all and those that consume at least some power are to be distinguished, passive elements can be classified further:

- Lossless versus lossy

Capacitors and inductors are lossless elements. Both of them are also *reactive* elements. Reactive elements can store energy received from a circuit. All reactive elements are lossless, but not all lossless elements are reactive. A three- or four-terminal element known as the *gyrator* is known to be lossless but is not reactive.

An *active* circuit element is a circuit element that may supply power to a circuit. The simplest examples are voltage and current sources. A more complicated example is the transistor. Typically, a transistor is powered by DC, and a portion of that power is used to amplify an AC signal applied to it. The transistor consumes net power as it amplifies AC components of the received signal, but if the DC component is ignored, it appears as if the transistor were supplying AC power to the circuit. Transistors are usually regarded as active elements in this sense.

The term "active" is sometimes used even when a transistor is performing purely passive operation, that is, dissipating power. For example, when a transistor is used as a load resistor in an amplifier circuit, it is sometimes called an "active load." However, this is essentially a misnomer and should really be called a "nonlinear" or "variable" load.
Multiterminal circuits can be classified according to:

- Reciprocal versus nonreciprocal

The gyrator is a lossless *nonreciprocal* element. *Reciprocal* elements can be composed of two-terminal elements, whereas nonreciprocal elements cannot be composed of two-terminal elements. In this sense, only nonreciprocal elements are genuine multiterminal elements. Controlled sources and transistors are nonreciprocal elements.

## 2.8 PROBLEMS

2.1 Draw the characteristics of an ideal rectifier on a plane with current $I$ on the horizontal axis and voltage $V$ on the vertical axis. Give the formula of the incremental conductance $G_{\text{inc}}(V)$.

2.2 The magnetic flux penetrating a nonlinear inductor is given by a single-valued function $\Phi(I)$ of current $I$. Find an equation relating the voltage $V(t)$ to the current $I(t)$ corresponding to (2.8) on p. 32 for a linear inductor. Hint: See (2.15) on p. 37.

2.3 Which of the following are applicable to nonlinear circuits?

- Synthesis formulas for series- and parallel-connected impedances and admittances
- Principle of superposition
- Equivalent source theorems (Thévenin's theorem and Norton's theorem)
- Reciprocity theorem
- Laplace transform
- Kirchhoff's voltage law (KVL)
- Kirchhoff's current law (KCL)

CHAPTER 3

# Waves in Periodic Structures

Electrons are regarded as particles in elementary physics. In this book, we present the physics of semiconductor devices adopting such a view. However, many of the physical properties of crystals are explained as a result of the interaction between electrons as waves and periodically arranged atoms. Treating electrons as waves requires quantum mechanics, on which modern solid-state physics is built. Since quantum mechanics is beyond the scope of this book, this chapter instead uses circuit theory to study the properties of periodic structures and the formation of energy bands, already mentioned in Chapter 1. This is because the formation of energy bands is not a phenomenon unique to quantum mechanics but a universal phenomenon that can be observed in electromagnetism and circuit theory, too, as long as there is a periodic structure.

## 3.1 ANALOGIES IN PHYSICS

### 3.1.1 Commonality of Mathematical Structures

Many problems that appear in science and engineering have a common or similar mathematical structure. For example, the motion of a weight hanging from a spring and the behavior of an LC resonator (Fig. 3.1) are described by equations of exactly the same form. If we can solve the problem of oscillations of the weight, we can obtain the answer for the LC resonator by replacing variables according to Table 3.1, and vice versa. As an example, the momentum for the mechanical system shown in Fig. 3.1 is $p = m\dot{x}$, whereas the corresponding equivalent "momentum" for the electrical system is $Q = C\dot{\Phi}$ (the superscript


DOI: 10.1201/ 9781003439417-3

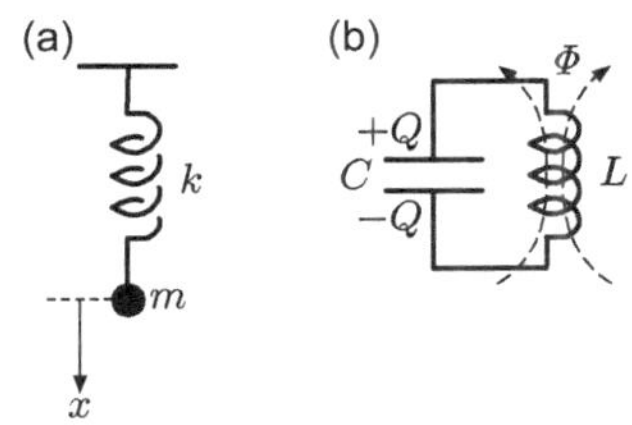


FIGURE 3.1 (a) A weight hanging from a spring. (b) An LC resonator.

TABLE 3.1 Correspondence between Mechanical and Electrical Oscillations

| | Mass | Spring constant | Coordinate | Momentum | Angular Frequency |
|---|---|---|---|---|---|
| Sprung mass point | $m$ | $k$ | $x$ | $p$ | $\sqrt{k/m}$ |
| LC resonator | $C$ | $1/L$ | $\Phi$ | $Q$ | $1/\sqrt{LC}$ |

TABLE 3.2 Forms of Differential Equation

| Field | Equation | Spatial derivative | Time derivative |
|---|---|---|---|
| Distributed circuits | Wave equation | 2nd order | 2nd order |
| Electromagnetism | Wave equation | 2nd order | 2nd order |
| Thermal physics | Diffusion/heat equation | 2nd order | 1st order |
| Quantum mechanics | Schrödinger equation | 2nd order | 1st order |

dot denotes the time derivative d/d$t$). The latter equation might appear unfamiliar, but from (2.7) on p. 32, $\dot{\Phi} = V$ is the voltage.

The methodology of treating seemingly different systems in a unified fashion has been systematized in physics and engineering [26, 36]. Correspondence relationships, as in Table 3.1, have been studied extensively and put to practical use. For example, problems in electromechanical systems could be transformed into problems in electric circuits and be solved by using circuit simulators.

In some cases, perfect correspondence may not be found, but still some similarity could be found between equations describing different systems. As an example, Table 3.2 shows the forms of differential equations that appear in different fields.

The differential equations in Table 3.2 all reduce to a differential equation with a second-order spatial derivative in the steady state,

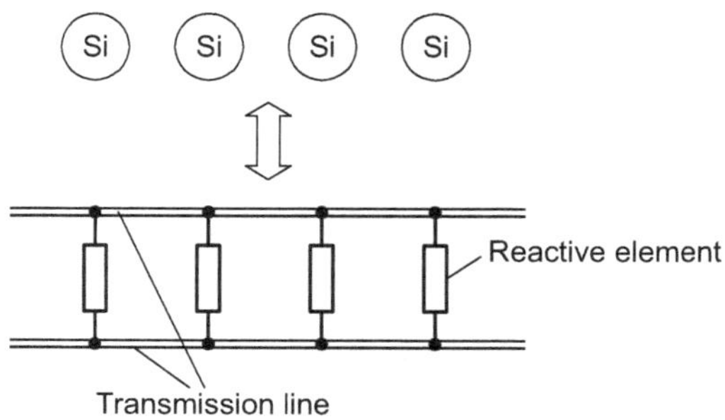


FIGURE 3.2 Similarity between one-dimensional crystals and the periodically loaded transmission line.

in which the time derivative equals zero. As a result, the following becomes possible, for example:

- Consider the properties of crystalline solids by looking at periodic networks as shown in Fig. 3.2.
- Solve problems in quantum mechanics as particle diffusion problems. In this case, the time derivative does not have to be zero.

### 3.1.2 Overview of the Chapter

In this chapter, we will look at energy band formation, an important concept in solid-state physics, by using the analogy described above. Although the band theory of solids is constructed using quantum mechanics, this book does not assume that the reader has learned quantum mechanics. Therefore, we investigate band formation using linear circuit theory (especially AC circuit theory), which the reader should have already learned. From §3.2.3 onward, transmission line theory or distributed circuit theory is also used. A basic explanation of transmission lines is included as well. It should be emphasized that this analogy is backed by equations having the same form. It should also be noted, however, that circuits do not enable us to understand every aspect of solid-state physics other than band formation.

Most solid crystals are three-dimensional, but for simplicity, we consider only one-dimensional periodic networks in this chapter (Fig. 3.2). This is because if you understand energy bands in one dimension, three-dimensional crystals are conceptually not much different (although mathematically much more complicated).

In Chapter 1, we saw that the electron energy $E$ has ranges of allowed values (i.e., allowed bands) and a range of disallowed values (i.e., the forbidden band) as shown in Fig. 1.6 (p. 12). In the same way, in this chapter, we will see that periodic networks have frequency ranges in which wave propagation is allowed (passband) and frequency ranges in which it is not allowed (stopband). Since frequency is related to energy, we see a connection to solid-state physics.

In what follows, we will first learn that the periodic networks called the *LC ladder* and the *CL ladder* have the characteristics shown in Figs. 3.3(a) and (b)—one forbidden band and one allowed band (§3.2.2). In the figure, $\omega_c$ is the angular frequency at the boundary between the stopband and the passband. At a certain limit, the LC ladder becomes a lossless transmission line and $\omega_c \to \infty$, as shown in Fig. 3.3(c) (§3.2.3). However, by introducing a new periodic structure to the lossless transmission line, as shown in Fig. 3.2, the characteristics shown in Figs. 3.3(d) and (e) emerge. The combination of "an allowed band on top of a forbidden band on top of another allowed band" in Figs. 3.3(d) and (e) corresponds to Fig. 1.6 (p. 12).

Incidentally, when we speak of "periodic structures and waves" in a narrow—or even ordinary—sense, we usually consider the case in which periodicity is introduced into a uniform medium (in this chapter, a lossless transmission line), taking the existence of such media for granted. The resulting band structures are shown in Figs. 3.3(d) and (e). However, in elementary circuit theory (i.e., lumped circuit theory), on which this chapter is initially based, the existence of a medium in which waves can propagate spatially is not obvious, given the fact that lumped circuits are spatially zero-dimensional. We, therefore, start by building a uniform one-dimensional medium by periodically arranging lumped circuit elements (§3.2.1 and §3.2.2). The resulting band structures are shown in Figs. 3.3(a) and (b). The physical origin of the forbidden bands in Figs. 3.3(a) and (b) may therefore be different from the origin of the forbidden bands in Figs. 3.3(d) and (e) (§3.5).

Furthermore, even in the cases of Figs. 3.3(d) and (e), what is happening physically is different between the cases in which

$$\text{(wavelength)} \lesssim \text{(period)} \tag{3.1}$$

and the cases in which

$$\text{(wavelength)} \gg \text{(period).} \tag{3.2}$$

In the former, Bragg diffraction occurs (§3.5). What plays an important role here is the "reduction of translational symmetry" due

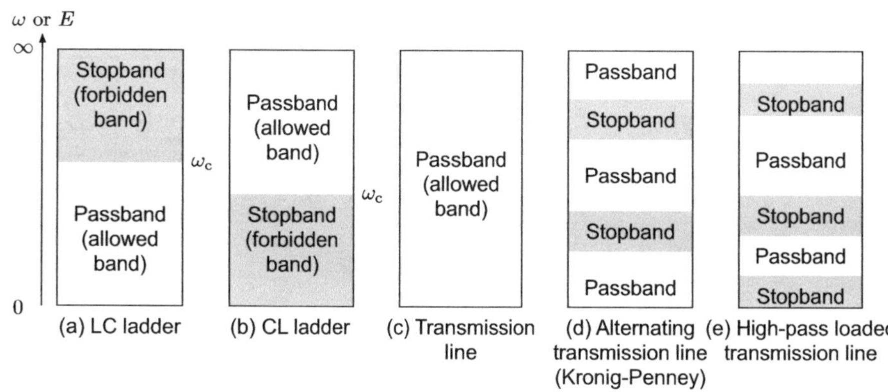


FIGURE 3.3 Energy band structures of periodic networks.

to the introduction of periodicity into a uniform medium. In the latter, on the other hand, the periodicity is not directly visible from the wave's point of view because the period is much smaller than the wavelength. However, this, of course, does not mean that nothing happens. The latter is related to the fact that macroscopic material parameters such as permittivity and magnetic permeability can be considered on a length scale (or for electromagnetic wavelengths) much larger than the distance between neighboring atoms in a crystal. On a subatomic scale, however, the space between the nucleus of an atom and the electrons surrounding it is a vacuum. In short, even when we are discussing seemingly the same thing for periodic structures, there can be different things happening physically.

## 3.2 PROPERTIES OF PERIODIC NETWORKS

In elementary solid-state physics, energy bands of infinitely large crystals are usually considered. There are actually two possible ways to look at the properties of a periodic one-dimensional (1D) network.

- Consider an infinitely long periodic network.
- Consider a periodic network consisting of $N(= 1, 2, 3, \cdots)$ unit cells, and make $N$ large.

In both cases, it turns out that (as $N$ becomes large) frequency region(s) in which waves can propagate and frequency region(s) in which waves cannot propagate appear.

### 3.2.1 Infinitely Long Ladder Networks

As a start to investigate the properties of periodic networks, let us examine the input impedance $Z_{\text{in}}$ of a (semi-)infinitely long ladder network, shown in Fig. 3.4. Note that both $Z$ and $Y$ are passive, and hence the real parts of the impedance $Z$ and the admittance $Y$ are assumed to be nonnegative.

$$\Re(Z) \geq 0, \tag{3.3}$$

$$\Re(Y) \geq 0. \tag{3.4}$$

Note that $\Re(Z)$ denotes the real part of the complex number $Z$, and $\Im(Z)$ is its imaginary part.

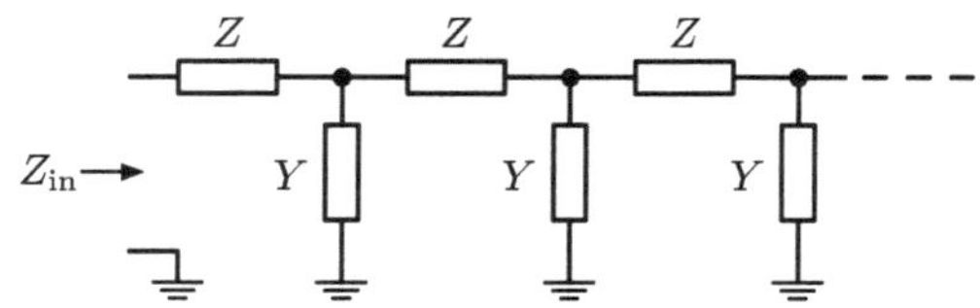


FIGURE 3.4 A ladder network consisting of series impedance $Z$ and shunt admittance $Y$.

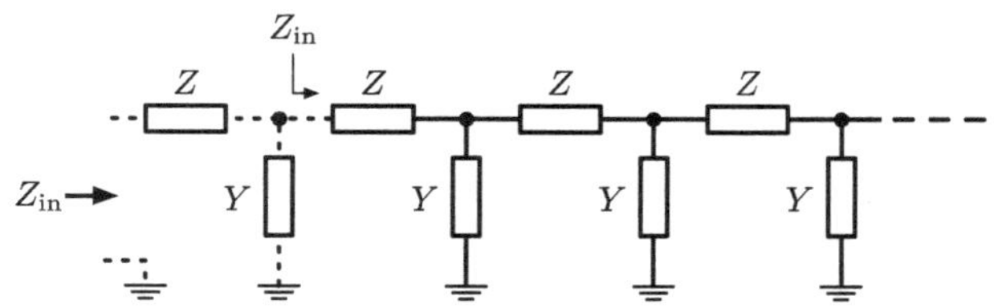


FIGURE 3.5 Shunt $Y$ and series $Z$ are added to the left of the ladder network in Fig. 3.4.

To find $Z_{in}$, note that when an L-shaped network consisting of $Y$ and $Z$ is added, as shown in Fig. 3.5, to the left of the infinite ladder network in Fig. 3.4, the input impedance $Z_{in}$ remains unchanged. This situation can be expressed as the following recurrence formula:

$$Z_{in} = Z + \left(\frac{1}{Y} \parallel Z_{in}\right) = Z + \frac{Z_{in}/Y}{1/Y + Z_{in}}, \quad \text{(Recurrence formula)} \quad (3.5)$$

where $\parallel$ represents parallel connection. From (3.5), $Z_{in}$ is found to be

$$\begin{aligned} Z_{in} &= \frac{Z \pm \sqrt{Z^2 + 4Z/Y}}{2} \\ &= \frac{Z}{2}\left(1 \pm \sqrt{1 + \frac{4}{ZY}}\right). \quad \text{(Iterative impedance of L-network)} \end{aligned} \quad (3.6)$$

By the passivity assumptions (3.3) and (3.4), the solution with $\Re(Z_{in}) \geq 0$ must be chosen. Equation (3.6) is known as the left-hand-side *iterative impedance* or *Bloch impedance* [9] of the L-network.[1]

[1] The right-hand-side iterative impedance of the L-network in Fig. 3.5 would be the input impedance looking leftward into the ladder network from its right end.

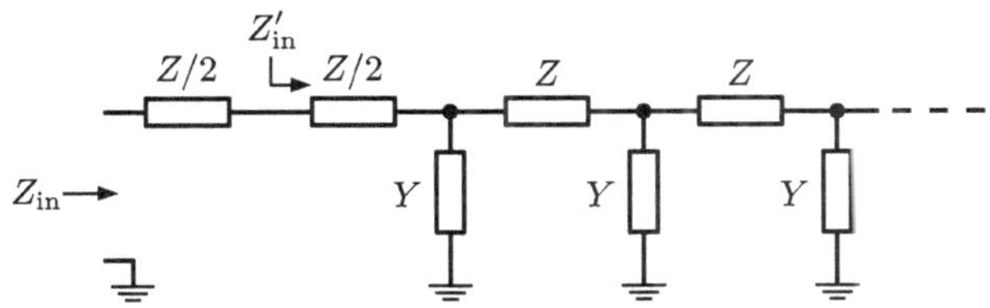


FIGURE 3.6 The leftmost $Z$ in the ladder network in Fig. 3.4 is split into $Z/2 + Z/2$.

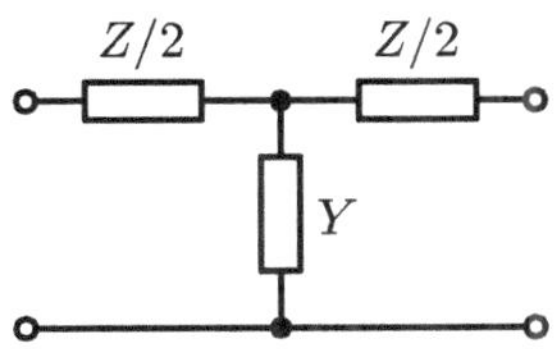


FIGURE 3.7 Symmetric T-network.

Before making the series impedance $Z$ and the shunt admittance $Y$ more specific, let us explain why the term $Z/2$ appears in (3.6). The network in Fig. 3.4 can be rewritten as shown in Fig. 3.6 by splitting the leftmost $Z$ into two series-connected $Z/2$. The first term of (3.6) corresponds to the leftmost $Z/2$ in Fig. 3.6, and it can be regarded as an extra element.

If the symmetric T-network shown in Fig. 3.7 had been infinitely cascaded, $Z/2$ in (3.6) would not have appeared. The essential term originating from the infinite periodic network, therefore, is only the second term of (3.6). Thus, let us consider from here on

$$Z'_{in} \equiv Z_{in} - \frac{Z}{2}, \quad \text{(Iterative impedance of symmetric T-network)} \tag{3.7}$$

shown in Fig. 3.6.

### 3.2.2 Infinitely Long LC Ladders

Let us consider more concretely the input impedance $Z'_{in}$ of the semi-infinitely long ladder network (Fig. 3.6), assuming that the series elements are linear inductors and the shunt elements are linear capacitors, as shown in Fig. 3.8.

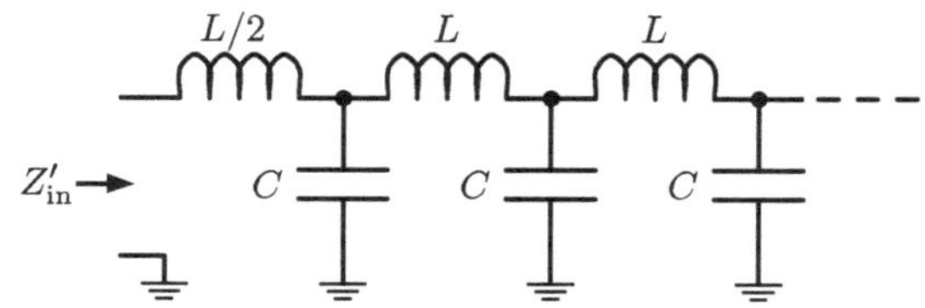


FIGURE 3.8 Infinitely long LC ladder.

By making the substitutions $Z = \mathrm{j}\omega L$ and $Y = \mathrm{j}\omega C$ in (3.7) and (3.6), we obtain

$$Z'_{\mathrm{in}} = \pm\frac{\mathrm{j}\omega L}{2}\sqrt{1-\frac{4}{\omega^2 LC}}. \tag{3.8}$$

Let the angular frequency at which the inside of the square root in (3.8) equals 0 be $\omega_{\mathrm{c}}$. From $1 - 4/\left(\omega_{\mathrm{c}}^2 LC\right) = 0$,

$$\omega_{\mathrm{c}} = \frac{2}{\sqrt{LC}}.\ \text{(Cutoff angular frequency of LC ladder)} \tag{3.9}$$

$\omega_{\mathrm{c}}$ is called the *cutoff angular frequency*. Equation (3.8) can be rewritten using $\omega_{\mathrm{c}}$ as follows.

$$Z'_{\mathrm{in}} = \pm\frac{\mathrm{j}\omega L}{2}\sqrt{1-\left(\frac{\omega_{\mathrm{c}}}{\omega}\right)^2}. \tag{3.10}$$

Since an imaginary unit j appears in (3.10), $Z'_{\mathrm{in}}$ is purely imaginary if the inside of the square root is positive.

$$\Re\left(Z'_{\mathrm{in}}\right) = 0\left(|\omega| > \omega_{\mathrm{c}}\right). \tag{3.11}$$

Conversely, if inside the square root of (3.10) is negative, $Z'_{\mathrm{in}}$ is real. As was noted below (3.6), a sign must be chosen from the double sign $\pm$ such that $\Re\left(Z'_{\mathrm{in}}\right) \geq 0$,

$$\Re\left(Z'_{\mathrm{in}}\right) > 0 \text{ and } \Im\left(Z'_{\mathrm{in}}\right) = 0\left(-\omega_{\mathrm{c}} < \omega < \omega_{\mathrm{c}}\right). \tag{3.12}$$

Here, we did not exclude the possibility of $\omega < 0$ in (3.11) and (3.12). In communication engineering and signal processing, both positive and negative frequencies are considered. However, we can forget about $\omega < 0$ in this book.

Let us now consider what (3.11) and (3.12) mean from the viewpoint of waves in periodic structures. Impedance in AC circuit theory

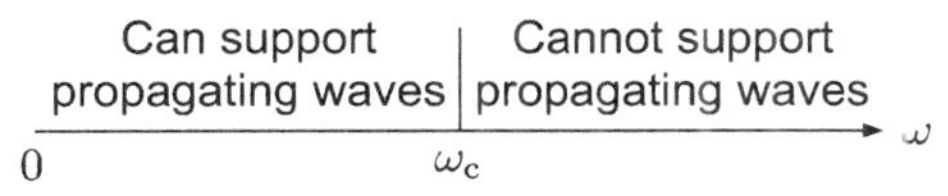


FIGURE 3.9 Frequency characteristics of an infinite LC ladder.

represents how difficult it is for the sinusoidal AC current of angular frequency $\omega$ to flow. $Z'_{in}$ being purely imaginary, as in (3.11), implies that the LC ladder shown in Fig. 3.8 appears reactive and lossless like an inductor or a capacitor (see the Box on p. 48). This may seem obvious because the LC ladder consists of inductors and capacitors. A reactive circuit element receives and emits energy from/to a circuit during a sinusoidal period but does not consume any energy. It only stores energy temporarily. The fact that $Z'_{in}$ is reactive suggests that no energy flows *steadily* into the semi-infinitely long LC ladder (Fig. 3.8). If a traveling wave with frequency $\omega > \omega_c$ is incident on the left end of the ladder, it will be completely reflected back. In other words, waves with $\omega > \omega_c$ cannot propagate along the LC ladder (Fig. 3.9).

In the frequency range given in (3.12), $Z'_{in}$ is positive real. Curiously, $Z'_{in}$ can actually have a nonzero real part, given the fact that the LC ladder consists only of inductors and capacitors. As a matter of fact, it is known to be impossible to produce an impedance with a nonzero real part from a *finite* number of reactive elements. The reason $\Re\left(Z'_{in}\right) > 0$ in (3.12) is that the LC ladder in Fig. 3.8 consists of *infinitely many* reactive elements. Although none of the inductors and capacitors consume energy, a sinusoidal wave with a frequency $(0 <)\omega < \omega_c$, impinging on the left end of the LC ladder, will enter the ladder and travel rightward indefinitely (Fig. 3.9). The energy associated with the wave is also carried away from the entrance, deep into the LC ladder. As a result, the input impedance of a semi-infinitely long LC ladder becomes $Z'_{in} = \Re\left(Z'_{in}\right) > 0$, as if the ladder were a linear resistor. Since only waves with $\omega < \omega_c$ can propagate, LC ladders are said to have *low-pass characteristics*. See also Fig. 3.3(a) (p. 54).

According to the quantum theory of electromagnetic fields, an electromagnetic wave of frequency $\nu$ consists of photons, each having energy $h\nu$, where $h \equiv 6.62607015 \times 10^{-34}$ J/Hz is the Planck constant. In quantum theory, $h$ divided by $2\pi$ rad ($\hbar \equiv h/2\pi$) is called the reduced Planck constant or Dirac constant. $\hbar$ is used more often than $h$. The relationship between $h$ and $\hbar$ is the same as that between angular frequency $\omega$ and frequency $\nu = f \equiv \omega/2\pi$. The photon energy can, therefore, be written as

$$h\nu = \hbar\omega. \qquad \text{(Photon energy)} \tag{3.13}$$

Voltage and current waves in electric circuits can be regarded as a form of electromagnetic waves. Rephrasing (3.11) and (3.12) in energy terms, waves that can propagate along an infinite LC ladder are only those consisting of photons with energy $\hbar\omega < \hbar\omega_c$.

In any periodic network, only waves of a certain frequency or energy range(s) can propagate. In the same way, electrons as waves can exist steadily in a crystal only if they have a certain range(s) of energy. Corresponding to the energy ranges in which electron waves can exist are the allowed bands (valence and conduction bands) mentioned in §1.3.4. However, the existence of an electron wave of a certain energy does not necessarily mean that electrons of that energy contribute to electrical conduction. The reasons for this are beyond the scope of this book, but most of the electrons in the valence band do not contribute to electrical conduction.

### 3.2.3 Lossless Transmission Lines

Those readers who are familiar with transmission lines should have noticed that the discussion in §3.2.2 was closely related to the *lossless transmission line*. For readers who have not yet studied transmission lines, a *transmission line* is simply a wiring consisting of two conducting wires in pairs. It is called a transmission line when the wiring length is not negligible compared with the wavelength. If the waveform is not sinusoidal and it is difficult to think of a wavelength, then if the time it takes for any wave front to pass through the wiring is not negligible in the sense that the two ends of the wiring may have different voltages or currents simultaneously (with a preferred time resolution), then the wiring must be considered a transmission line.

Circuit theoretic treatment of transmission lines in transmission line theory is also known as *distributed circuit theory*. The ordinary elementary circuit theory is also known as *lumped circuit theory*. In a lumped circuit, every conductor is considered to have an equipotential surface and, therefore, is treated as a single *node*. The dimensions of any node in a lumped circuit are assumed infinitesimal. In contrast, since a transmission line has a length not negligible compared

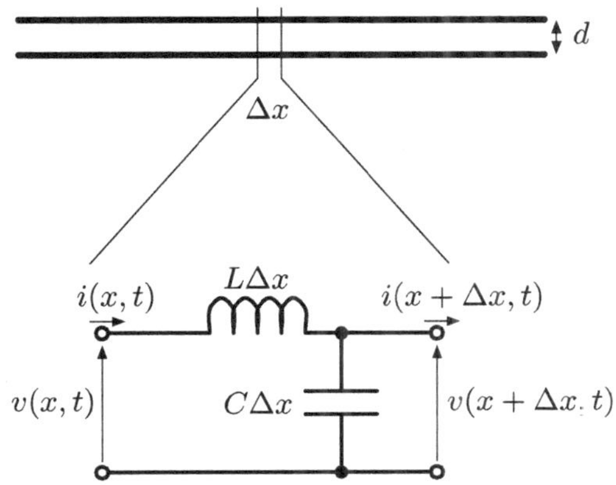


FIGURE 3.10 Model of a short section, Δx, of lossless transmission line.

with the wavelength, the surface of each of the conducting wires constituting the line is not necessarily equipotential. Thus, the voltage difference between the two conducting wires that constitute a transmission line may depend on the position. Kirchhoff's laws can be applied to lumped circuits, but these are generally not applicable to distributed circuits, including transmission lines. In the following, we will derive the differential equations that govern the waves on a lossless transmission line.

An ordinary lossless transmission line can be modeled as an LC ladder as shown in Fig. 3.10. Note that $L$ and $C$ in Fig. 3.10 are per-unit-length inductance and capacitance, respectively. $L\Delta x$ and $C\Delta x$ are the series inductance and the shunt capacitance for the *short* section of length $\Delta x$. Let us suppose the short section in Fig. 3.10 is so small in size that it can be regarded as a lumped circuit, and therefore that Kirchhoff's laws are applicable.

Applying Kirchhoff's voltage law to the circuit in Fig. 3.10, we obtain

$$v(x,t) - L\Delta x\frac{\partial i(x,t)}{\partial t} - v(x+\Delta x,t) = 0. \tag{3.14}$$

This can be rearranged as follows:

$$\frac{v(x+\Delta x,t) - v(x,t)}{\Delta x} = -L\frac{\partial i(x,t)}{\partial t}. \tag{3.15}$$

Similarly, applying Kirchhoff's current law, we obtain

$$i(x,t) - C\Delta x \frac{\partial v(x+\Delta x,t)}{\partial t} - i(x+\Delta x,t) = 0. \tag{3.16}$$

Rearrangement gives

$$\frac{i(x+\Delta x,t) - i(x,t)}{\Delta x} = -C\frac{\partial v(x+\Delta x,t)}{\partial t}. \tag{3.17}$$

Taking the limit of $\Delta x \to 0$ in (3.15) and (3.17), the following simultaneous differential equations are obtained.

$$\frac{\partial v(x,t)}{\partial x} = -L\frac{\partial i(x,t)}{\partial t}, \tag{3.18}$$

$$\frac{\partial i(x,t)}{\partial x} = -C\frac{\partial v(x,t)}{\partial t}. \tag{3.19}$$

Equations (3.18) and (3.19), taken together, are known as *telegrapher's equations*. Both $v(x,t)$ and $i(x,t)$ appear in (3.18) and (3.19), and these differential equations are not convenient as they are. It is, actually, possible to derive separate differential equations for $v(x,t)$ and $i(x,t)$ as follows (see Problem 3.2 on p. 85):

$$\frac{\partial^2 v(x,t)}{\partial x^2} = LC\frac{\partial^2 v(x,t)}{\partial t^2}, \quad \text{(Wave equation for voltage)} \tag{3.20}$$

$$\frac{\partial^2 i(x,t)}{\partial x^2} = LC\frac{\partial^2 i(x,t)}{\partial t^2}. \quad \text{(Wave equation for current)} \tag{3.21}$$

Equations (3.20) and (3.21) are the wave equations for voltage $v(x,t)$and current $i(x,t)$, respectively (see Table 3.2 on p. 51).

All the equations above are *time-domain* equations. These equations are applicable regardless of the waveform. However, if the waveform is limited to a sinusoid, then we can derive *frequency-domain* equations by simply replacing the time derivative $\partial/\partial t$ with $\mathrm{j}\omega$. Let us use capital letters $V$ and $I$ to represent frequency-domain voltage and current *phasors*, respectively. Note that phasors are complex-valued. The frequency-domain telegrapher's equations are given by

$$\frac{\mathrm{d}V(x)}{\mathrm{d}x} = -\mathrm{j}\omega L I(x), \tag{3.22}$$

$$\frac{dI(x)}{dx} = -j\omega CV(x). \tag{3.23}$$

There are no longer are any partial differential operators in (3.22) and (3.23) because $t$ disappeared. Noting that $\partial^2/\partial t^2$ in (3.20) and (3.21) are to be replaced with $(j\omega)^2$, the frequency-domain wave equations are

$$\frac{d^2V(x)}{dx^2} = \beta^2 V(x), \qquad \text{(Wave equation for voltage)} \tag{3.24}$$

$$\frac{d^2I(x)}{dx^2} = \beta^2 I(x), \qquad \text{(Wave equation for current)} \tag{3.25}$$

where

$$\beta \equiv \omega\sqrt{LC} \qquad \text{(Phase constant)} \tag{3.26}$$

is the *phase constant*. We will see shortly that $\beta$ represents the phase rotation per unit length of a sinusoid of angular frequency $\omega$. $\beta$ is related to wavelength $\lambda$ as follows.

$$\lambda = \frac{2\pi}{\beta} = \frac{2\pi}{\omega\sqrt{LC}}. \qquad \text{(Wavelength)} \tag{3.27}$$

In physics, per-unit-length phase rotation is called the *wave number* or *angular wave number* and is usually denoted by $k(= 2\pi/\lambda)$, instead of $\beta$. The difference between the wave number and the phase constant is that the former may be a vector: $\mathbf{k} = (k_x, k_y, k_z)$.

The wave equations (3.24) and (3.25), respectively, have solutions of the form

$$V(x) = V^+e^{-j\beta x} + V^-e^{j\beta x}, \tag{3.28}$$

$$I(x) = \frac{V^+}{Z_0}e^{-j\beta x} - \frac{V^-}{Z_0}e^{j\beta x}. \tag{3.29}$$

The first term of (3.28) represents a *voltage traveling wave phasor* propagating in the direction of positive $x$ (rightward in Fig. 3.10). The second term of (3.28) represents a voltage traveling wave phasor propagating in the opposite direction. $V^+$ represents the amplitude phasor of the rightward voltage traveling wave and $V^-$ represents that of the leftward voltage traveling wave.[2] $V^+$ and $V^-$ are independent of the position $x$.

In (3.29),

$$Z_0 = \sqrt{\frac{L}{C}} \qquad \text{(Characteristic impedance of lossless line)} \qquad (3.30)$$

is called the *characteristic impedance*. The first term of (3.29) is a *current traveling wave phasor* propagating in the positive direction of $x$, and $I^+ \equiv V^+/Z_0$ is its position-independent amplitude. The second term of (3.29) is a current traveling wave phasor propagating in the opposite direction and $I^- \equiv -V^-/Z_0$ is its position-independent amplitude. Note that $V^\pm$ and $I^\pm$ are complex-valued phasors, although their arguments, $\angle V^\pm$ and $\angle I^\pm$, may well be 0.

Taking the ratios of voltage and current traveling wave phasors from (3.28) and (3.29) propagating in the same direction, we see

$$Z_0 \equiv \frac{V^+}{I^+} = -\frac{V^-}{I^-}. \qquad \text{(Characteristic impedance)} \qquad (3.31)$$

Equation (3.31) indicates that the characteristic impedance $Z_0$ represents the input impedance of a semi-infinitely long transmission line. In other words, $Z_0$ equals the iterative impedance of a section of the transmission line. In spite of the term "impedance" in its name, the characteristic impedance $Z_0$ in (3.30) is a positive real number. This is similar to the situation where the input impedance $Z'_{\text{in}}$ of an LC ladder ((3.10) on p. 58) assumed a positive real value as in (3.12). It follows that the voltage and current traveling wave phasors propagating in the same direction on a lossless transmission line have the same phase.[3] Thus, a lossless transmission line is a one-dimensional medium that can support the propagation of sinusoidal waves without attenuation.

---

[2] Here we defined $V^\pm$ as amplitude phasors rather than rms (root-mean-square) phasors, but it is also permissible to define $V^\pm$ as rms phasors, provided necessary modifications are made to some equations.

[3] In general, the characteristic impedance of a lossy transmission line, defined by (3.31), is complex-valued, and the voltage and current traveling wave phasors propagating in the same direction have a phase difference.

Now let us consider the relationship between the LC ladder in §3.2.2 and the lossless transmission line. Replacing $L$ and $C$ of (3.10) on p. 58 with $L\Delta x$ and $C\Delta x$, respectively, we get

$$Z'_{\text{in}} = \pm \frac{\mathrm{j}\omega L\Delta x}{2}\sqrt{1 - \frac{4}{\omega^2 L\Delta x \cdot C\Delta x}} = \pm\sqrt{\left(\frac{\mathrm{j}\omega L\Delta x}{2}\right)^2 + \frac{L}{C}}. \qquad (3.32)$$

From the condition of the inside the square root of (3.32) becoming 0, the cutoff angular frequency is given by

$$\omega_{\text{c}} = \frac{2}{\sqrt{LC}\Delta x}. \qquad (3.33)$$

Equation (3.33) is the cutoff angular frequency of an LC ladder where the unit cell is as shown in Fig. 3.10 (p. 61). Taking the limit of $\Delta x \to 0$ in (3.32),

$$\lim_{\Delta x \to 0} Z'_{\text{in}} = Z_0. \qquad (3.34)$$

That is to say, the input impedance of a semi-infinite lossless transmission line equals the characteristic impedance $Z_0$. Likewise, taking the limit of $\Delta x \to 0$ in (3.33), the cutoff angular frequency of a lossless transmission line is

$$\lim_{\Delta x \to 0} \omega_{\text{c}} = \infty. \qquad (3.35)$$

An ideal lossless transmission line, therefore, has no cutoff frequency (see Fig. 3.3(c) (p. 54).

**HIDDEN ASSUMPTIONS IN TRANSMISSION LINE THEORY**

Contrary to the conclusion (3.35) about the cutoff angular frequency, actual transmission lines have upper-frequency limits of operation. The smaller the cross-sectional dimensions of a given transmission line, the higher its highest usable frequency. We will not discuss in detail how such upper-frequency limits come about, but it is worth pointing out the sloppiness in the above standard development of transmission line theory.

It is probably not valid to take the limit $\Delta x \to 0$, which we did on p. 62 to derive the telegrapher's equations (3.18) and (3.19), and in (3.34) and (3.35). In Fig. 3.10 (p. 61), $d$ denotes the distance between the two wires. When we applied Kirchhoff's laws to the short section $\Delta x$, it was necessary that $\Delta x \ll \lambda$. But at the same time, we also implicitly assumed that $d \ll \lambda$. Otherwise, Kirchhoff's laws would not have held.

As the frequency becomes higher, the wavelength $\lambda$ becomes shorter, and $d \ll \lambda$ may not hold if the frequency is too high. Then, the development based on Kirchhoff's laws becomes questionable, however small $\Delta x$ might be. Thus, $d \ll \lambda$ must hold to begin with, and the lower limit of $\Delta x$ should be comparable to $d$. This also explains why we referred to a *short* section $\Delta x$ on p. 61, not an *infinitesimal* section. If (3.35) is to hold, $d$ must also become smaller such that $d \leq \Delta x$ as $\Delta x$ approaches 0.

### 3.2.4 Periodic Networks with a Finite Number of Repetitions

The ladder networks and transmission lines we considered above were infinitely long. In this section, let us start over in a different way. Let us consider what happens when a unit *two-port*[4] is on its own, cascaded twice, three times, ⋯, and $N$ times, as shown in Fig. 3.11.

The matrix representation of a two-port suitable for cascade connection is the ABCD-matrix (see §A.1.1):

$$\mathbf{F} = \begin{bmatrix} A & B \\ C & D \end{bmatrix}. \tag{3.36}$$

If the characteristics of a unit two-port are given by (3.36), the characteristics of a periodic network built by cascading $N$ unit two-ports (Fig. 3.11) are given by $\mathbf{F}^N$. The common method of calculating the $N$-th power of a square matrix is to diagonalize it. Here we use an alternative method that is applicable when the determinant of F satisfies $\det\mathbf{F} = 1$. The details are given in §A.2. Using (A.16) on p. 273,

$$\mathbf{F}^N = \mathbf{F}U_{N-1}(\xi) - \mathbf{1}_2 U_{N-2}(\xi) \qquad (N \geq 2), \tag{3.37}$$

[4] See §A.1.1 for what *port* and *two-port* are.

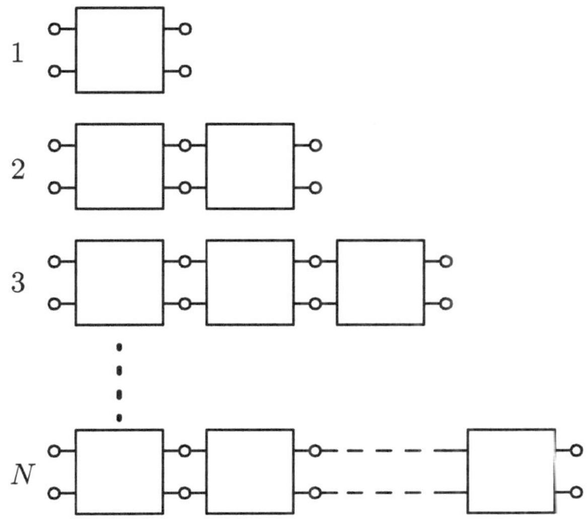


FIGURE 3.11 $N$-times cascade of unit two-ports.

where, from (A.37), (A.20), and (A.24),

$$U_N(\xi) = \frac{\sin[(N+1)\arccos\xi]}{\sin(\arccos\xi)}, \tag{3.38}$$

$$\xi \equiv \frac{A+D}{2}, \tag{3.39}$$

$$\mathbf{1}_2 \equiv \begin{bmatrix} 1 & 0 \\ 0 & 1 \end{bmatrix}. \qquad (2 \times 2\,\text{identity matrix}) \tag{3.40}$$

In (3.38), “arccos” (arc cosine) is the inverse function of “cos” (cosine).

*Example: Alternating Transmission Lines*

Let us look at the two-port shown in Fig. 3.12, consisting of two types of lossless transmission lines. Their phase constants are $\beta_A$ and $\beta_B$, respectively, and the characteristic impedances are $Z_{0A}$ and $Z_{0B}$. Let the lengths be $\ell_A = \ell_B = 0.5\,\text{mm}$.

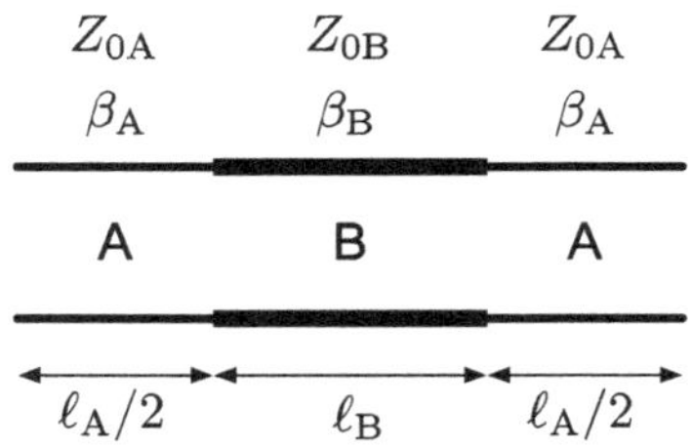


FIGURE 3.12 Unit two-port consisting of type A and type B lossless transmission lines.

TABLE 3.3 Specifications of Type A and Type B Transmission Lines

| | $L$ | $C$ | $Z_0$ | $\omega/\beta$ | $\ell$ |
|---|---|---|---|---|---|
| Type A | 0.4 nH/mm | 1 pF/mm | 20 Ω | $0.5 \times 10^8$ m/s | 0.5 mm |
| Type B | 0.5 nH/mm | 0.2 pF/mm | 50 Ω | $1 \times 10^8$ m/s | 0.5 mm |

The ABCD-matrix of a length, $\ell$, of a lossless transmission line is given by

$$\begin{bmatrix} \cos\beta\ell & Z_0 \sin\beta\ell \\ Z_0^{-1}\sin\beta\ell & \cos\beta\ell \end{bmatrix} \quad \text{(ABCD-matrix of lossless line)}, \quad (3.41)$$

where $\beta$ is given by (3.26) (p. 63), and $Z_0$ by (3.30) (p. 64). Using (3.41), the ABCD-matrix of the unit two-port (Fig. 3.12) is given by

$$\mathbf{F} = \begin{bmatrix} \cos\frac{\beta_A\ell_A}{2} & Z_{0A}\sin\frac{\beta_A\ell_A}{2} \\ Z_{0A}^{-1}\sin\frac{\beta_A\ell_A}{2} & \cos\frac{\beta_A\ell_A}{2} \end{bmatrix} \begin{bmatrix} \cos\beta_B\ell_B & Z_{0B}\sin\beta_B\ell_B \\ Z_{0B}^{-1}\sin\beta_B\ell_B & \cos\beta_B\ell_B \end{bmatrix}$$
$$\times \begin{bmatrix} \cos\frac{\beta_A\ell_A}{2} & Z_{0A}\sin\frac{\beta_A\ell_A}{2} \\ Z_{0A}^{-1}\sin\frac{\beta_A\ell_A}{2} & \cos\frac{\beta_A\ell_A}{2} \end{bmatrix}. \quad (3.42)$$

Since the determinant of (3.41) equals unity ($\because \sin^2\beta\ell + \cos^2\beta\ell = 1$), **F** given by (3.2.4) also satisfies det**F** = 1. Thus, we can apply (3.37) to calculate the *N*-th power of F. Specifications of the type A and type B transmission lines are given in Table 3.3.

Fig. 3.13 shows the magnitude of the *transmission coefficient* $S_{21}$ (see §A.1.2) for different values of repetition count ($1 \le N \le 32$)

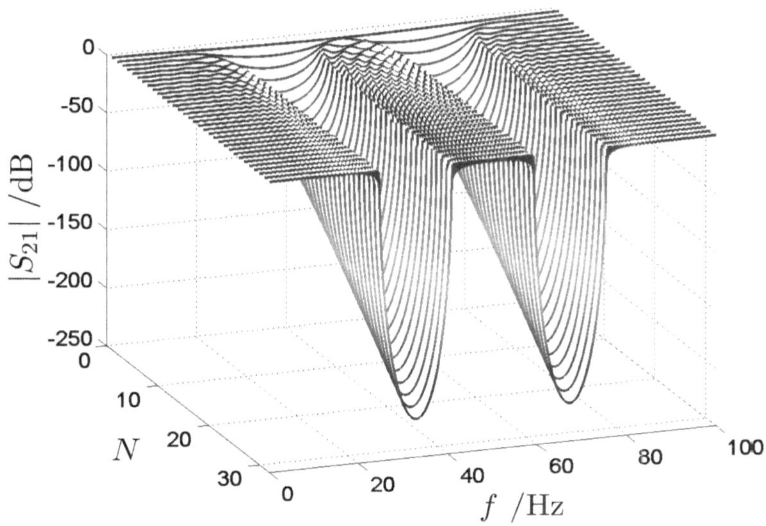


FIGURE 3.13 The transmission coefficient of $N$ unit two-ports, shown in Fig. 3.12, in cascade ($1 \leq N \leq 32$).

up to 100 GHz, with the reference resistance being 50 Ω. The vertical axis of Fig. 3.13 is in decibels (dB), meaning that $10 \log_{10}|S_{21}|^2$ is plotted. Low-attenuation *passbands* and high-attenuation *stopbands* can clearly be seen to appear in turn (see Fig. 3.3(d) on p. 54) even for small values of $N$. It should be clear what happens when $N \to \infty$. The passbands correspond to the allowed bands in solid-state physics, and stopbands correspond to forbidden bands. Usually, in elementary solid-state physics, only infinitely large crystals are considered, and it is not so clear how to handle crystals of finite size. However, Fig. 3.13 suggests that small crystals should have quite similar properties to large crystals. ■

### 3.2.5 Kronig–Penney Model

Many introductory solid-state physics books discuss the *Kronig–Penney model*, which explains energy band formation—the emergence of allowed and forbidden bands—due to a periodic potential field for electrons. The periodic "battlements" potential, $U(x)$, of the Kronig–Penney model is shown in Fig. 3.14. The meaning of $U(x)$ is that an electron at a position $x$ will have potential energy $U(x)$. In quantum

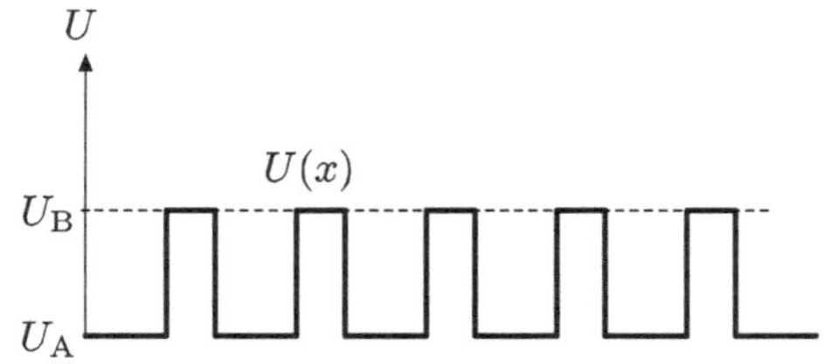


FIGURE 3.14 Battlements potential for electrons.

mechanics, an electron exhibiting wave nature is described by a mathematical function, $\phi(x)$, known as the *wave function*. It is a solution of the governing differential equation called the *Schrödinger equation*.

$\phi(x)$ can be written in the following form for the potential given in Fig. 3.14:

$$\phi(x) = \begin{cases} D_A^+ e^{ik_A x} + D_A^- e^{-ik_A x} & \text{(in regions where} U(x) = U_A) \\ D_B^+ e^{ik_B x} + D_B^- e^{-ik_B x} & \text{(in regions where} U(x) = U_B) \end{cases} \quad (3.43)$$

where $k_A$ and $k_B$ are wave numbers, and $D_A^{\pm}$ and $D_B^{\pm}$ are constants that are to be determined so that boundary conditions are satisfied.

Actually, the "alternating transmission lines" example that we considered on p. 67 can be regarded as a transmission line version of the Kronig–Penney model. This follows from the fact that in a steady state (or more precisely in this case, a *periodic steady state*; see Fig. 5.1 on p. 116), the wave equation, (3.24) on p. 63, for voltage and the one-dimensional *time-independent* Schrödinger equation have the same form (see Table 3.2 on p. 51). Fig. 3.15 (p. 71) contrasts transmission line and quantum mechanical versions of the Kronig–Penney model. Note that j denotes the imaginary unit in electrical engineering, whereas i does so in physics (see the Box on p. 72 for further discussion). The battlements are given by $U(x)$ in the quantum mechanical version, whereas they are given by alternating, piecewise-constant per-unit-length inductance $L(x)$ and capacitance $C(x)$ in the transmission line version (Table 3.3 on p. 68). The only marked distinction between the two versions is the shapes of the dispersion curves (see §3.3): piecewise-linear $\beta(\omega)$ versus parabolic $k(E)$. The bottom left graph in Fig. 3.15 is an $E$-$k$ diagram mentioned in the Box on p. 15.

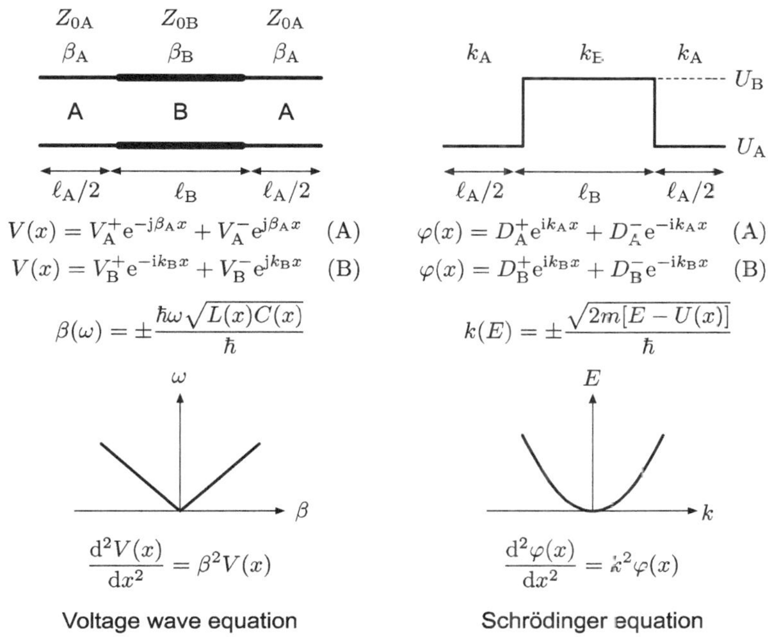


FIGURE 3.15 Transmission line and quantum mechanical versions of the Kronig–Penney model.

## 3.3 DISPERSION RELATION AND PHASE AND GROUP VELOCITIES

### 3.3.1 Dispersion Relation

As is clear from (3.26) (p. 63), the phase constant $\beta$ of a transmission line depends on the frequency. The relation between $\beta$ and $\omega$ is called the *dispersion relation* and is often plotted on a graph as shown in Fig. 3.16 (p. 72), with angular frequency as the vertical axis and phase constant as the horizontal axis. $\beta < 0$ corresponds to propagation in the direction of $-x$. In the case of a lossless transmission line, the $\omega$-$\beta$ diagram exhibits a piecewise-linear *dispersion curve*, as shown in Fig. 3.16 (p. 72).

More generally, the dispersion relation is the frequency dependence of wave number (may be a vector; see p. 63) or of phase velocity (§3.3.2). The abscissa of a dispersion diagram may be wave number,

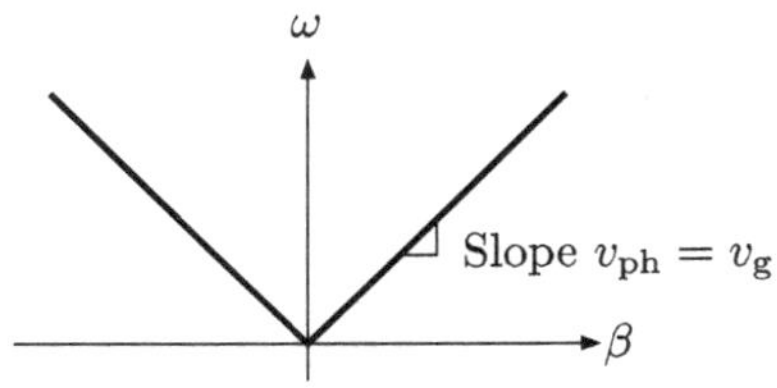


FIGURE 3.16 The dispersion relation of a lossless transmission line.

which is the per-unit-length phase rotation, or phase rotation itself. The ordinate may be angular frequency $\omega$ or energy ($\hbar\omega$ or $E$).

If the dispersion relation of a given medium satisfies

$$\frac{\omega}{\beta} = \pm\text{const.} \qquad \text{(Dispersion relation of nondispersive medium)} \tag{3.44}$$

as in Fig. 3.16 (p. 72), the medium is said to be *nondispersive*. Lossless transmission lines are nondispersive.

If (3.43) on p. 70 does not hold, the medium is said to be *dispersive*. Fig. 3.17 (p. 74) shows an example of a dispersion relation of a dispersive medium. A stopband is seen in Fig. 3.17, wherein waves cannot propagate (see Fig. 3.3(d) on p. 54). The range of the horizontal axis is limited to $\pm\pi/\ell$ in Fig. 3.17. Although the absolute value of $\beta$ can be arbitrarily large, $\beta$ is usually passed to a trigonometric function as an angle in a form like $\beta\ell$ or $\beta x$, as we saw in (3.41) (p. 68) and Fig. 3.15 (p. 71) (see also (3.44) on p. 72). Since an angle "rotates back" every $2\pi$ rad, Fig. 3.17 limits the horizontal axis range so that the corresponding angle is limited to $\pm\pi$ rad. As a result, $\omega(\beta)$ in Fig. 3.17 is multi-valued. This format of showing the dispersion relation $\omega(\beta)$ is known as the *reduced zone scheme*. A different format can be found in Fig. 3.20 (p. 77).

## IMAGINARY UNIT IN ELECTRICAL ENGINEERING AND PHYSICS

In mathematics and physics, "i" denotes the *imaginary unit*. In electrical engineering, "j" denotes the imaginary unit, presumably to avoid possible confusion with $i$, denoting electrical current. But the difference is not just the choice of symbol, although

both of them, of course, satisfy $j^2 = i^2 = -1$. j and i are usually used in subtly different ways in the respective fields.
In electrical engineering, sinusoidal time dependence is often expressed, using an exponential function, as $e^{j\omega t}$. This really means that the sinusoidal time dependence is given by the real part (or at times the imaginary part, but this is not recommended as discussed below) of

$$e^{j\omega t} = \cos\omega t + j \sin \omega t. \qquad \text{(Euler's formula)} \qquad (3.45)$$

At the heart of electrical engineering is lumped circuit theory. Since a lumped circuit has no spatial extent, there is no need to consider spatial coordinates. Moving on to distributed circuit theory, we need to add position dependence to mathematical formulas. At a given time instant $t$, the phase of our sinusoidal wave at a distance $x$ from the origin is $(\omega t - \beta x)$, because there is a *phase lag* at $x$ (and hence the minus sign before $\beta x$). The wave front at the origin will reach $x$ after a duration of $\beta x/\omega$. Therefore, the time- and position-dependence is written as $e^{j(\omega t - \beta x)}$.

On the other hand, physicists presumably began with $x$-dependence (because there is no time in statics) and wrote a sinusoid, using an exponential function, as $e^{ikx}$. Recall that the wave number $k$ is the same thing as the phase constant $\beta$ in one-dimensional space (p. 63). If we add time dependence to $e^{ikx}$, the result will be $e^{i(kx-\omega t)}$. Since we fix the position $x$ first and then consider a time instant after a duration $t$ has passed, the $\omega t$ term in the exponent needs a minus sign. The wave front at $x$ was a distance $\omega t/k$ away at $t = 0$. The real part of

$$e^{-i\omega t} = \cos\omega t - i \sin \omega t \qquad \text{(Euler's formula)} \qquad (3.46)$$

gives the time dependence. Note that $\Re\left(e^{j\omega t}\right) = \Re\left(e^{-i\omega t}\right)$. We also see that it does not seem like a very good idea to take the imaginary part because $\Im\left(e^{j\omega t}\right) \neq \Im\left(e^{-i\omega t}\right)$.

To switch between the two notational conventions, replacing j with $-i$ or vice versa will do. Stratton pointed it out in the preface of his famous book on electromagnetism as follows [29]:

> The use of the factor $e^{-i\omega t}$ instead of $e^{+i\omega t}$ is another point of mild controversy. This has been done because the time factor is invariably discarded, and it is somewhat more convenient to retain the

positive exponent $e^{+ikR}$ for a positive traveling wave. To reconcile any formula with its engineering counterpart, one need only replace −i by +j.

## IMAGINARY UNIT IN ELECTRICAL ENGINEERING AND PHYSICS (CONT.)

So the impedance of an inductor, for example, is $j\omega L$ in electrical engineering and $-i\omega L$ in physics. Unfortunately, not all authors adhere to this established convention, thereby causing confusion, especially when dissipative components, such as lossy transmission lines or *complex permittivity* (with a nonzero imaginary part), are involved [11]. A dissipative system can end up erroneously exhibiting exponential divergence, rather than exponential decay, in the time domain.

The choice between i and j is yours, but we ask the reader to follow the convention that the time dependence is $e^{-i\omega t}$ and $e^{j\omega t}$, respectively.

### 3.3.2 Phase Velocity and Group Velocity

For a given medium, important information that can be read from dispersion curves ($\omega$-$\beta$ diagrams) like Fig. 3.16 (p. 72) and Fig. 3.17

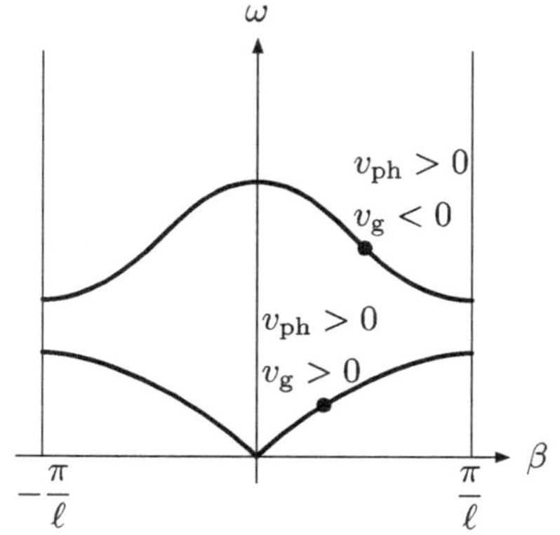


FIGURE 3.17 Example of a dispersion relation of a dispersive medium.

is the *phase velocity*, given by

$$v_{\mathrm{ph}} \equiv \frac{\omega(\beta)}{\beta}. \qquad \text{(Phase velocity)} \tag{3.47}$$

On the right-hand side of (3.46), the angular frequency $\omega$ is regarded as a function of the phase constant $\beta$ and is written as $\omega(\beta)$. Equation (3.46) has the same form as the expression of the chord resistance, (2.10) (p. 33). The phase velocity $v_{\mathrm{ph}}$ at a given value of $\omega$ equals the slope of a straight line joining a point on the dispersion curve and the origin. It represents the (signed) velocity of propagation of a wave front of a sinusoid of angular frequency $\omega$. Table 3.3 (p. 68) gave phase velocities of type A and type B transmission lines.

Another important quantity that can be read from an $\omega$-$\beta$ diagram is the *group velocity*[5]. It is given as the local slope of a dispersion curve as follows:

$$v_{\mathrm{g}} \equiv \frac{\mathrm{d}\omega(\beta)}{\mathrm{d}\beta}. \qquad \text{(Group velocity)} \tag{3.48}$$

Equation (3.47) has the same form as the expression of the incremental resistance, (2.10) (p. 33). The group velocity represents the (signed) velocity of propagation of the *envelope* of a waveform made up by the superposition of multiple sinusoids. For example, if the wave form is an amplitude-modulated carrier wave (Fig. 3.18) given by

$$v(t) = v_{\mathrm{s}}(t)\cos\omega_{\mathrm{c}}t, \qquad \text{(Amplitude-modulated carrier wave)} \tag{3.49}$$

$$v_{\mathrm{s}}(t) = V_0 \sin\omega_{\mathrm{s}} \quad (\omega_{\mathrm{s}} \ll \omega_{\mathrm{c}}), \qquad \text{(Modulating signal or envelope)} \tag{3.50}$$

where $\omega_{\mathrm{c}}$ is the carrier angular frequency, and $\omega_{\mathrm{s}}$ is the signal angular frequency, the group velocity represents the velocity of a wave front of $v_{\mathrm{s}}(t)$.

Fig. 3.16 (p. 72) shows the dispersion relation of a nondispersive medium, and therefore $v_{\mathrm{ph}} = v_{\mathrm{g}}$ holds. However, this is a special case. In general, phase velocity and group velocity have different values, as is clear from Fig. 3.17 (p. 74). This is analogous to the fact that the chord resistance $R_{\mathrm{ch}}$ and the incremental resistance $R_{\mathrm{inc}}$ of a resistor coincide only if the resistor is a linear resistor.

---

[5] For a more detailed account of the group velocity, see [4].

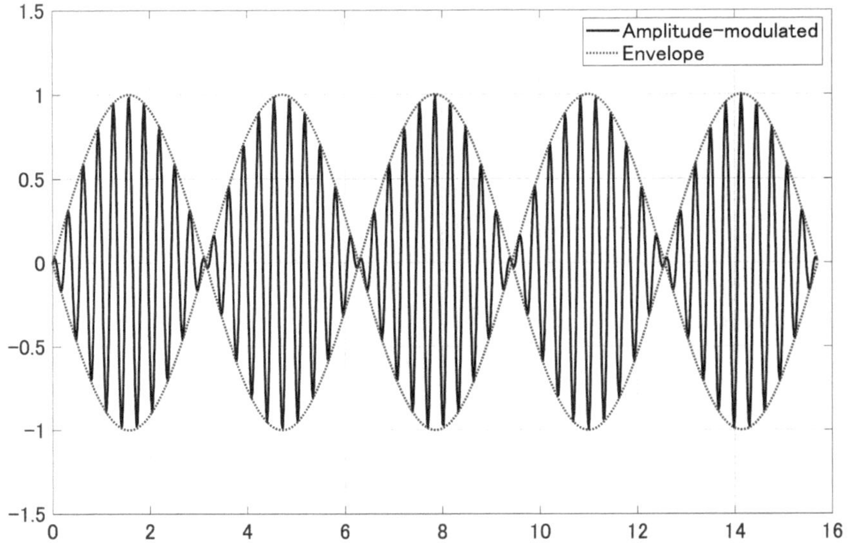


FIGURE 3.18 Amplitude-modulated carrier wave and its envelope.

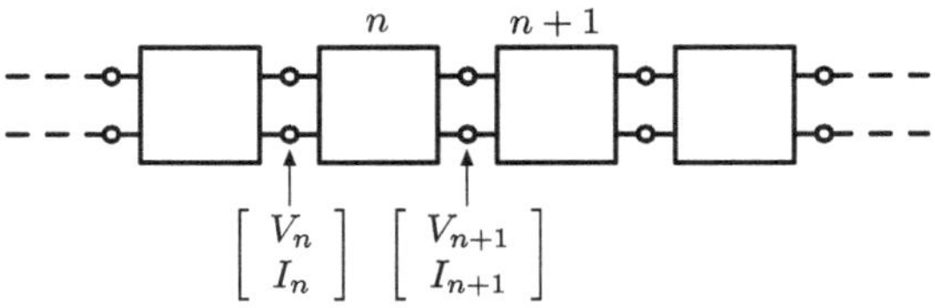


FIGURE 3.19 An infinite cascade of phase-shifting unit two-ports.

### 3.3.3 Calculation of the Dispersion Relation

Let us consider an infinitely long cascade of lossless unit two-ports, as shown in Fig. 3.19. Let us suppose further that the unit two-port affects only the phase of the wave that passes through it and does not affect the amplitude. Such a network is called a *phase shifter*. The dispersion relation of the one-dimensional periodic network can be calculated as follows.

Let the ABCD-matrix of the unit two-port be

$$\mathbf{F} = \begin{bmatrix} A(\omega) & B(\omega) \\ C(\omega) & D(\omega) \end{bmatrix}. \tag{3.51}$$

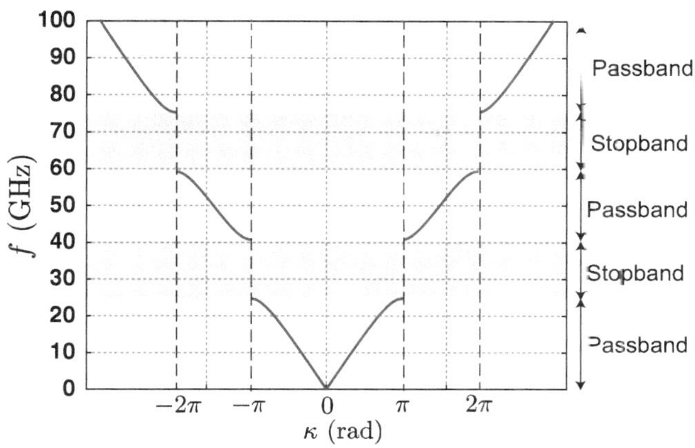


FIGURE 3.20 *f*-κ diagram of an infinite cascade of a unit two-port shown in Fig. 3.12 (p. 68).

Let us suppose that the unit two-port is *reciprocal* (see the Box on p. 48) and therefore satisfies (p. 273)

$$\det\mathbf{F} = A(\omega)D(\omega) - B(\omega)C(\omega) = 1. \tag{3.52}$$

The left and right port voltages and currents of the $n$-th unit two-port are related to each other by

$$\begin{bmatrix} V_n \\ I_n \end{bmatrix} = \mathbf{F}\begin{bmatrix} V_{n+1} \\ I_{n+1} \end{bmatrix} = \begin{bmatrix} A & B \\ C & D \end{bmatrix}\begin{bmatrix} V_{n+1} \\ I_{n+1} \end{bmatrix}. \tag{3.53}$$

Since the unit two-port, by assumption, only introduces phase rotation, (3.52) can also be written as follows:

$$\begin{bmatrix} V_n \\ I_n \end{bmatrix} = \begin{bmatrix} A & B \\ C & D \end{bmatrix}\begin{bmatrix} V_{n+1} \\ I_{n+1} \end{bmatrix} = \begin{bmatrix} V_{n+1}\mathrm{e}^{\mathrm{j}\kappa} \\ I_{n+1}\mathrm{e}^{\mathrm{j}\kappa} \end{bmatrix}, \tag{3.54}$$

where $\kappa$ is the phase rotation. The right-hand side of (3.53) can be rewritten as

$$\begin{bmatrix} V_{n+1}\mathrm{e}^{\mathrm{j}\kappa} \\ I_{n+1}\mathrm{e}^{\mathrm{j}\kappa} \end{bmatrix} = \begin{bmatrix} \mathrm{e}^{\mathrm{j}\kappa} & 0 \\ 0 & \mathrm{e}^{\mathrm{j}\kappa} \end{bmatrix}\begin{bmatrix} V_{n+1} \\ I_{n+1} \end{bmatrix}. \tag{3.55}$$

By subtracting (3.54) from (3.52), we obtain

$$\begin{bmatrix} A - \mathrm{e}^{\mathrm{j}\kappa} & B \\ C & D - \mathrm{e}^{\mathrm{j}\kappa} \end{bmatrix}\begin{bmatrix} V_{n+1} \\ I_{n+1} \end{bmatrix} = \begin{bmatrix} 0 \\ 0 \end{bmatrix}. \tag{3.56}$$

For the equality (3.55) to hold, the determinant of the matrix on the left-hand side must be 0:

$$\left(A - e^{j\kappa}\right)\left(D - e^{j\kappa}\right) - BC = AD + e^{2j\kappa} - (A + D)\, e^{j\kappa} - BC = 0. \quad (3.57)$$

Using (3.51), (3.56) simplifies to

$$1 + e^{2j\kappa} - (A + D)\, e^{j\kappa} = 0. \quad (3.58)$$

Dividing both sides of (3.57) by $e^{j\kappa}$ and rearranging leads to

$$e^{j\kappa} + e^{-j\kappa} = 2\cosh(j\kappa) = 2\cos\kappa = A + D, \quad (3.59)$$

where we used the following relation:

$$\cosh(x + jy) = \cosh x \cos y + j \sinh x \sin y. \quad (3.60)$$

From (3.58), the dispersion relation is given by

$$\cos\kappa(\omega) = \frac{A(\omega) + D(\omega)}{2}, \quad (3.61)$$

or equivalently,

$$\kappa(\omega) = \arccos\frac{A(\omega) + D(\omega)}{2}. \quad (3.62)$$

An $\omega$-$\kappa$ diagram could be plotted by sweeping $\omega$. If the horizontal axis needs to be the wave number $k$ (or $\beta$) instead of $\kappa$, then let the length of the unit two-port be $\ell$, and $k = \kappa/\ell$ can be the horizontal axis.

*Example: Dispersion Relation for the Kronig–Penney Model*

Draw a dispersion diagram for the transmission line version of the Kronig–Penney model shown in Fig. 3.12 (p. 68). Use the values given in Table 3.3, (p. 68), the ABCD-matrix of the unit two-port (3.42) (p. 68), and the dispersion relation (3.61).

A result is shown in Fig. 3.20. Note that the vertical axis is the frequency $f = \omega/2\pi$. Fig. 3.20 does not limit the range of the horizontal axis to $\pm\pi$. This format of displaying the dispersion relation is known as the *extended zone scheme*. If the range is limited to $\pm\pi$ by horizontally shifting inward the curves lying outside this range by $\pm 2n\pi$ with $n$ being an integer, we will get a similar plot to Fig. 3.17 (p. 74). Check that Fig. 3.20 is consistent with Fig. 3.13 (p. 69). ■

**LUMPED VERSUS DISTRIBUTED, FINITE VERSUS INFINITE**

In §3.2.3, we explained that the distributed circuit theory is the theory for transmission lines. We might have given the impression that a distributed circuit is a circuit that is nonnegligible in size compared to the wavelength.

However, there are some distributed circuits that are not quite comparable with the wavelength. The infinite LC ladder in §3.2.2 (Fig. 3.8 on p. 58) was such an example. We may as well say that we simply forgot to mention the dimensions of the inductors and the capacitors constituting the LC ladder, so we cannot compare it with the wavelength. However, circuit theory draws a line between lumped and distributed circuits based on the number of elements. A circuit consisting of a finite number of lumped circuit elements is a *lumped circuit*. A circuit consisting of infinitely many lumped circuit elements is a *distributed circuit*.

As we have noted on p. 59, it is possible to synthesize an impedance with a positive real part if infinitely many reactive elements are available. There is a fundamental difference between a finite number and infinity. Problem 3.3 (p. 85) also gives such an example.

## 3.4 DISPERSION RELATION AND PROPERTIES OF SEMICONDUCTORS

In solid-state physics, the dispersion relation for electron waves in a solid is plotted with the electron energy $E$ as the vertical axis and the wave number $k$ as the horizontal axis. Such an "$E$-$k$ diagram" is also called an *energy band diagram* (see the box on p. 15). The dispersion relation of a semiconductor typically looks like Fig. 3.21. The energy gap $E_g$ is the distance between "the bottom of the dispersion curve for the conduction band" and "the top of the dispersion curve for the valence band."

In solid-state physics, the group velocity $v_g$ that can be read from the dispersion relation $E(k)$ is interpreted as the velocity of electrons. In other words, the velocity of the envelope of the wave function is associated with the motion of electrons.

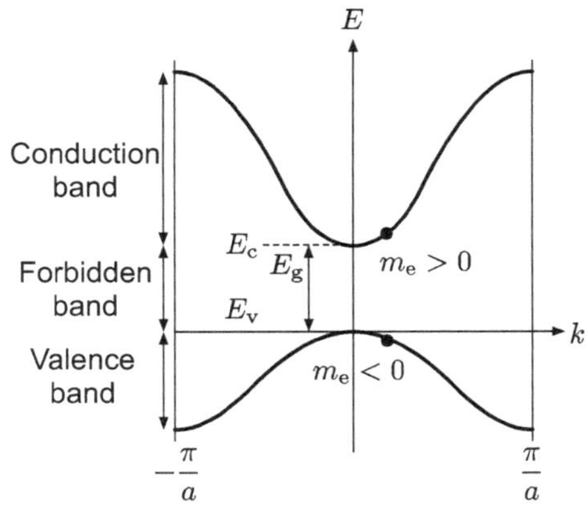


FIGURE 3.21 Sketch of an *E*-*k* diagram of a semiconductor. *a* is the spacing between adjacent atoms.

From (3.47) (p. 75),

$$v_g = \frac{d\omega}{dk} = \frac{d(\hbar\omega)}{d(\hbar k)} = \frac{1}{\hbar}\frac{dE(k)}{dk}, \tag{3.63}$$

where we used the relationship $E(k) = \hbar\omega$ between energy and angular frequency. If (3.62) gives the electron velocity, its time derivative $\dot{v}_g = dv_g/dt$ should represent acceleration. If force $F$ acts on an electron, the equation of motion is

$$m_e\dot{v}_g = F, \tag{3.64}$$

where $m_e$ is called the *effective mass* and assumes a different value from the electron rest mass, $m_0$, in free space. If $F$ is known, $m_e$ is given by

$$m_e = \frac{F}{\dot{v}_g}. \tag{3.65}$$

The electron effective mass depends on the material and crystal orientation. Its value is typically between a hundredth of and several times the free space rest mass $m_0$. It can be shown that $m_e$ is inversely proportional to the second derivative of $E(k)$ as follows.

$$\frac{1}{m_e} \propto \frac{d^2E(k)}{dk^2}. \tag{3.66}$$

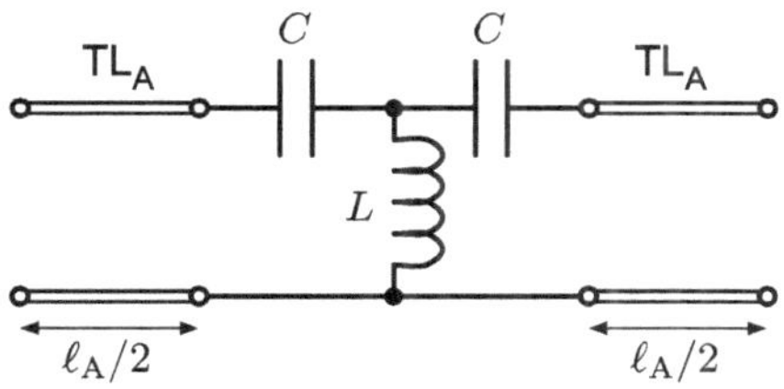


FIGURE 3.22 Transmission line loaded with a high-pass two-port.

From the shape of the dispersion curves in Fig. 3.21, we see that $m_e > 0$ near the bottom of the conduction band, and $m_e < 0$ near the top of the valence band. Electrons with negative effective mass have to do with holes (see the Box on p. 136).

Incidentally, Fig. 3.21 looks quite different from the dispersion relation for the Kronig–Penney model, shown in Fig. 3.20 or Fig. 3.17 (p. 74). In Fig. 3.17, the bottommost allowed band has the smallest energy at the center, but the bottommost allowed band in Fig. 3.21 peaks at the center. Similarly, the second allowed bands from the bottom look like upside-down versions of each other. The quantum mechanical versions of the Kronig–Penney model are essentially the same in this regard. The only difference from the transmission line version is that the dispersion curve is not sharp-cornered as in Fig. 3.20.

One way to reconcile the discrepancy is to consider that the middle band in Fig. 3.20 corresponds to the valence band in Fig. 3.21 and that the top band in Fig. 3.20 corresponds to the conduction band in Fig. 3.21. Another possibility is to consider a unit two-port that exhibits a different dispersion relation than that in Fig. 3.20. In the following example, the bottom two allowed bands exhibit a semiconductor-like dispersion relation.

### *Example: Periodically High-Pass Loaded Transmission Line*

The unit two-port shown in Fig. 3.22 consists of a lumped reactive network sandwiched between transmission line sections. Let us find the dispersion relation of a one-dimensional periodic network built of it. Unlike the unit two-port for the Kronig–Penney model (Fig. 3.12 on p. 68), this two-port has *high-pass characteristics* and blocks DC voltage and current.

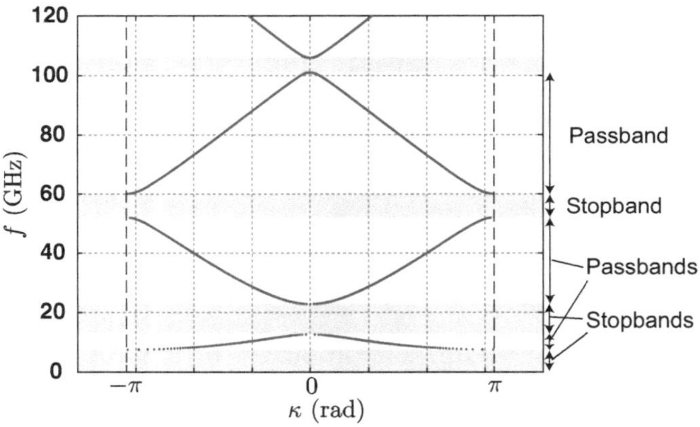


FIGURE 3.23 $f$-$\kappa$ diagram of the one-dimensional periodic network composed of the unit two-port shown in Fig. 3.22.

Let the inductance be $L = 0.5\,\text{nH}$, the capacitance be $C = 0.2\,\text{pF}$, and the length be $\ell_A = 0.5\,\text{mm}$. Suppose that the characteristics of the transmission line are given by Table 3.3 (p. 68). The resulting $f$-$\kappa$ diagram is shown in Fig. 3.23. The lower two bands look semiconductor-like as in Fig. 3.21. See also Fig. 3.3(e) (p. 54). ■

## LIMITATIONS OF CLASSICAL ANALOGUES OF QUANTUM THEORY

The analogy shown in Fig. 3.15 (p. 71) between the time-independent Schrödinger equation and the frequency-domain wave equation for the transmission line is powerful. The quantum mechanical *tunneling* effect (p. 174), which is often described as "peculiar to quantum mechanics," can also be described by the transmission line or electromagnetic wave equation. In electromagnetism, the wave corresponding to the tunneling effect is named an evanescent wave, and in microwave engineering, the equivalent phenomenon is at times used in attenuators. The above correspondence between quantum theory and transmission line theory holds only in periodic steady states (p. 115). If one wants to track the temporal evolution of a system, the diffusion equation could be used (see Table 3.2 on p. 51).

However, not all quantum mechanical phenomena can be described or understood by using classical analogues. A perfect example is the *quantum computer*. Some computational problems are known to be practically intractable for any classical computer but are, in principle, solvable by quantum computers with sufficient computational capacity, although there are not so many problems that fall into this category of computational complexity. Many other difficult problems are intractable even for quantum computers. Anyway, the speed-up originates intrinsically from the properties of quantum mechanics. This is where the peculiarity of quantum mechanics reveals itself. It is generally believed that no classical system (or analogue) can emulate a quantum computer efficiently. Note that inefficient emulation gives no speed-up. If it were possible to emulate a quantum computer efficiently by a classical analogue, then there would be no point in trying to implement a quantum computer.

Then, where is the boundary between cases where classical analogues exist and cases where they do not? In this book, we do not deprive the reader of the pleasure of finding the answer to this intriguing question.

## 3.5 BRAGG REFLECTION

The reader might have some knowledge of X-ray diffraction in crystals, including the *Bragg condition*, under which diffracted X-rays add up constructively. If X-rays impinge on a cleavage plane perpendicularly, the rays are just reflected under the Bragg condition.

Consider X-rays with wavelength $\lambda$, incident from the left on a one-dimensional crystal, as shown in Fig. 3.24 (p. 84). Ignore the amplitude in Fig. 3.24, and focus only on the phase relationship. Assume that the wavelength does not change inside or outside the crystal. The crystal lattice spacing is $a$. The wave traveling in the crystal is partially reflected by each atom as it travels through the crystal. If the Bragg condition, $\lambda = 2a$, is satisfied, reflected waves are in phase with each other as shown in Fig. 3.24, resulting in strong outgoing reflected waves. This is the *Bragg reflection*.

The above was the case of a wave entering the crystal from the outside. If the one-dimensional crystal were infinitely long, what would happen if we tried to excite a wave with wavelength $\lambda = 2a$ inside

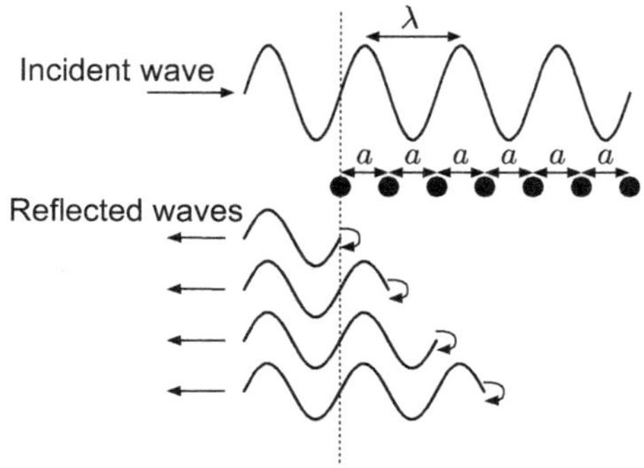


FIGURE 3.24 The Bragg reflection is due to an array of atoms.

the crystal? Since reflections by atoms occur for waves traveling in either direction, waves propagating in either direction are canceled out by the Bragg reflection, and waves of this wavelength cannot exist persistently in the crystal.

In Fig. 3.17 (p. 74) and Fig. 3.21 (p. 80), $k = \pm 2\pi/\lambda = \pm\pi/\ell$ or $k = \pm 2\pi/\lambda = \pm\pi/a$ holds on the left and right edges. These are equivalent to the Bragg condition $\lambda = 2n\ell$ or $\lambda = 2na$ with $n\,(\neq 0)$ being an integer. In Fig. 3.17, two dispersion curves come closer to each other but flatten out at $k = \pm\pi/\ell$. An interpretation of this is that the wave with an angular frequency corresponding to the center of the stopband has a wavelength $\lambda = 2n\ell$ and, therefore, cannot exist persistently in the periodic network. The center frequency of the stopband is known as the *Bragg (angular) frequency* and the corresponding wavelength is called the *Bragg wavelength*.

The first example that we considered in this chapter was the infinite LC ladder shown in Fig. 3.8 (p. 58). The infinitely long LC ladder and CL ladder (Problem 3.1 on p. 85) have one passband and one stopband, as shown in Figs. 3.3(a) and (b) (p. 54). It does not seem possible to think of "the center frequency of the stopband." This might imply that the origin of the stopband in these ladder networks is not the Bragg reflection (§3.1.2). We started our discussion with the infinite LC ladder because it is more elementary and easier to understand.

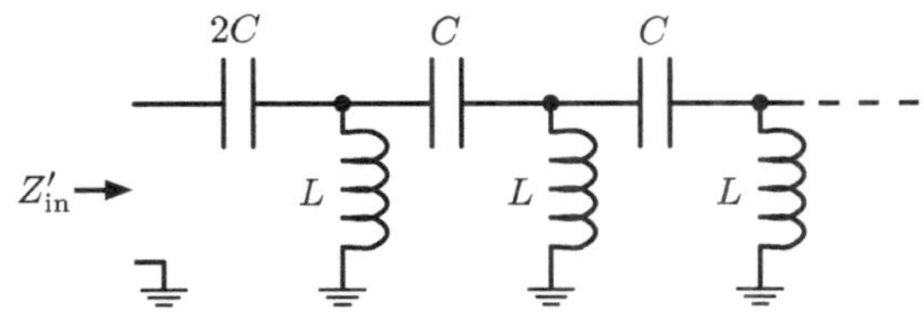


FIGURE 3.25 Infinite CL ladder.

## 3.6 SUMMARY

Quantum mechanics and band theory are quite difficult for beginners. In this chapter, instead of dealing squarely with band theory in solid-state physics, which requires quantum mechanics, we investigated energy band formation using circuit theory.

- The frequency-domain wave equations for a transmission line and the time-independent Schrödinger equation describing an electron have the same mathematical form.
- In a periodic structure, only waves with a certain range of energy (or frequency) can exist persistently.
- The Kronig–Penney model can be formulated as both a quantum mechanics problem and a transmission line problem, and the results are almost the same.
- The energy band formation in periodic structures is a universal phenomenon common to various systems that can support wave propagation and is not exclusive to quantum mechanics.

## 3.7 PROBLEMS

3.1 Consider the infinite CL ladder, shown in Fig. 3.25, which is in a dual relationship to the infinite LC ladder that we discussed in §3.2.2. Find its cutoff angular frequency $\omega_c$ and the input impedance $Z'_{\text{in}}$.

3.2 Derive the wave equations (3.20) and (3.21) on p. 62 from the telegrapher's equations (3.18) and (3.19).

3.3 Plot the magnitudes of the reflection coefficient $S_{11}$ and the transmission coefficient $S_{21}$ (see §A.1.2) of an LC ladder (Fig. 3.8, p. 58) composed of $N$ unit two-ports versus the frequency. Number of repetitions: $1 \leq N \leq 32$. Series inductance: $L = 0.5\,\text{nH}$. Parallel capacitance: $C = 0.2\,\text{pF}$. Reference resistance: $R_{\text{ref}} = 50\,\Omega$. Use a computer, etc. for plotting as appropriate.

3.4 Do a search on phase velocity and group velocity to better understand them.

3.5 Examples of applications of periodic structures described by the wave equation for electromagnetic waves (Table 3.2, p. 51) include "photonic crystals" for visible light, "electromagnetic bandgap (EBG) structures" in the microwave band, and "metamaterials" related to both light and microwaves. Do a search on these.

CHAPTER 4

# Physics of Semiconductors in Equilibrium

We have already briefly discussed the properties of semiconductors in Chapter 1, but the discussion was largely qualitative. In this chapter, we will look at the physics of semiconductors in more detail, using mathematical formulas as well. The contents of this chapter are the basis for the discussion of electrical conduction and the analysis of semiconductor devices in the following chapters. This chapter deals only with the equilibrium states of spatially uniform semiconductors. Nonequilibrium states, including net current flow, and cases where the equilibrium state is not spatially uniform are considered in Chapter 5 and thereafter.

## 4.1 DENSITY OF STATES IN ENERGY BAND AND DISTRIBUTION FUNCTION

As mentioned in §1.3.4, the valence and conduction bands of semiconductors consist of densely distributed orbitals that electrons may populate (Fig. 4.1). The orbitals in the conduction band are almost empty, but orbitals near the bottom of the band are filled with electrons. These electrons are responsible for electrical conduction. The orbitals in the valence band are mostly filled with electrons, but there are some unoccupied orbitals near the top of the band. These unoccupied orbitals contribute to electrical conduction as holes

What we have called "orbitals" are more often referred to as *states* in solid-state physics, so we will follow the convention from here on. Let the energy of an electron be $E$ as before (§1.3.5). The number of states per unit energy range per unit volume is called the *density of states*. The density of states as a function of $E$, written as $N(E)$, is called

DOI: 10.1201/ 9781003439417-4

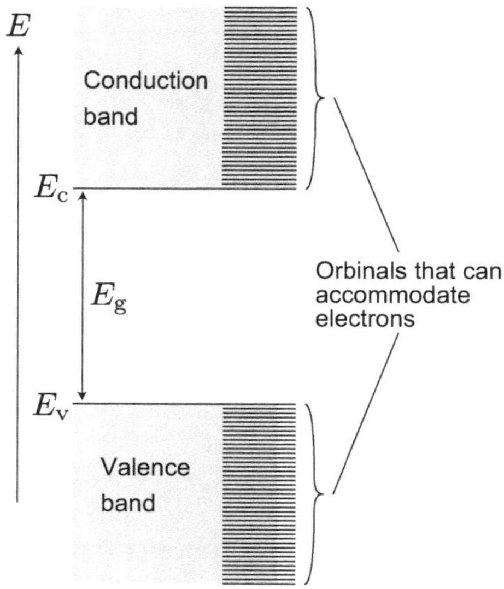


FIGURE 4.1 Allowed bands consist of a dense collection of electron orbitals.

the density-of-states function. The number of states in an infinitesimal energy interval $dE$ is $N(E)\,dE$ per unit volume.

Here is a concrete form of the density-of-states function for the conduction band of a three-dimensional semiconductor crystal, although we will leave its derivation to solid-state physics books [24, 30]:

$$N(E) = \frac{1}{\pi^2}\frac{m_c\sqrt{2m_c}}{\hbar^3}(E - E_c)^{1/2}, \tag{4.1}$$

where $m_c$ is called the *density-of-states effective mass* of an electron. $m_c$ is somewhat different in value and meaning from the effective mass on p. 80. $N(E)$ has a parabolic functional form. Similarly, the density-of-states function for the valence band of a three-dimensional semiconductor crystal is given by

$$N(E) = \frac{1}{\pi^2}\frac{m_v\sqrt{2m_v}}{\hbar^3}|E - E_v|^{1/2}, \tag{4.2}$$

where $m_v$ is the density-of-states effective mass of a hole. Note that in the valence band $E < E_v$.

Incidentally, you might have thought it is strange to treat the number of states as if it were a continuous quantity, as in (4.1) and (4.2). Since the number of states is an integer and $E$ is inherently discrete (see Fig. 1.4 on p. 10 and Fig. 1.5 on p. 11), such a question is valid. If the number of atoms in the crystal under consideration is small, the treatment assuming such a continuous quantity is not correct. However, crystals we deal with usually contain a huge number of atoms and states, so in such cases, the above treatment is a very good approximation. In the derivations of (4.1) and (4.2), the discrete quantities are approximated as continuous quantities on this basis.

Now, the number of states in an infinitesimal energy interval $\mathrm{d}E$ is $N(E)\,\mathrm{d}E$, but the conduction band states are not fully filled except near the bottom of the band, and the valence band states have vacant states near the top of the band (p. 11). Thus, the number of states actually occupied by electrons is smaller than $N(E)\,\mathrm{d}E$. We, therefore, bring in a function $f(E)$ such that $0 \le f(E) \le 1$ and express the number of states per unit volume actually occupied by electrons as

$$N(E)f(E)\,\mathrm{d}E. \qquad \text{(Electron density within } \mathrm{d}E\text{)} \qquad (4.3)$$

$f(E)$ represents the probability that an electron actually occupies a state with energy $E$ and is called the *distribution function*. The functional form of the distribution function depends on the system under consideration. The distribution function for the conduction band and valence band states is called the *Fermi–Dirac distribution function*:

$$f(E) = \frac{1}{1 + \exp\left(\frac{E-\zeta}{kT}\right)}, \qquad \text{(Fermi-Dirac distribution function)} \qquad (4.4)$$

where $k$ is the Boltzmann constant, and $T$ is the absolute temperature. $\zeta$ is called the *Fermi level* and has the dimensions of energy. The Fermi level is the *electrochemical potential* of electrons in solid-state physics, but this statement may not make much sense at this point. Electrochemical potential is an extremely important concept in understanding the physics of semiconductor devices and is discussed in detail in §4.4.

The value of $\zeta$ depends on impurity doping, but it often falls roughly within the following range:

$$E_{\mathrm{v}} \lesssim \zeta \lesssim E_{\mathrm{c}}, \qquad \text{(Range of Fermi level)} \qquad (4.5)$$

where the symbol "$\lesssim$" means that the inequality sign is approximately satisfied. In words, (4.5) says "Fermi level $\zeta$ of a semiconductor often falls within the forbidden band, but it may at times step into the conduction or valence band."

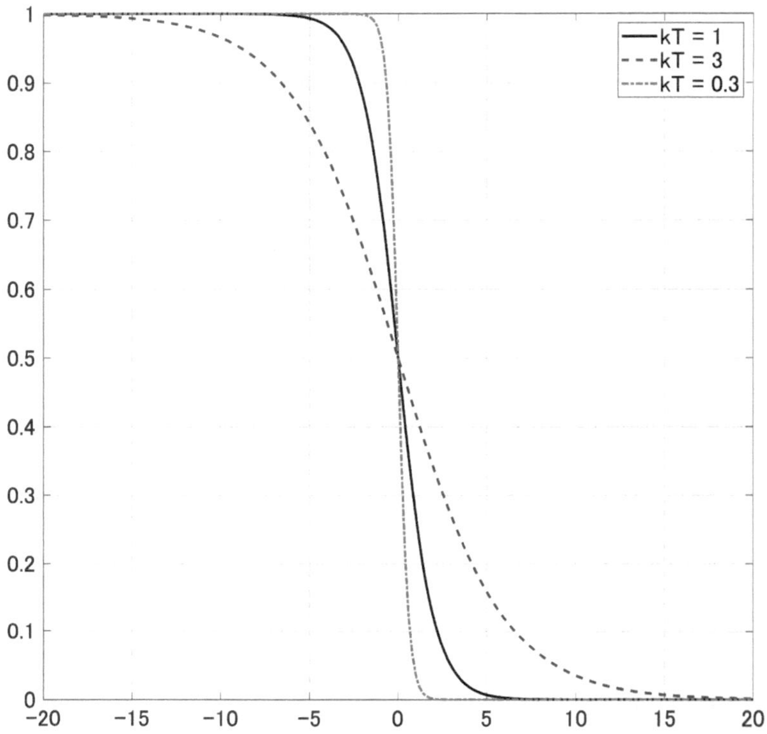


FIGURE 4.2 Fermi–Dirac distribution function (4.4) with $(E-\zeta)$ on the horizontal axis.

Fig. 4.2 (p. 90) plots the Fermi–Dirac distribution function (4.4) with $(E-\zeta)$ on the horizontal axis. The shape of the curve depends on temperature $T$, but always $f(E) = 1/2$ at $E = \zeta$. At $T = 0\,\text{K}$, $f(E)$ looks like a left-right reversed unit step function as follows:

$$f(E) = \begin{cases} 1 & (E < \zeta) \\ 1/2 & (E = \zeta) \\ 0 & (E > \zeta) \end{cases} \qquad \text{(Fermi-Dirac function at} T = 0\text{K)} \quad (4.6)$$

Although it is not clear from Fig. 4.2 because the horizontal axis is $(E-\zeta)$, the Fermi level $\zeta$ itself depends on the temperature (Problem 4.1 on p. 112). The Fermi level at absolute zero is called the *Fermi energy* [3, 14].

Some authors use the term "Fermi energy" synonymously with "Fermi level." In fact, the Fermi level of room temperature metals

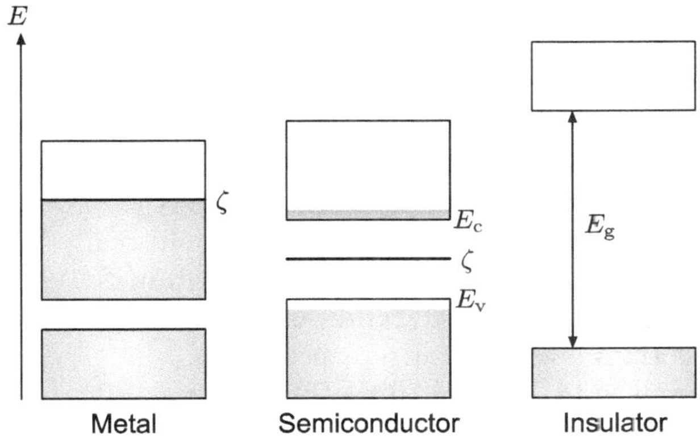


FIGURE 4.3 Energy band structures of metal, semiconductor, and insulator.

is almost equal to the Fermi energy. $f(E)$ for room temperature metals is almost like a reversed step function, (4.6). We can also assume, for most practical purposes, that metals do not have the forbidden band, that there are many electrons in the allowed band corresponding to the conduction band, and that the Fermi energy is located in the allowed band. Therefore, almost all states below the Fermi energy are filled with electrons (Fig. 4.3 on p. 91). This makes the Fermi energy so important in the physics of metals. The value of the Fermi energy $E_F$ is also important when considering the properties of metals.

On the other hand, as can be inferred from (4.5), the Fermi energy of a semiconductor lies in the forbidden band where no state exists, so it is not the boundary between occupied and unoccupied states [3]. Also, the value of the Fermi energy or the Fermi level itself is rarely an issue. Usually, we consider only relative values such as $(E_c - \zeta)$ and $(\zeta - E_v)$. The significance and importance of the Fermi level in semiconductors is discussed further in §4.4.

## 4.2 CARRIER DENSITIES OF NONDEGENERATE SEMICONDUCTORS

### 4.2.1 Electron Density

As we saw in (4.3) on p. 89, the electron density per energy interval d$E$ is $N(E)f(E)\,\mathrm{d}E$, so integrating it over $E$ gives the number of electrons per unit volume. The electron density in the conduction band is given by

$$n = \int_{E_c}^{\infty} N(E) f(E)\, \mathrm{d}E. \qquad \text{(Electron density)} \tag{4.7}$$

The upper end of the integral in (4.7) should really be the top of the conduction band, but considering the form of $f(E)$ given in (4.4) and Fig. 4.2, the result of integration will not change if the upper end is above a certain level, so it is replaced by $\infty$.

The integral in (4.7) is called the *Fermi–Dirac integral* of one-half order and is difficult to evaluate analytically. We make the following approximations. Noting that $\mathrm{e}^3 \simeq 20 \gg 1$, the Fermi–Dirac distribution function can be approximated as follows, provided $(E - \zeta)/kT \gtrsim 3$:[1]

$$\begin{aligned} f(E) &= \frac{1}{1 + \exp\left(\frac{E-\zeta}{kT}\right)} \approx \frac{1}{\exp\left(\frac{E-\zeta}{kT}\right)} \\ &= \exp\left(-\frac{E-\zeta}{kT}\right) \quad (\ll 1). \end{aligned} \tag{4.8}$$

Considering the fact that the range of integration of (4.7) is $E \geq E_c$ (i.e., within the conduction band) and that the range of the Fermi level $\zeta$ is given by (4.5) on p. 89, the condition $(E - \zeta)/kT \gtrsim 3$ implies $\zeta \lesssim E_c - 3kT$. The exponential distribution function in (4.8) is called the *Maxwell–Boltzmann distribution function*. Semiconductors to which this approximation can be applied are called *nondegenerate semiconductors*. Simply put, a nondegenerate semiconductor is one in which the doping density is not very high and, consequently, the majority carrier density is not very high. Electrons and holes in nondegenerate semiconductors can be regarded as ideal gases. Conversely, a semiconductor with a very high doping density and a very high majority carrier density is called a *degenerate semiconductor* (§4.3.2). *We will consider nondegenerate semiconductors from here onward unless otherwise specified.*

If $f(E)$ in (4.7) is given by the Maxwell–Boltzmann distribution (4.8), then the integral can be performed analytically. The result is

[1] 3 here is a rough guide.

(see Problem 4.2 on p. 113)

$$n = N_c \exp\left(-\frac{E_c - \zeta}{kT}\right), \qquad \text{(Nondegenerate electron density)} \quad (4.9)$$

where $N_c$ is called the *effective density of states* of the conduction band, and $E_c$ is the electron energy at the bottom of the conduction band (Fig. 4.1 on p. 88). The effective density of states is a constant determined for each material. Numerical examples are given in Table 1.3. (p. 5).

The exponential factor in (4.9) can be written as $f(E_c)$ using the Maxwell–Boltzmann distribution function, (4.8). The inequality in (4.8) holds for the energy range for electrons in the conduction band, that is, for $E \geq E_c$, because $f(E)$ is to be used within the range of integration of (4.7). Thus,

$$f(E_c) = \exp\left(\frac{-E_c - \zeta}{kT}\right) \ll 1 \qquad (4.10)$$

holds for the exponential factor in (4.9), too. From the above, the electron density of a nondegenerate semiconductor satisfies $n \ll N_c$. To be more specific, if $n$ is not greater than about a tenth of $N_c$, the semiconductor can be regarded as nondegenerate.

The meaning of the term "effective density of states" is that, if all the states ($N_c$ per unit volume) were concentrated at the bottom of the conduction band ($E = E_c$), then only $f(E_c)(\ll 1)$ times that amount would actually be occupied by electrons, as suggested by (4.9). That is to say, $N_c$ is the number of states per unit volume if the states in the conduction band were considered to be concentrated at the bottom of the conduction band, and (4.10) is the occupancy of those states. Note also that $N_c$ has the same dimensions as the electron density.

### 4.2.2 Hole Density

The hole density can be found in basically the same way as for electrons, but note that "the probability that a hole occupies a state" equals "the probability that an electron does not occupy a state." Thus, using (4.4) on p. 89, the distribution function $f_h(E)$ for holes is given by

$$f_h(E) \equiv 1 - f(E) = \frac{1}{\exp\left(\frac{\zeta - E}{kT}\right) + 1}. \qquad (4.11)$$

Using the density-of-states function (4.2) on p. 88, the hole density in the valence band is given by

$$p = \int_{E_v}^{-\infty} N(E) f_h(E)\,\mathrm{d}E. \tag{4.12}$$

The upper end of the integral range in (4.12) should really be the bottom of the valence band, but it is replaced with $-\infty$.

If $(\zeta - E)/kT \gtrsim 3$ is satisfied, that is, if the semiconductor is nondegenerate, (4.11) can be approximated as

$$f_h(E) = \frac{1}{\exp\left(\frac{\zeta - E}{kT}\right) + 1} \approx \frac{1}{\exp\left(\frac{\zeta - E}{kT}\right)}$$
$$= \exp\left(-\frac{\zeta - E}{kT}\right) \quad (\ll 1). \tag{4.13}$$

This is the Maxwell–Boltzmann distribution function for holes. Considering the facts that the range of integration of (4.12) is $E \leq E_v$ (i.e., within the valence band) and that the range of the Fermi level $\zeta$ is given by (4.5), $(\zeta - E)/kT \gtrsim 3$ implies $\zeta \gtrsim E_v + 3kT$. Note that this inequality can hold concurrently with the inequality $\zeta \lesssim E_c - 3kT$ considered below (4.8) on p. 92. Those semiconductors in which these inequalities hold are degenerate semiconductors.

The hole density is given by

$$p = N_v \exp\left(-\frac{\zeta - E_v}{kT}\right), \qquad \text{(Nondegenerate hole density)} \tag{4.14}$$

where $N_v$ is the effective density of states of the valence band, and $E_v$ is the electron energy at the top of the valence band (Fig. 4.1 on p. 88). By the same argument as for the electron density, it can be shown that $p \ll N_v$ for nondegenerate semiconductors. More specifically, if $p$ is not greater than about a tenth of $N_v$, the semiconductor can be regarded as nondegenerate.

### 4.2.3 Product of Hole and Electron Densities

Based on the above discussion, we can summarize the range of the Fermi level $\zeta$ for nondegenerate semiconductors as follows:

$$E_v + 3kT \lesssim \zeta \lesssim E_c - 3kT. \tag{4.15}$$

Again, the factor 3 in (4.15) is a rough guide.

Since both electron density $n$ and hole density $p$ are equal to the intrinsic carrier density $n_i$ in intrinsic semiconductors (p. 7), $np = n_i^2$ clearly holds. In fact, this is a more general relationship that holds in uniform nondegenerate semiconductors in equilibrium. This can be confirmed by multiplying (4.9) and (4.14) as follows:

$$pn = N_c N_v \exp\left(-\frac{E_g}{kT}\right) = n_i^2. \qquad \text{(Equilibrium } pn \text{ product)} \qquad (4.16)$$

If you have any doubts about the second equality in (4.16), try to find the $pn$ product using (4.24) on p. 98 and (4.25), which will be derived later. Note that (4.16) does not hold in degenerate semiconductors.

From (4.16), the intrinsic carrier density can be written as

$$n_i = \sqrt{N_c N_v} \exp\left(-\frac{E_g}{2kT}\right). \qquad \text{(Intrinsic carrier density)} \qquad (4.17)$$

It can be seen from (4.17) that the larger the energy gap $E_g$, the smaller the intrinsic carrier density $n_i$. $n_i$ depends also on the effective densities of states, $N_c$ and $N_v$, but the exponential dependence on $E_g$ has a bigger impact. Table 1.3 (p. 5) shows that GaAs with $E_g = 1.4\,\text{eV}$ has a smaller value of intrinsic carrier density ($n_i = 2.1 \times 10^6$) than Si with $E_g = 1.1\,\text{eV}$ and $n_i = 1.0 \times 10^{10}$ (see Problem 4.3 on p. 113).

### 4.2.4 Insulators

Let us now discuss the difference between semiconductors and insulators. Insulators can be considered as materials with a very large energy gap. For example, the $E_g$ of silicon dioxide ($SiO_2$) is as high as 8 to 9 eV. If the $E_g$ is very large, $n_i$ becomes very small by (4.17). As a result, carriers can hardly exist in the allowed bands above and below the energy gap, resulting in extremely large resistivity (§5.4.2).

Fig. 4.3 summarizes the energy band structures of metals (p. 91), semiconductors, and insulators. The rectangles in Fig. 4.3 represent allowed bands, and the shaded regions indicate the states that are actually occupied by electrons. Note that the drawings are not to scale. Specifically, the energy gap $E_g$ of the insulator in Fig. 4.3 is drawn much smaller than it actually is. Completely empty allowed bands, such as in the right band diagram in Fig. 4.3, do not contribute to electrical conduction. Allowed bands that are completely filled with electrons, such as those in the left and right band diagrams in Fig. 4.3, do not contribute to electrical conduction either. For an allowed band to contribute to electrical conduction, unoccupied states are required. The Fermi level of insulators, like that of semiconductors, is located in the forbidden band, but it is not shown in Fig. 4.3 because it is usually unnecessary to consider it explicitly.

### 4.2.5 Fermi Level of Intrinsic Semiconductors

So far, we have postponed the discussion of how the value of the Fermi level $\zeta$ is determined. As mentioned in §1.3.7, uniform semiconductor crystals are usually electrically neutral. The value of $\zeta$ is determined such that the *charge neutrality condition* is satisfied. Here we use (4.9) on p. 93 for the electron density and (4.14) on p. 94 for the density to find an expression for the Fermi level of an intrinsic semiconductor, also known as the *intrinsic Fermi level* $E_{\mathrm{i}}$.

Intrinsic semiconductors are undoped, and there are no donor or acceptor ions in them. The charge neutrality condition can be written as $n = p$, which is the same as (1.1) on p. 8. Equating (4.9) on p. 93 and (4.14) on p. 94, we obtain

$$N_{\mathrm{c}} \exp\left(-\frac{E_{\mathrm{c}} - \zeta}{kT}\right) = N_{\mathrm{v}} \exp\left(-\frac{\zeta - E_{\mathrm{v}}}{kT}\right). \tag{4.18}$$

Solving (4.18) for $\zeta$, the intrinsic Fermi level, $\zeta = E_{\mathrm{i}}$, is given as follows (see Problem 4.4 on p. 113):

$$E_{\mathrm{i}} = \frac{E_{\mathrm{c}} + E_{\mathrm{v}}}{2} - \frac{kT}{2} \ln\left(\frac{N_{\mathrm{c}}}{N_{\mathrm{v}}}\right), \qquad \text{(Intrinsic Fermi level)} \tag{4.19}$$

where ln is the natural logarithm ($\log_e$). The first term of (4.19) represents the middle of the energy gap (or midgap), and the second term represents the deviation from the middle.

#### *Example: Deviation of $E_{\mathrm{i}}$ from the Midgap*

Let us look at a numerical example to see how far the intrinsic Fermi level $E_{\mathrm{i}}$ gets from the middle of the forbidden band. First, note that at room temperature $T = 300\,\mathrm{K}$, $kT \simeq 26\,\mathrm{meV}$ (see Problem 1.3 on p. 26). According to Table 1.3 (p. 5), the effective densities of states of silicon are $N_{\mathrm{c}} \simeq 2.8 \times 10^{19}\,\mathrm{cm}^{-3}$ and $N_{\mathrm{v}} \simeq 2.6 \times 10^{19}\,\mathrm{cm}^{-3}$. Since

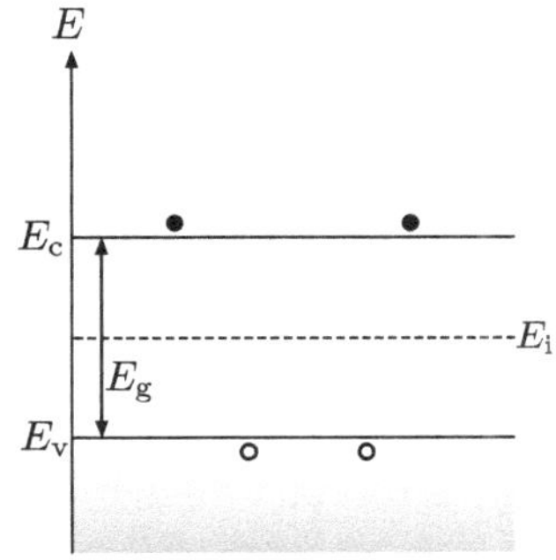


FIGURE 4.4 Intrinsic Fermi level $E_i$ lies almost in the middle of the energy gap.

$N_c/N_v \simeq 1$, $\ln(N_c/N_v)$ should be close to 0. Actual calculations show that $E_i$ lies only about 1 meV below the midgap. ■

In general, the intrinsic Fermi level $E_i$ lies almost in the middle of the forbidden band, as shown in Fig. 4.4 (see Problem 4.5 on p. 113). Similar to (1.5) and (1.6) on p. 15, $E_i$ is related to the electrostatic potential $\psi$ as follows:

$$E_i(x) = -q\psi(x) + \text{const.} \tag{4.20}$$

Note that the constant term in (1.5) and (1.6) and that in (4.20) differs by about $E_g/2$. In this connection, Fermi level $\zeta$ can also be expressed as follows:

$$\zeta = -q\psi_F + \text{const.}, \tag{4.21}$$

where $\psi_F$ is called the *Fermi potential*.

### 4.2.6 Carrier Density in Terms of Intrinsic Carrier Density

Equation (4.9) on p. 93 for electron density expresses $n$ as "an exponential factor times $N_c$." That is to say, the conduction band effective density of states, $N_c$, serves as the *reference density* in (4.9). Likewise, equation (4.14) on p. 94 for hole density expresses $p$ as "an exponential factor times the reference density $N_v$." In the following, we will show that it is possible to express $n$ and $p$ using the intrinsic carrier density, $n_i$, as the reference density in place of $N_c$ or $N_v$.

Let us first consider the electron density $n$. In an intrinsic semiconductor, the electron density is given by $n = n_i$ and the Fermi level is given by $\zeta = E_i$. Putting these in (4.9), we obtain

$$n_i = N_c \exp\left(-\frac{E_c - E_i}{kT}\right). \tag{4.22}$$

Solving (4.22) for $N_c$, we get

$$N_c = n_i \exp\left(\frac{E_c - E_i}{kT}\right). \tag{4.23}$$

$N_c$ has now been written in terms of $n_i$. $N_c$ in (4.9) can be eliminated by using (4.23):

$$n = n_i \exp\left(\frac{\zeta - E_i}{kT}\right). \qquad \text{(Electron density in terms of } n_i\text{)} \tag{4.24}$$

This is the new expression of the electron density $n$ we were looking for.

Similarly, the following equation can be derived from (4.14) on p. 94 for the hole density $p$ (see Problem 4.6 on p. 113):

$$p = n_i \exp\left(\frac{E_i - \zeta}{kT}\right). \qquad \text{(Hole density in terms of} n_i\text{)} \tag{4.25}$$

## 4.3 FERMI LEVEL OF DOPED SEMICONDUCTORS

Doped semiconductors share the same distribution functions (§4.1) and carrier density expressions (§4.2.1, §4.2.2, §4.2.6) with intrinsic semiconductors. However, the value of the Fermi level $\zeta$ is not equal to the intrinsic Fermi level ($\zeta \neq E_i$). In this section, we investigate what happens to the Fermi level of doped semiconductors. The Fermi level $\zeta$ is determined based on the charge neutrality condition, just as in the case of intrinsic semiconductors (§4.2.5).

### 4.3.1 Nondegenerately Doped Semiconductors

#### *4.3.1.1 Dopant Density and Carrier Density*

The charge neutrality condition for a doped semiconductor in equilibrium can be written as

$$n + N_A^- = p + N_D^+, \qquad \text{(Charge neutrality condition)} \tag{4.26}$$

where $N_A^-$ is the ionized acceptor density, and $N_D^+$ is the ionized donor density. $N_A^-$ and $N_D^+$ are lower than the acceptor density $N_A$ and the donor density $N_D$, respectively. That is, $N_A^- < N_A$ and $N_D^- < N_D$. However, since the ionization rate of dopants is quite high at room temperature, ionized dopant densities are often approximated to be equal to dopant densities: $N_A^- \approx N_A$ and $N_D^+ \approx N_D$. In this book, we use $N_A^-$ and $N_D^+$ to emphasize the facts that acceptor ions are anions and donor ions are cations.

Let us define $\Delta n$ as the difference between the electron density and the hole density:

$$\Delta n \equiv n - p. \tag{4.27}$$

$\Delta n$ represents the "signed net carrier density" corresponding to the "net carrier charge density" $-q(n-p)$. $\Delta n = 0$ in an intrinsic semiconductor and $\Delta n \neq 0$ in a doped semiconductor. Let us investigate the carrier density of a doped semiconductor using $\Delta n$ and then consider what happens to the Fermi level $\zeta$. Equation (4.26) can be rewritten using $\Delta n$ as follows:

$$\Delta n = N_D^+ - N_A^-. \qquad \text{(Charge neutrality condition)} \tag{4.28}$$

Equation (4.28) says that the net carrier density (left-hand side) is determined by the net ionized dopant density (right-hand side).

Incidentally, it was implicitly assumed in §1.3.7 that a doped semiconductor is only doped with either donors or acceptors. However, it is, of course, possible to dope a semiconductor with both donors and acceptors, in which case whichever is doped more determines the polarity of the semiconductor (n-type or p-type). That is the meaning of (4.28). The polarity of an n- or p-type semiconductor can be reversed by injecting dopants of the opposite polarity. This operation is called *compensation*. Depending on $\Delta n \gtrless 0$, the right-hand side of (4.28) may be written as $N_D^{+\prime}$ (effective ionized donor density) or $N_A^{-\prime}$ (effective ionized acceptor density). These may further be simplified to $N_D^+$ or $N_A^-$, without a prime.

Now, let us substitute (4.27) for $p$ in (4.16) on p. 95.

$$n(n - \Delta n) = n_i^2,$$
$$n^2 - n\Delta n - n_i^2 = 0. \tag{4.29}$$

From this quadratic equation, the electron density is found to be

$$n = \frac{\Delta n + \sqrt{\Delta n^2 + 4n_i^2}}{2}. \tag{4.30}$$

The hole density is given by

$$p = n - \Delta n = \frac{-\Delta n + \sqrt{\Delta n^2 + 4n_i^2}}{2}. \tag{4.31}$$

Note that we used the facts that $n > 0$ and $p > 0$ to get (4.30) and (4.31).

Usually, doped semiconductors are doped such that $|\Delta n| \gg n_i$ (p. 16). So let us apply the following approximate formula:

$$(1+x)^{1/2} \approx 1 + \frac{x}{2} \qquad (\text{for } |x| \ll 1) \tag{4.32}$$

to the square root terms in (4.30) and (4.31). Then,

$$\begin{aligned} \sqrt{\Delta n^2 + 4n_i^2} &= |\Delta n| \left[1 + \left(\frac{2n_i}{\Delta n}\right)^2\right]^{1/2} \approx |\Delta n| \left(1 + \frac{2n_i^2}{|\Delta n|^2}\right) \\ &= |\Delta n| + \frac{2n_i^2}{|\Delta n|}. \end{aligned} \tag{4.33}$$

Substitute this result into (4.30) and (4.31).

Since $\Delta n = |\Delta n| > 0$ holds in an n-type semiconductor, we obtain

$$n_N \approx \Delta n + \frac{n_i^2}{\Delta n} \approx \Delta n, \qquad \text{(Electron density in n-type)} \tag{4.34}$$

$$p_N \approx \frac{n_i^2}{\Delta n} (\ll n_i). \qquad \text{(Hole density in n-type)} \tag{4.35}$$

The subscript "N" in (4.34) and (4.35) indicates that the quantity is that in an n-type semiconductor. From (4.34) and (4.28), we see that the majority carrier (electron) density $n_N$ is determined by the net ionized impurity density, (4.28). The minority carrier density can be found from the *pn* product (4.16) on p. 95. From (4.35), the following inequality holds in an n-type semiconductor:

$$n_N > n_i > p_N. \tag{4.36}$$

This was actually implied by (4.16), too. Note, however, that (4.36) is only for equilibrium and uniform n-type semiconductors without band bending. We will see cases in which (4.36) does not hold in Chapter 5 onward.

Since $\Delta n = -|\Delta n| < 0$ in a p-type semiconductor, we get from (4.31) and (4.33)

$$p_\mathrm{P} \approx |\Delta n| + \frac{n_\mathrm{i}^2}{|\Delta n|} \approx |\Delta n|, \qquad \text{(Hole density in p-type)} \qquad (4.37)$$

$$n_\mathrm{P} \approx \frac{n_\mathrm{i}^2}{|\Delta n|}(\ll n_\mathrm{i}). \qquad \text{(Electron density in p-type)} \qquad (4.38)$$

The subscript "P" in (4.37) and (4.38) indicates that the quantity is that in a p-type semiconductor. The following inequality holds in a p-type semiconductor:

$$p_\mathrm{P} > n_\mathrm{i} > n_\mathrm{P}. \qquad (4.39)$$

Equation (4.39) too, is for equilibrium and uniform p-type semiconductors without band bending.

#### *4.3.1.2 Fermi Level*

Let us derive the Fermi level $\zeta$. First, let us consider n-type semiconductors. Solving (4.9) on p. 93 for $\zeta$, we obtain

$$\zeta = E_\mathrm{c} - kT \ln\left(\frac{N_\mathrm{c}}{n}\right) \qquad \text{(Fermi level of n-type relative to } E_\mathrm{c}) \quad (4.40)$$

$$\approx E_\mathrm{c} - kT \ln\left(\frac{N_\mathrm{c}}{\Delta n}\right) \approx E_\mathrm{c} - kT \ln\left(\frac{N_\mathrm{c}}{N_\mathrm{D}^+}\right). \qquad (4.41)$$

Equation (4.33) was used for the first approximation in (4.41). Equation (4.28) and the fact that $N_\mathrm{D}^+ \gg N_\mathrm{A}^-$ in n-type semiconductors were used in the last approximation. Recall that $n \ll N_\mathrm{c}$ in nondegenerate semiconductors (p. 94), and therefore the logarithm in the second term of (4.40) is positive. So (4.40) says that the Fermi level $\zeta$ of a nondegenerate n-type semiconductor is below the conduction band bottom, $E_\mathrm{c}$. This was already seen in (4.15) on p. 94.

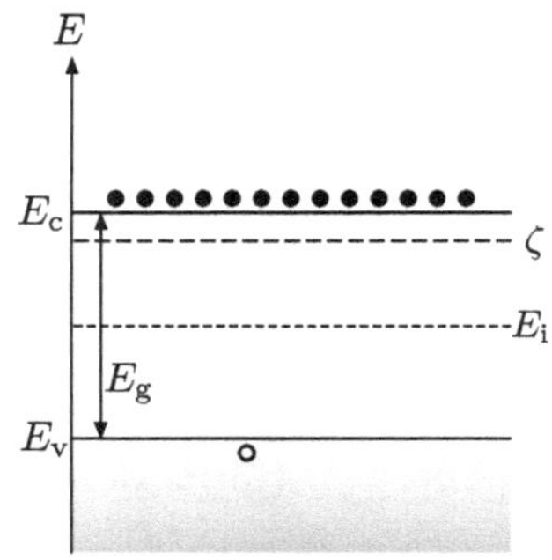


FIGURE 4.5 Fermi level of an n-type semiconductor lies between $E_c$ and $E_i$.

Let us derive yet another expression that will tell us a little more about the n-type Fermi level. Solving (4.24) on p. 98 for $\zeta$, we get

$$\zeta = E_i + kT \ln\left(\frac{n}{n_i}\right) \qquad \text{(Fermi level of n-type relative to } E_i) \quad (4.42)$$

$$\approx E_i + kT \ln\left(\frac{N_D^+}{n_i}\right). \quad (4.43)$$

Equation (4.42) says that the Fermi level $\zeta$ of a nondegenerate n-type semiconductor is above $E_i$. This is new information that could not be read from (4.15) on p. 94. From the above, the range of the Fermi level of the nondegenerate n-type semiconductor is given by

$$E_i < \zeta \lesssim E_c - 3kT. \quad (4.44)$$

This situation is shown in the band diagram in Fig. 4.5.

The Fermi level of p-type semiconductors can be considered in the same way. Solving (4.14) on p. 94 for $\zeta$ gives

$$\zeta = E_v + kT \ln\left(\frac{N_v}{p}\right) \qquad \text{(Fermi level of p-type relative to } E_v) \quad (4.45)$$

$$\approx E_v + kT \ln\left(\frac{N_v}{|\Delta n|}\right) \approx E_v + kT \ln\left(\frac{N_v}{N_A^-}\right). \quad (4.46)$$

$N_A^- \gg N_D^+$ in p-type semiconductors was used in (4.46). Equation (4.45) says that the Fermi level $\zeta$ of a nondegenerate p-type

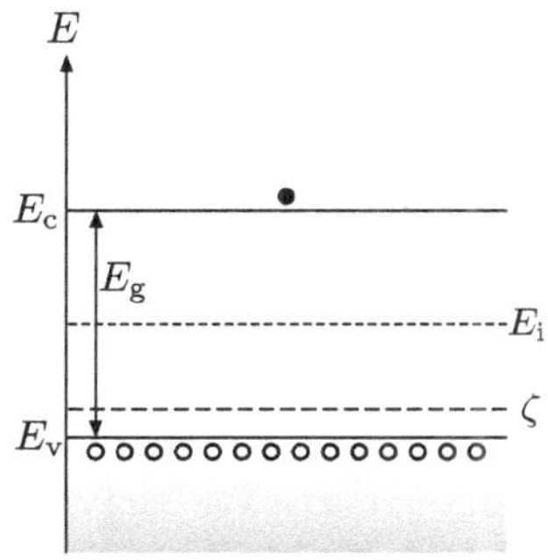


FIGURE 4.6 Fermi level of a p-type semiconductor lies between $E_v$ and $E_i$.

semiconductor is above $E_v$. By solving (4.25) on p. 98 for $\zeta$, we obtain

$$\zeta = E_i - kT \ln\left(\frac{p}{n_i}\right) \qquad \text{(Fermi level of p-type relative to } E_i\text{)} \qquad (4.47)$$

$$\approx E_i - kT \ln\left(\frac{N_A^-}{n_i}\right). \qquad (4.48)$$

Equation (4.47) says that the Fermi level $\zeta$ of a nondegenerate p-type semiconductor is below $E_i$. From the above and (4.15) on p. 94, the range of the Fermi level of the nondegenerate p-type semiconductor is given by

$$E_v + 3kT \lesssim \zeta < E_i. \qquad (4.49)$$

This situation is shown in the band diagram in Fig. 4.6.

### *Example: Nondegenerate Doping Density*

Let us look at a numerical example. As noted on p. 92, nondegenerate semiconductors are semiconductors with low doping density. From (4.34) and (4.37) on p. 100, the majority carrier density equals the ionized dopant density (after compensation), $N_D^{+\prime}$ or $N_A^{-\prime}$. In n-type semiconductors, this means

$$N_D^{+\prime} \ll N_c. \qquad \text{(Condition for nondegenerate n-type doping)} \qquad (4.50)$$

In p-type semiconductors,

$$N_A^{-\prime} \ll N_v. \qquad \text{(Condition for nondegenerate p-type doping)} \qquad (4.51)$$

Typical doping densities for nondegenerate silicon are $10^{16}$ to $10^{18}\,\text{cm}^{-3}$. A comparison of these values with the values of $N_c$ and $N_v$ given in Table 1.3 (p. 5) shows that (4.50) and (4.51) are satisfied. ■

### 4.3.2 Degenerate Semiconductors

Nondegenerate semiconductors include intrinsic semiconductors, but degenerate semiconductors are always heavily doped. The previously derived equations (4.9) and (4.24) for electron density and (4.14) and (4.25) for hole density are not applicable to degenerate semiconductors, making analytical treatment of these materials difficult. Qualitatively, when the Fermi level $\zeta$ goes outside the nondegenerate range (4.15) on p. 94, the increase in majority carrier density becomes slower than is suggested by a nondegenerate expression, (4.24) or (4.25). When the doping density is high, it is not enough to simply perform the numerical integration in (4.7) on p. 92 or (4.12) on p. 94 to obtain the carrier density. As the dopant density increases, the ionization rate of the dopants decreases, and the presence of many dopants in a crystal (a kind of periodic structure) results in the formation of the so-called "impurity bands," which, in effect, reduces the energy gap. In short, rigorous theoretical treatment of degenerate semiconductors is cumbersome.

However, it is possible to infer the general properties of degenerate semiconductors by carrying over the theory for nondegenerate semiconductors. In degenerate n-type semiconductors, (4.50) does not hold, and hence $N_D^+ \approx N_c$. From (4.41) on p. 101,

$$\zeta \approx E_c. \qquad \text{(Fermi level of degenerate n-type semiconductor)} \quad (4.52)$$

The above derivation using (4.41) intended for nondegenerate semiconductors is, of course, suspect, but (4.52) is often used as a rough estimate for the Fermi level of degenerate n-type semiconductors. Equation (4.52) also appears in (4.5) on p. 89. Substituting (4.52) into (4.9) yields

$$n \approx N_c. \quad (4.53)$$

Similarly, since (4.51) does not hold in degenerate p-type semiconductors, from $N_A^- \approx N_v$ and (4.46),

$$\zeta \approx E_v. \qquad \text{(Fermi level of degenerate p-type semiconductor)} \quad (4.54)$$

Substituting (4.54) into (4.14) yields

$$p \approx N_v. \tag{4.55}$$

The above results were obtained by applying the equations for non-degenerate semiconductors to degenerate semiconductors. This is the crudest approximate treatment of degenerate semiconductors. The next level of approximation will involve similar formulas with some corrections for degenerate semiconductors.

**CHICKEN-AND-EGG QUESTION: CARRIER DENSITY AND FERMI LEVEL**

The expressions (4.9) on p. 93 and (4.24) on p. 98 for the electron density $n$ are written in the form $n(\zeta)$, as a function of the Fermi level $\zeta$ (likewise for the hole density expressions). Equation (4.7) on p. 92, from which these equations were derived, also contained $\zeta$ via the distribution function. So, it might appear that some physical (possibly quantum-mechanical) mechanism first determines $\zeta$, which in turn determines $n$. But in the discussion of §4.3.1, (4.40) on p. 101, for example, shows that $\zeta$ is a function, $\zeta(n)$, of $n$. Astute readers might have wondered which really is determined first—the Fermi level $\zeta$ or the electron density $n$.

To make a long story short, carrier density and Fermi level are determined simultaneously. $n$ and $\zeta$ are, actually, two ways of expressing essentially the same thing. The equations for the electron density, (4.9) and (4.24), and those for the Fermi level, (4.40) and (4.42), are conversion equations for converting $n$ and $\zeta$ into each other. This fact is closely related to the importance of the energy band diagram (§5.2.4).

When performing numerical calculations, the Poisson equation and the continuity equations are solved simultaneously, often using the electron and hole densities (as well as the electrostatic potential) as the unknowns (§5.7). The results ($n$ and $p$) are used to determine the Fermi level (or more precisely, the quasi-Fermi levels, $\zeta_n$ and $\zeta_p$ (see §5.2)) via the conversion equation. However, it is, in principle, perfectly fine to calculate the (quasi)-Fermi level as an unknown, and then determine the carrier densities.

## 4.4 FERMI LEVEL AND CHEMICAL POTENTIAL

In §4.1, we introduced a quantity called the Fermi level, but we have postponed a detailed explanation of its meaning. The Fermi level $\zeta$ is often described as the value of electron energy $E$ at which the value of the Fermi–Dirac distribution function $f(E)$, given in (4.4) on p. 89, equals 1/2. This statement is technically correct, as is clear from Fig. 4.2 on p. 88, except that it is of little help in understanding the operation of semiconductor devices. Given the fact that the Fermi level of nondegenerate semiconductors lies in the forbidden band (see (4.15) on p. 94) *where there are no states*, it does not make much sense to say that the probability of the state at $E = \zeta$ being occupied is 1/2, does it? Then, how can we make sense of the Fermi level in a way conducive to understanding electrical conduction and device physics? The answer is to recognize that the Fermi level is the (electro)chemical potential of conduction electrons. Let us take a look at what exactly chemical potential is.

### 4.4.1 Properties of Chemical Potential

The *chemical potential* is a quantity defined for freely moving particles that make up a gas. From this point on, a *system* in this section refers to a box containing such gas particles, as shown in Fig. 4.7. For example, a small piece of semiconductor containing many conduction electrons is an example of such a system. The particles we are considering here are microscopic gas particles in thermal motion (§5.4.1), not like marbles that can stay still in a box. Electrons and holes in solids can be considered gases (p. 92).

Chemical potential is a quantity that is mathematically formulated in statistical mechanics or thermodynamics. But for the sake of simplicity, we will try to explain it without using mathematical expressions as much as possible. The chemical potential has the following properties:

(i) Chemical potential $\zeta$ is a quantity with the dimensions of energy.

(ii) Chemical potential is a quantity defined only in equilibrium (p. 115).

(iii) In a system in equilibrium, $\zeta$ takes a constant value throughout the system.

(iv) $\zeta$ is the energy or work required to add one additional particle to a system kept at a constant temperature and volume.

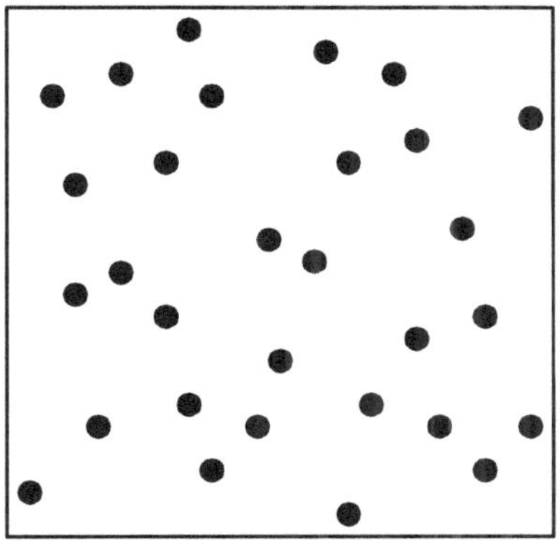

FIGURE 4.7 Conduction electrons in a semiconductor can be regarded as a gas.

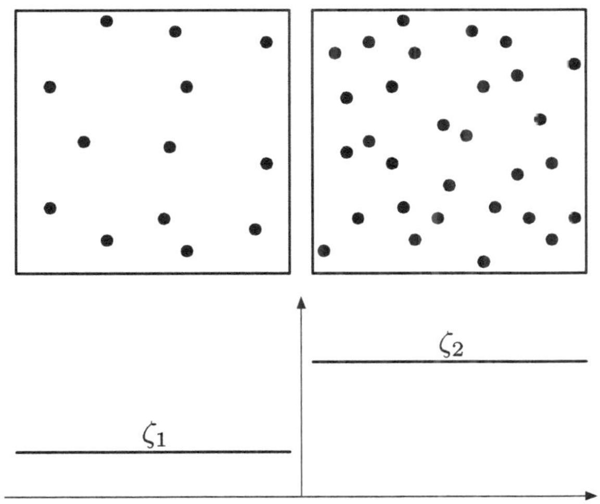


FIGURE 4.8 Two systems with different values of chemical potential.

(v) A larger $\zeta$ means that the particles have a stronger tendency to leave the system.

(vi) Thus, if two systems (system 1 and system 2) contain the same kind of particles and $\zeta_1 < \zeta_2$ (Fig. 4.8), then particles will tend to go from system 2 to system 1.

(vii) The work required to move one particle from system 1 to system 2 equals $\zeta_2 - \zeta_1$.

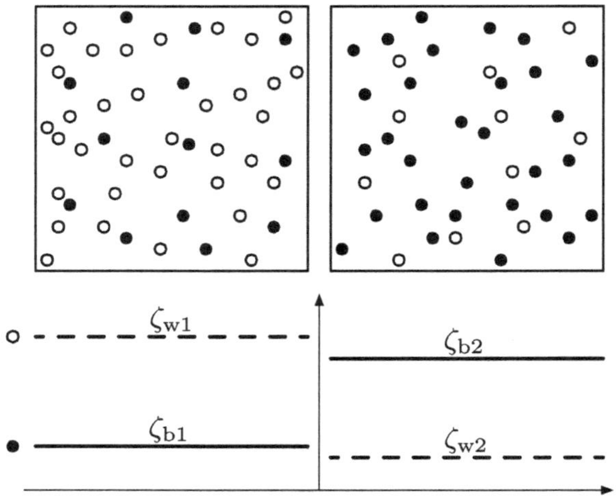


FIGURE 4.9 Chemical potentials for two kinds of gases.

(viii) When the two systems with $\zeta_1 < \zeta_2$ are brought into contact and are allowed to exchange particles for a long time until equilibrium is reached, the overall chemical potential will eventually be $\zeta_3$, where $\zeta_1 < \zeta_3 < \zeta_2$.

(ix) $\zeta$ is related to the density of particles; the higher the density, the larger the value of $\zeta$ (Fig. 4.8).

(x) Chemical potential is an *intensive* quantity.

(xi) If there is more than one particle species, the chemical potential is defined for each particle species (Fig. 4.9).

(xii) If an external force (electromagnetic force, gravity, etc.) is acting on particles, the chemical potential (in the broad sense) is also related to the external force.

"Intensive" in (x) means that the value does not change when the size of the system is changed. Such a quantity is called an *intensive quantity* or *intensive variable*. Temperature, for example, is an intensive quantity. The antonym is the *extensive quantity* or *extensive variable*, which is a variable related to the quantity or amount of something. For example, the number of particles and volume are extensive quantities.

From the above, we can see that the chemical potential is related to particle density, and from (iii), (vi), and (viii), we can also see that it is somewhat similar to temperature. Of course, in the case of temperature, it is not particles that move between systems, but thermal energy.

Point (xi) becomes relevant when considering electrons and holes (§5.2.2).

From (v) through (viii), the chemical potential is related to the inclination of particles to diffuse, so if an external force, such as electrostatic or gravitational force, is acting on particles, it should also affect the chemical potential. This is what (xii) suggests. We will discuss the chemical potential in a broad sense in the presence of external force in §4.4.2.

### 4.4.2 Chemical Potential in the Presence of an External Force

When an external force is acting on particles, the *total chemical potential*, $\zeta_{tot}$, which includes the effect of the external force, is given by [15]:

$$\zeta_{tot} = \zeta_{int} + \zeta_{ext}, \qquad \text{(Total chemical potential)} \qquad (4.56)$$

where $\zeta_{int}$ is the *internal chemical potential*, which is the chemical potential that does not depend on external force. $\zeta_{ext}$ is the *external chemical potential*, which accounts for the effect of the external force, as mentioned in (xii) on p. 108. If the force is a *conservative force* such as electrostatic force or gravity, then $\zeta_{ext}$ equals the potential energy per particle due to the conservative field. If the particles under consideration are charged particles and also if an electrostatic field exists, the total chemical potential is called the *electrochemical potential*. In solid-state physics, the electrochemical potential of electrons is also called the *Fermi level*.

#### *Example: Fermi Levels of Two Metals*

Consider two dissimilar metal pieces. There is no electrostatic potential gradient ($\zeta_{ext}$ gradient) inside either piece of metal. As shown in Fig. 4.10, the initial Fermi levels of the metal pieces are assumed to be different ($\zeta_{1,initial} < \zeta_{2,initial}$). However, both metal pieces contain cations that neutralize the negative charge of electrons, so they are electrically neutral. If the two metal pieces were connected by a conductor, electrons would flow from right to left (current would flow from left to right) due to (vi) on p. 107.

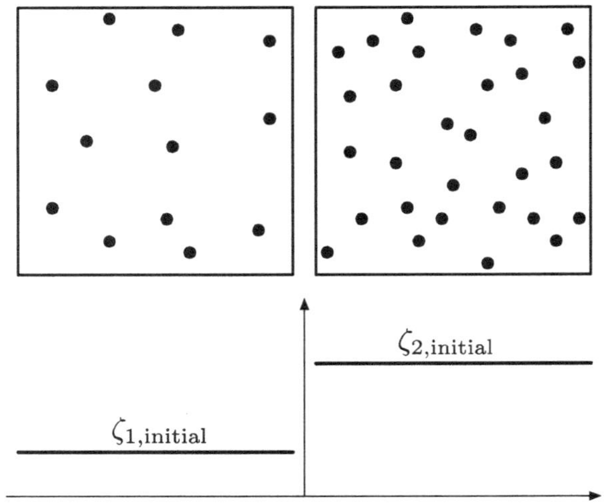


FIGURE 4.10 Two dissimilar metal pieces with electrons.

Next, choose a voltage $\Delta V$ so that $q\Delta V = \zeta_{2,\text{initial}} - \zeta_{1,\text{initial}}$ ($> 0$), and connect a voltage source as shown in Fig. 4.11. The voltage source is connected so that the Fermi level of the metal piece on the right is lowered. The result is $\zeta_{2,\text{final}} = \zeta_{1,\text{final}}$. Since the incremental resistance of the voltage source is 0, electrons are allowed to move between the two metal pieces in the state shown in Fig. 4.11. Nevertheless, there is no net flow of electrons. That is, the two metal pieces are in diffusive equilibrium (Problem 4.9 on p. 114).■

Internal chemical potential does not include the effects of external force. Thus, if a system is in equilibrium with an external force acting and if there is no net flow of particles, there may be a gradient in particle density, and $\zeta_{\text{int}}$ will not necessarily be constant in the system. This means that the $\zeta$ in (iii) on p. 106 is not the internal chemical potential $\zeta_{\text{int}}$ but the total chemical potential $\zeta_{\text{tot}}$.

### *Example: Atmosphere on Earth's Surface*

Gravity is a force that acts on all particles with mass, whether or not they are charged. Earth's atmosphere is in thermal motion and tends to diffuse into space (vacuum). Earth's gravity acts on the atmosphere and counters its attempt to escape. The density or the existence of the atmosphere on a celestial body is determined by the balance between diffusion due to the thermal motion of gas particles and gravity.

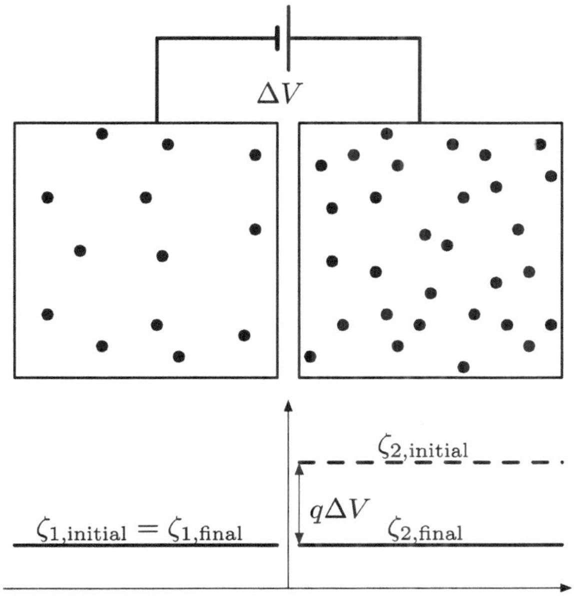


FIGURE 4.11 Two metal pieces are connected via a voltage source.

Gravity depends on the distance from Earth's center of gravity, but for simplicity, let us consider the region near the Earth's surface where the gravitational field can be assumed uniform with the acceleration of gravity, $g$. Suppose there is a gas particle of mass $m$ at height $h$. The potential energy (i.e., external chemical potential) of this gas particle is given by

$$\zeta_{\text{ext}} = mgh. \qquad \text{(Potential energy of gas particle)} \tag{4.57}$$

In equilibrium,

$$\zeta_{\text{tot}} = \text{const.} \tag{4.58}$$

holds, and therefore the internal chemical potential is given by

$$\zeta_{\text{int}} = \zeta_{\text{tot}} - \zeta_{\text{ext}} = -mgh + \text{const.} \tag{4.59}$$

This is depicted in Fig. 4.12.

Needless to say, the real atmosphere is not in equilibrium. ■

Chapter 5 extends the concept of chemical potential to nonequilibrium states and links it to the operation of semiconductor devices.

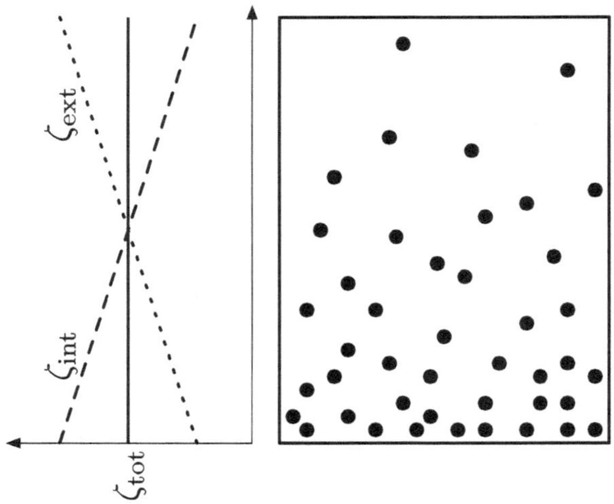


FIGURE 4.12 Chemical potential of the atmosphere near the ground surface.

## 4.5 SUMMARY

In this chapter, we discussed the physics of spatially uniform semiconductors in equilibrium.

- Electron orbitals in an atom have discrete energy levels, whereas those in a large crystal are band-like with continuously distributed energy levels.
- Electron and hole densities of semiconductors are determined by the effective densities of states, $N_c$ and $N_v$, which are material constants, and the Fermi level $\zeta$, which depends on impurity doping.
- The Fermi level $E_i$ of an intrinsic semiconductor is located around the middle of the forbidden band. For n-type semiconductors, $\zeta > E_i$, and for p-type semiconductors, $\zeta < E_i$.
- The Fermi level is the electrochemical potential of conduction electrons in solids.
- The electrochemical potential is determined by the density of charged gas particles and the electrostatic field acting on them.

## 4.6 PROBLEMS

4.1 If the Fermi energy is $E_F$ and the Fermi level at a finite temperature is $\zeta$, then $E_F \neq \zeta$ in general (§4.1). Now, which of the following holds, $E_F > \zeta$ or $E_F < \zeta$? Consider the fact that the Fermi level is the electrochemical potential of electrons (§4.4) as a clue.

4.2 Substitute the Maxwell–Boltzmann distribution function (4.8) into the expression (4.7) on p. 92 for the electron density expressed as an integral, and derive the equation (4.9) for the electron density of a nondegenerate semiconductor. Also, find the expression for the effective density of states, $N_c$, of the conduction band. Note that

$$\int_0^{\infty} x^{1/2} e^{-x} dx = \Gamma(3/2) = \frac{\sqrt{\pi}}{2}, \tag{4.60}$$

where $\Gamma(x)$ is a special function called the *gamma function*.

4.3 Search the Internet for lattice constants and energy band gaps of group IV and III–V semiconductors. Does any trend in the relationship between lattice constants and energy band gaps exist? If so, discuss qualitatively why such trends exist.

4.4 Derive (4.19) on p. 96 for the intrinsic Fermi level.

4.5 Using the effective densities of states, $N_c$ and $N_v$, of GaAs in Table 1.3 (p. 5), find how far the intrinsic Fermi level $E_i$ is from the middle of the forbidden band at room temperature.

4.6 From the hole density expression (4.14) on p. 94 in terms of $N_v$, derive the other hole density expression (4.25) on p. 98 in terms of the intrinsic carrier density $n_i$.

4.7 Let us consider the temperature dependence of the Fermi level of a doped semiconductor. Plot the Fermi level as a function of the absolute temperature. Show that, in the case of an n-type semiconductor, the Fermi level goes to the midpoint between the conduction band bottom $E_c$, and the donor level $E_D$ (see Fig. 1.13 on p. 20) at the low-temperature limit, as schematically shown in Fig. 4.13. What does this mean physically? On the other hand, at the high-temperature limit, the Fermi level approaches the midgap. Why? Also plot the logarithm of the majority carrier density as a function of the inverse temperature, $1/T$, for a few different dopant densities. The result for one dopant density should look roughly like Fig. 4.14. Show that the dopant level ($E_D$ or $E_A$) can be found from the slope at low temperatures.

4.8 The energy band diagrams in Fig. 4.5 on p. 102 and Fig. 4.6 on p. 103 do not show dopant atoms. Draw an energy band diagram

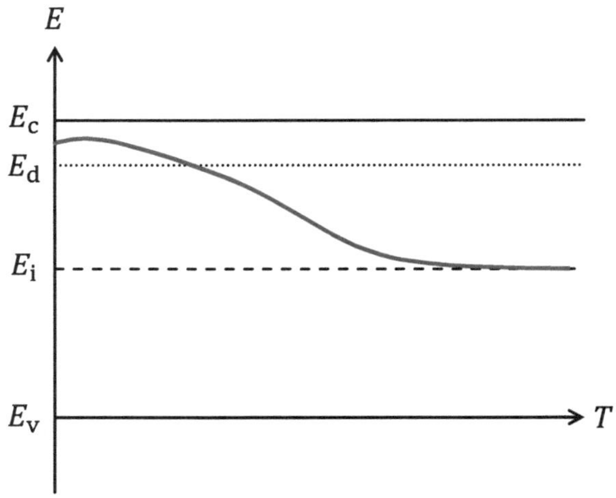


FIGURE 4.13 Temperature dependence of the Fermi level of an n-type semiconductor.

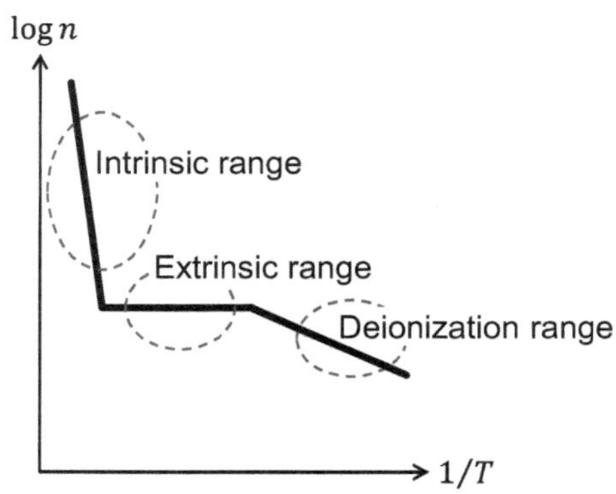


FIGURE 4.14 Approximate log carrier density versus inverse temperature. Three temperature ranges can be identified: intrinsic temperature range, intrinsic temperature range (also known as saturation range), and deionization temperature range (also known as freeze-out range).

including electrons, holes, and dopant atoms (both ionized and non-ionized).

4.9 Is the orientation of the voltage source in Fig. 4.11 (p. 93) correct? Explain in simple terms why it is correct (or wrong).

CHAPTER 5

# Carrier Dynamics in Semiconductors

In Chapter 4, we learned about semiconductors that are in equilibrium and have no net current flow in them. In this chapter, we consider electrical conduction and related phenomena. The presence of detectable current flow means there is net carrier transfer, and the state is a nonequilibrium state. The generation and recombination of carriers, phenomena closely related to electrical conduction, are also discussed.

## 5.1 EQUILIBRIUM AND NONEQUILIBRIUM STATES, STEADY AND NONSTEADY STATES

We have used the terms *equilibrium (state)* and *nonequilibrium (state)* without any particular explanation. There are also two similar terms, *steady state* and *nonsteady state*, and it may be difficult to distinguish between equilibrium state and steady state and between nonequilibrium state and nonsteady state. Fig. 5.1 summarizes the relationship between these terms.

An *equilibrium state* is a state in which temperature does not change in time or space (i.e., thermal equilibrium), there is no net flow of particles (i.e., diffusive equilibrium), and chemical reactions in the broad sense are balanced (i.e., chemical equilibrium). Chemical reactions include, for example, the ionization of neutral dopants and the reverse processes. Thermodynamics and elementary statistical mechanics deal with relations between equilibrium states. Nevertheless, thermodynamics, in particular, yields a variety of conclusions of practical importance because transitions between states may occur via nonequilibrium states.

DOI: 10.1201/ 9781003439417-5

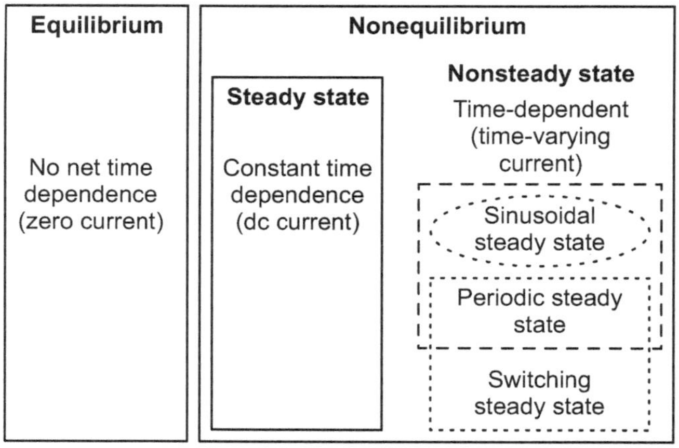


FIGURE 5.1 Equilibrium versus nonequilibrium, steady state versus nonsteady state.

States that are not in equilibrium are *nonequilibrium states*, but these can be classified into some subcategories. A *steady state* corresponds to a state in which there is a temperature gradient or a net flow of particles, but it does not change with time. For example, if a DC current is flowing through a device and its temperature is not changing, the device is in a steady state. Note that in this book we will not consider temperature gradients. Steady states can often be treated in a way similar to equilibrium.

States that are not steady states are *nonsteady states*, and these can also be divided into some subclasses. If the time variation of the system is periodic, it is called a *periodic steady state* or a *cyclostationary state*. In particular, if the periodic change is sinusoidal, it is called a *sinusoidal steady state*, and it can be easily handled in the frequency domain; that is, it can be treated like a steady state (p. 63). The AC circuit theory is a theory for sinusoidal steady states with angular frequency $\omega$. Even if the time variation is not necessarily periodic, it is possible to consider the statistical average behavior over a long period. If the average behavior does not change with time, then we can regard the state as a kind of steady state. For example, it is possible to consider the average behavior of a transistor that is repeatedly turned on and off, seemingly randomly, in a digital circuit. A state like this may be called a *switching steady state*. The switching may be periodic or aperiodic.

Actually, there are two more relevant terms, as suggested by the term “cyclostationary state”: *stationary states* and *nonstationary states*. The term “stationary state” may refer to an equilibrium state or a steady state, depending on the context. And, of course, nonstationary states are states that are not stationary.

In this book, we will mainly consider equilibrium states, steady states (at a fixed temperature), and sinusoidal steady states.

*Example: Differential Equations for Periodic State States*

The frequency-domain wave equation for lossless transmission lines and the time-independent Schrödinger equation, both shown in Fig. 3.15 (p. 71), are, mathematically, differential equations for periodic steady states.

Note, however, that in quantum mechanics, sinusoidally oscillating quantum states, called energy eigenstates, are not regarded as changing with time. Rather, sinusoidally oscillating quantum states are considered to be (and called) *stationary states* for reasons beyond the scope of this book. ■

## 5.2 QUASI-FERMI LEVELS AND CARRIER DENSITIES

The Fermi level discussed in §4.4 can only be defined in equilibrium (see (ii) on p. 106). Therefore, it is not possible to consider the Fermi level in situations where a net current is flowing, even if it is a DC current. In nonequilibrium states, *quasi-Fermi levels* are introduced and used as substitutes for the Fermi level.

### 5.2.1 Quasi-Chemical Potential

As mentioned in §4.4.1, the chemical potential is a quantity that can only be defined for equilibrium states. However, given that the difference between the chemical potentials of two systems with particles indicates which way the particles want to move, it seems convenient if we can define a similar quantity even when there is a flow of particles, that is, in nonequilibrium states.

Let us consider it more concretely using the example of a gas particle near the surface of the earth, shown in Fig. 5.2(a). The potential energy of the particle is given by (4.60) on p. 113. Since gravity is

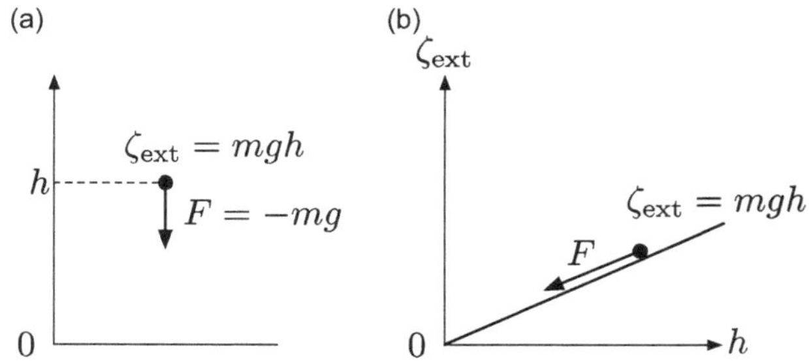


FIGURE 5.2 (a) A particle in uniform gravitational field. (b) Gravity acts on the particle such that it is pulled down the slope of $\zeta_{ext}$.

a conservative force, the force acting on this particle is given by the derivative of the potential energy as follows:

$$F = -\frac{\mathrm{d}\zeta_{ext}}{\mathrm{d}h} = -mg, \qquad \text{(Force acting on particle)} \qquad (5.1)$$

where $h$ is the height of the particle, $m$ is its mass, and $g$ is the acceleration of gravity. The minus sign before the derivative in (5.1) indicates that the force is directed toward a smaller value of the potential energy $\zeta_{ext}$. In other words, as shown in Fig. 5.2(b), the particle tries to go down the "slope" of $\zeta_{ext}$.

So far we have considered a single gas particle, but if we have an ensemble of particles of the same kind in thermal motion, and if we can define a quantity similar to the total chemical potential $\zeta_{tot}$ (p. 112) even in nonequilibrium, we expect that the *effective force* acting on each particle can be expressed in the same form as (5.1).

$$F = -\frac{\mathrm{d}\zeta'_{tot}}{\mathrm{d}h} = -\left(\frac{\mathrm{d}\zeta_{ext}}{\mathrm{d}h} + \frac{\mathrm{d}\zeta'_{int}}{\mathrm{d}h}\right), \qquad \text{(Effective force)} \qquad (5.2)$$

where $\zeta'_{tot}$ and $\zeta'_{int}$ are the nonequilibrium counterparts of $\zeta_{tot}$ and $\zeta_{int}$, respectively. The effective force $F$ in (5.2) includes the component originating from the external force (first term) and the component originating from the density gradient of the particles (second term) (see Fig. 4.12 on p. 112). The implications of (5.2) are that

- If there is a gradient in $\zeta'_{tot}$, then a net flow of gas particles exists.
- Conversely, if $\zeta'_{tot}$ is constant, there is no net flow of gas particles. That is, the system is in equilibrium.

- The steeper the gradient of $\zeta'_{tot}$, the stronger the effective force will be, so the gas particles will have a stronger tendency to go down the slope of $\zeta'_{tot}$.

The "chemical potential counterpart of a (weak) nonequilibrium state" introduced on the basis of the above ideas is called the *quasi-chemical potential* [3]. The electrochemical potential extended to a weak nonequilibrium state is called the *quasi-electrochemical potential*. Naturally, if there is no net flow of particles (i.e., in equilibrium), the total quasi-chemical potential $\zeta'_{tot}$ matches the normal total chemical potential.

If $F \neq 0$, *at least one of the signs of the first and second terms on the right-hand side of (5.2) must match the sign of the left-hand side.* If the first and second terms on the right-hand side of (5.2) have different signs, then the term with the same sign as the left-hand side has a larger absolute value and is dominant.

### 5.2.2 Electron and Hole Quasi-Fermi Levels

The Fermi level is the electrochemical potential of electrons in solid-state physics (§4.4.2). Similarly, the *quasi-Fermi level* is the Fermi level counterpart in weak nonequilibrium. Note, however, that there is a significant difference between the Fermi level and the quasi-Fermi level. Although it was not explicitly stated in Chapter 4, it is sufficient to consider only the electron Fermi level, $\zeta$, and there is no need to consider the hole Fermi level. If the latter were needed, $-\zeta$ could be regarded as the hole Fermi level. This is also evident in the carrier density equations (4.27) and (4.28) on p. 99, or in the $pn$ product (4.19) on p. 96. That is, in equilibrium, if either the electron density $n$ or the hole density $p$ is known, the other is also known. In contrast, for nonequilibrium states, it is necessary to consider the quasi-Fermi levels for electrons and holes separately. Let us consider why this is so in the following.

First, in an equilibrium semiconductor, both electrons and holes are in equilibrium. Specifically, the following (and other, see §5.6) "chemical reactions" of carrier generation and recombination are all in equilibrium:

$$\text{(covalent bond)} \rightleftharpoons \text{(electron)} + \text{(hole)} \qquad \text{(Fig. 1.7 (p. 13))} \qquad (5.3)$$

$$\text{(neutral donor)} \rightleftharpoons \text{(donor ion)} + \text{(electron)} \qquad \text{(Fig. 1.13 (p. 20))} \tag{5.4}$$

$$\text{(neutral acceptor)} \rightleftharpoons \text{(acceptor ion)} + \text{(hole)} \qquad \text{(Fig. 1.18 (p. 22))} \tag{5.5}$$

In equilibrium, rightward and leftward reactions occur at the same rate, and it appears as if no reaction is occurring.

It might seem to contradict (xi) on p. 108 that a single electrochemical potential $\zeta$ is sufficient for both electrons and holes, which are clearly different particles. The reason only one electrochemical potential is sufficient is because all of the above chemical reactions have reached chemical equilibrium. In such a state, $pn = n_i^2$, given in (4.19), holds, and if we know the electron Fermi level $\zeta$, we know the hole density immediately from (4.17) on p. 95 or (4.28) on p. 99. So *(xi) on p. 108 applies to particle species that are independent of each other*. Electrons and holes in equilibrium are interdependent as in (5.3) through (5.5).

However, when the system is in a nonequilibrium state, the equality $pn = n_i^2$ no longer holds (§5.2.3). This means that even if $n$ is known, $p$ is not. The physical reason for this is related to a couple of time constants involved in the dynamics of carriers in semiconductors.

Let us consider a p-type semiconductor as an example. Suppose that equilibrium is disturbed by some cause, such as a current flow, and that $pn \neq n_i^2$ results in a small region of the semiconductor. What can disturb the equilibrium condition, $pn = n_i^2$, with a *small change* should be the deviation, $\Delta n$, of the minority carrier (electron) density $n$ from the equilibrium value.[1] If the electron density increases and becomes $\Delta n > 0$, the excess electrons can *recombine* with holes—the majority carriers—and disappear (§5.6). Such recombination processes occur and try to bring the system back to equilibrium. The leftward reaction in (5.3) is an example of a recombination process. The approximate time that it takes for the excess carrier density $\Delta n$ to decrease appreciably is the time constant $\tau_n$, called the *electron lifetime* (§5.6.4).

---

[1] This $\Delta n$ is different from the net carrier density in (4.30) on p. 100. $\Delta n$ here is the *excess electron density* in (5.79) on p. 149. $\Delta n = 0$ in equilibrium.

TABLE 5.1 Correspondence between Drift of Electrons Due to Electric Field and Drift of Gas Due to Gravity

| | Conduction electrons | Atmospheric gas |
|---|---|---|
| External field | Electric field $\mathcal{E}$ | Acceleration of gravity $g$ |
| Force acting on particles $F$ | $-q\mathcal{E}$ | $-mg$ |
| Drift velocity $\lvert v_{\mathrm{drift}} \rvert$ | $\mu_{\mathrm{n}}\mathcal{E}$ | $\tau_{\mathrm{g}} g$ |
| Transport coefficient | Mobility $\mu_{\mathrm{n}}$ | Mean free time $\tau_{\mathrm{g}}$ |
| Einstein's relation | $D_{\mathrm{n}}/\mu_{\mathrm{n}} = kT/q$ | $D_{\mathrm{g}}/\tau_{\mathrm{g}} = kT/m$ |

But recombination is not the only process triggered by $\Delta n > 0$. Immediately after becoming $\Delta n > 0$, that small region may well be negatively charged (i.e., breakdown of the charge neutrality condition). Then, holes will be attracted to the negatively charged region, which eventually will neutralize the region. The phenomenon of majority carriers gathering around and neutralizing opposite charges is called *dielectric relaxation*, and the associated time constant, $\tau_{\mathrm{drp}}$, is called the *dielectric relaxation time* (§5.8).

Now the question is the relation between the electron (or minority carrier) lifetime $\tau_{\mathrm{n}}$ and the dielectric relaxation time $\tau_{\mathrm{drp}}$. What actually happens depends on which time constant is shorter. If $\tau_{\mathrm{n}} \ll \tau_{\mathrm{drp}}$, then the excess electrons disappear by recombination before the dielectric relaxation occurs. But at the same time, an equal number of holes disappear ($\Delta p = -\Delta n$, where $\Delta p$ is the excess hole density), so this region is still negatively charged (the charge conservation law). Then, holes are supplied from the surroundings to compensate for the recombined and vanished holes (dielectric relaxation). In this scenario, if $\tau_{\mathrm{n}}$ is sufficiently small, $\Delta n$ goes to zero ($\Delta n \to 0$) instantaneously, and since the majority carrier density $p$ is huge, we can safely assume that $|\Delta p|/p = 0$ at all times. Therefore, $pn = n_{\mathrm{i}}^2$, too, continues to hold all the time.

But in reality, it is known that $\tau_{\mathrm{n}} \gg \tau_{\mathrm{drp}}$ (see Table 5.2 on p. 122), and majority carriers spatially redistribute (i.e., dielectric relaxation) before recombination occurs. As a result, neutralization of the charged region occurs very quickly, but the nonequilibrium condition manifested by $pn \neq n_{\mathrm{i}}^2$ persists. Consequently, *electrons and holes become practically independent particles*. Then, even if one of $n$ and $p$ is known, the other is not.

If, however, the degree of deviation from the equilibrium state is small (i.e., weak nonequilibrium), the electrons can be considered to

TABLE 5.2 Time Constants Related to Carriers in Silicon [31]

| Name | Symbol | Typical value | Page |
|---|---|---|---|
| Mean free time | $\tau_e$, $\tau_h$ | $10^{-13} \sim 10^{-12}$ s | 135 |
| Minority carrier lifetime | $\tau_n$, $\tau_p$ | $10^{-9} \sim 10^{-4}$ s | 149 |
| Dielectric relaxation time | $\tau_{drn}$, $\tau_{drp}$ | $10^{-12}$ s | 155 |

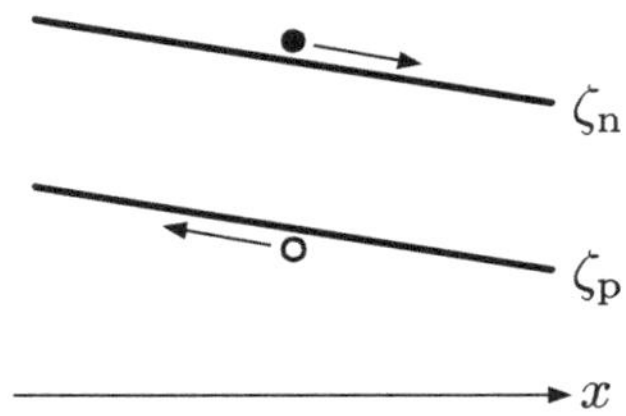


FIGURE 5.3 Directions of movement of electrons and holes with respect to quasi-Fermi level gradients.

be locally close to equilibrium by themselves, without involvement of holes via (5.3) through (5.5). Similarly, we may consider holes to be locally close to equilibrium by themselves. Being close to equilibrium, we can expect the functional form of the carrier density equation to remain the same as in the equilibrium case (§5.2.3). Now we need separate quasi-electrochemical potentials for electrons and holes, according to (xi) on p. 108. So we introduce electron quasi-Fermi level $\zeta_n$ and hole quasi-Fermi level $\zeta_p$ at each location in the semiconductor. The quasi-Fermi levels, $\zeta_n$ and $\zeta_p$, should be drawn in energy band diagrams for nonequilibrium semiconductors instead of the Fermi level $\zeta$. Since holes flow from the smaller value of $\zeta_p$ to the larger value of $\zeta_p$ as shown in Fig. 5.3, the electrochemical potential for holes would be $-\zeta_p$ (p. 119).

In equilibrium, the following equality holds.

$$\zeta_n = \zeta_p = \zeta. \qquad \text{(Quasi Fermi levels in equilibrium)} \qquad (5.6)$$

We will consider how the values of $\zeta_n$ and $\zeta_p$ are determined in nonequilibrium states in §5.7.

The quasi-Fermi level is sometimes called the *imref*, coming from "Fermi" spelled backward. William Shockley, the inventor of the quasi-Fermi level, told Fermi that "quasi-Fermi level" was too long and asked if there was a shorter alternative. Then Fermi suggested "imref" [28].

Quasi-Fermi levels have the dimensions of energy. But if one wants to consider them in the dimensions of voltage or electrostatic potential, we can use the *quasi-Fermi potentials* defined as follows:

$$\psi_{\mathrm{n}} \equiv \frac{\zeta_{\mathrm{n}}}{-q} + \mathrm{const.} \qquad \text{(Electron quasi Fermi potential)} \qquad (5.7)$$

$$\psi_{\mathrm{p}} \equiv \frac{\zeta_{\mathrm{p}}}{-q} + \mathrm{const.} \qquad \text{(Hole quasi Fermi potential)} \qquad (5.8)$$

The constant terms in (5.7) and (5.8) may be determined as appropriate for convenience.

### 5.2.3 Nonequilibrium Carrier Densities

The previous section suggested that, given the electron quasi-Fermi level $\zeta_{\mathrm{n}}$, the equations for the electron density are obtained by replacing $\zeta$ with $\zeta_{\mathrm{n}}$ in the equilibrium electron density equations (4.12) on p. 94 and (4.27) on p. 99.

$$n = N_{\mathrm{c}} \exp\left(-\frac{E_{\mathrm{c}} - \zeta_{\mathrm{n}}}{kT}\right) \qquad (N_{\mathrm{c}}\text{-referenced electron density}) \qquad (5.9)$$

$$= n_{\mathrm{i}} \exp\left(\frac{\zeta_{\mathrm{n}} - E_{\mathrm{i}}}{kT}\right) \qquad (n_{\mathrm{i}}\text{-referenced electron density}) \qquad (5.10)$$

$E_{\mathrm{c}} - \zeta_{\mathrm{n}}$ in (5.9) and $\zeta_{\mathrm{n}} - E_{\mathrm{i}}$ in (5.10) are "lengths" that can be read from an energy band diagram. By reading either of these lengths from an energy band diagram, we can tell the electron density. If any of the relevant quantities depend on the position $x$ as $\zeta_{\mathrm{n}}(x)$, $E_{\mathrm{c}}(x)$, and $E_{\mathrm{i}}(x)$, we can also see the gradient of the electron density by following $E_{\mathrm{c}}(x) - \zeta_{\mathrm{n}}(x)$ or $\zeta_{\mathrm{n}}(x) - E_{\mathrm{i}}(x)$ as a function of $x$.

Likewise, if the hole quasi-Fermi level $\zeta_{\mathrm{p}}$ is given, the equations for the hole density are obtained by replacing $\zeta$ with $\zeta_{\mathrm{p}}$ in the equilibrium hole density equations (4.17) on p. 95 and (4.28) on p. 99.

$$p = N_{\mathrm{v}} \exp\left(-\frac{\zeta_{\mathrm{p}} - E_{\mathrm{v}}}{kT}\right) \qquad (N_{\mathrm{v}}\text{-referenced hole density}) \qquad (5.11)$$

$$= n_{\mathrm{i}} \exp\left(\frac{E_{\mathrm{i}} - \zeta_{\mathrm{p}}}{kT}\right) \qquad (n_{\mathrm{i}}\text{-referenced hole density}) \qquad (5.12)$$

$\zeta_p - E_v$ in (5.11) and $E_i - \zeta_p$ in (5.12) are also "lengths" that can be read from an energy band diagram. The hole density can be found by reading either of these. $x$-dependent $\zeta_p(x) - E_v(x)$ and $E_i(x) - \zeta_p(x)$ provide information about the gradient of hole density, too.

*Example: Translating "Vertical Length" in Band Diagram into Carrier Density*

Let us check a numerical example of the relationship between the "vertical lengths" on the energy band diagram and carrier densities for silicon. The Fermi level $\zeta$ of the nondegenerate semiconductor falls within the range given by (4.18) on p. 96. The electron and hole quasi-Fermi levels, $\zeta_n$ and $\zeta_p$, can also be assumed to fall within the same range. From the values of the effective densities of states in Table 1.3 (p. 5), the majority carrier density of nondegenerate silicon is $n_N \simeq 10^{18}\,\mathrm{cm}^{-3}$ when $\zeta$ lies near the upper end of (4.18) (the silicon is n-type), and $p_P \simeq 10^{18}\,\mathrm{cm}^{-3}$ when $\zeta$ lies near the lower end of (4.18) (the silicon is p-type). Using the intrinsic carrier density $n_i \simeq 10^{10}\,\mathrm{cm}^{-3}$ in Table 1.3 and the $pn$ product expression (4.19) on p. 96, the corresponding minority carrier densities are $p_N \simeq 10^2\,\mathrm{cm}^{-3}$ and $n_P \simeq 10^2\,\mathrm{cm}^{-3}$, respectively. So if the Fermi level $\zeta$ shifts by about 1 eV on a band diagram, which is a little smaller than the bandgap energy $E_g \simeq 1.1$ eV, the electron and hole densities change by about 16 orders of magnitude. Even if we restrict the range of change of $\zeta$ to the n-type range ((4.47) on p. 103) or the p-type range ((4.52) on p. 104), the carrier densities change by eight orders of magnitude. ■

Now, multiplying (5.10) and (5.12) together gives

$$pn = n_i^2 \exp\left(\frac{\zeta_n - \zeta_p}{kT}\right). \qquad \text{(Non equilibrium } pn \text{ product)} \qquad (5.13)$$

This is the nonequilibrium $pn$ product. The extent of splitting of electron and hole quasi-Fermi levels, $|\zeta_n - \zeta_p|$, can be interpreted as indicating the degree of deviation from equilibrium (see Problem 5.1 on p. 164). This can also be read from an energy band diagram.

Taking the square root of (5.13),

$$n_i' \equiv n_i \exp\left(\frac{\zeta_n - \zeta_p}{2kT}\right) \qquad \text{(Effective intrinsic carrier density)} \quad (5.14)$$

can be regarded as the *effective intrinsic carrier density* in nonequilibrium states. Depending on $\zeta_n - \zeta_p \gtrless 0$, $n_i'$ may be larger or smaller than the intrinsic carrier density $n_\mathrm{i}$, that is, $n_\mathrm{i}' \gtrless n_\mathrm{i}$. If $n_\mathrm{i}' > n_\mathrm{i}$, then at least one of the carrier densities, $n$ and $p$ (actually, almost always the minority carrier density), is larger than at equilibrium, and vice versa (the minority carrier density is smaller than at equilibrium) if $n_\mathrm{i}' < n_\mathrm{i}$. We will see specific examples where $\zeta_\mathrm{n} \neq \zeta_\mathrm{p}$ in Chapter 6 onward. As will be explained in §5.6, the sign of $\zeta_\mathrm{n} - \zeta_\mathrm{p}$ tells us whether carrier generation or recombination is dominant.

The Fermi level $\zeta$ has the property that the value of the Fermi–Dirac distribution function (4.4) on p. 89 at $E = \zeta$ is invariably $f(\zeta) = 1/2$, but the quasi-Fermi levels $\zeta_\mathrm{n}$ and $\zeta_\mathrm{p}$ have no such property. Since (4.4) is a monotonically decreasing function, if $\zeta_\mathrm{n} \neq \zeta_\mathrm{p}$, then both $f(\zeta_\mathrm{n})$ and $f(\zeta_\mathrm{p})$ cannot be equal to 1/2. In the first place, we cannot define the Fermi level $\zeta$ contained in (4.4) in nonequilibrium.

However, it is possible to consider the occupation probability in nonequilibrium states, too. For example, since the exponential factor of (4.12) for the electron density $n$ on p. 93 represents the occupancy of $N_\mathrm{c}$ states per unit volume in the conduction band, the exponential factor of (5.9) on p. 123 can also be considered to represent the occupancy of conduction band states at nonequilibrium. The exponential factor in (5.11) on p. 123 can likewise be regarded as the occupancy of valence band states at nonequilibrium. In other words, separate distribution functions are required for the conduction band and the valence band at nonequilibrium. If one prefers to avoid the use of the effective densities of states, we can use the following distribution functions, which have the same form as the Fermi–Dirac distribution function, (4.4) (see Problem 5.2 on p. 164):

$$f_\mathrm{c}(E) = \frac{1}{1 + \exp\left(\frac{E-\zeta_\mathrm{n}}{kT}\right)} \qquad (E \geq E_\mathrm{c}) \tag{5.15}$$

$$f_\mathrm{v}(E) = \frac{1}{1 + \exp\left(\frac{E-\zeta_\mathrm{p}}{kT}\right)} \qquad (E \leq E_\mathrm{v}) \tag{5.16}$$

### 5.2.4 Logarithmic Transform of Carrier Densities

Solving the electron density expression (5.9) on p. 123 for $\zeta_\mathrm{n}$ yields

$$\zeta_\mathrm{n} = E_\mathrm{c} - kT \ln\left(\frac{N_\mathrm{c}}{n}\right). \qquad \text{(Electron quasi Fermi level)} \tag{5.17}$$

By the discussion in §4.4.2 (p. 109), $\zeta_n$ on the left-hand side corresponds to the total chemical potential $\zeta_{tot}$ in (4.59) on p. 111, and for the right-hand side, the correspondence is as follows:

$$\zeta_{ext} = E_c, \qquad \zeta_{int} = kT \ln\left(\frac{n}{N_c}\right). \tag{5.18}$$

Similarly, solving (5.10) on p. 123 for $\zeta_n$, we obtain

$$\zeta_n = E_i + kT \ln\left(\frac{n}{n_i}\right). \qquad \text{(Electron quasi Fermi level)} \tag{5.19}$$

In this case, external and internal chemical potentials are shifted by a constant ($\simeq E_g/2$, see §4.2.5) compared to (5.18) as follows:

$$\zeta_{ext} = E_i, \qquad \zeta_{int} = kT \ln\left(\frac{n}{n_i}\right). \tag{5.20}$$

The quasi-Fermi level $\zeta_n$ on the left-hand side of (5.17) and (5.19) is a quantity that can be plotted on an energy band diagram. All the quantities on the right-hand side other than the electron density $n$ can be regarded as constants (at a given position $x$). Therefore, *(5.17) and (5.19) can be regarded as logarithmic transforms of the electron density*. Put differently, (5.17) and (5.19) are "change of variables" from $n$ to $\zeta_n$ (see the Box on p. 15). Since $N_c > n$ in nondegenerate semiconductors, (5.17) implies $\zeta_n < E_c$. Also, since $n > n_i$ in n-type semiconductors and $n < n_i$ in p-type semiconductors, (5.19) implies that $\zeta_n > E_i$ in n-type and $\zeta_n < E_i$ in p-type. The above discussion is mostly a repetition of the discussion on p. 101, but the position of $\zeta_n$ may be different from the position of the equilibrium Fermi level $\zeta$. Equations (5.9) and (5.10) on p. 123 can be understood as expressions for reading the electron density $n$ from the relative position of $\zeta_n$ (measured from $E_c$ or $E_i$) in the energy band diagram.

Basically, the same is true for the hole density. Solving the hole density expressions (5.11) and (5.12) for $\zeta_p$, we obtain

$$\zeta_p = E_v + kT \ln\left(\frac{N_v}{p}\right), \qquad \text{(Hole quasi Fermi level)} \tag{5.21}$$

$$\zeta_p = E_i - kT \ln\left(\frac{p}{n_i}\right). \qquad \text{(Hole quasi Fermi level)} \tag{5.22}$$

Equations (5.21) and (5.22) transform the hole density $p$ into $\zeta_p$, which can be plotted on an energy band diagram. As a result, by using

(5.11) or (5.12), we can read the hole density $p$ from the relative position of $\zeta_p$ (measured from $E_v$ or $E_i$) on the energy band diagram. From (5.22), we see that $\zeta_p < E_i$ in p-type semiconductors and $\zeta_p > E_i$ in n-type semiconductors.

In Chapter 6 onward, we will see cases where $\zeta_n > E_i$ (i.e., $n > n_i$) in p-type semiconductors and $\zeta_p < E_i$ (i.e., $p > n_i$) in n-type semiconductors. In semiconductor devices, such situations can also be realized by electrical means (p. 3).

Note that the change of variables from carrier density to quasi-Fermi level is a logarithmic transform, which makes it easier to plot carrier densities that may change by many orders of magnitude on an energy band diagram. However, as a result, a small difference in carrier density becomes difficult to discern.

### THE JOY OF READING ENERGY BAND DIAGRAMS

Herbert Kroemer, who was awarded the Nobel Prize in Physics in 2000 for his research on semiconductor heterostructures, stated the following about energy band diagrams in his Lecture [16].

> **Kroemer's Lemma of Proven Ignorance** If, in discussing a semiconductor problem, you cannot draw an energy band diagram, this shows that you don't know what you are talking about, with the corollary: If you can draw one, but don't, your audience won't know what you are talking about.The "energy band diagram" here is the *E-x* diagram (see the Box on p. 15). It seems very important that we draw energy band diagrams!

It should, however, be noted that it is often quite difficult to draw energy band diagrams including the quasi-Fermi levels by hand. This is because determining the electrostatic potential and quasi-Fermi levels of a device, which inevitably is spatially nonuniform, generally requires numerical simulation (§5.7). We, therefore, propose that the reader first try to develop the ability to properly *read* energy band diagrams that are drawn using a device simulator (also known as "TCAD," see p. 154). Even if you have an energy band diagram of a device, nicely drawn using a simulator, it is useless unless you can read physics from it. We hope that you will discover and experience the joy of reading energy band diagrams, especially TCAD-drawn ones, by going through this book.

### 5.2.5 General Form of Nondegenerate Carrier Density Expressions

The expressions of electron density, both (5.9) and (5.10) on p. 123, had the following form:

$$n = (\text{reference density}) \times (\text{exponential factor})$$
$$= (\text{reference density}) \times \exp\left[\frac{\zeta_\text{n} - (\text{reference energy})}{kT}\right]. \tag{5.23}$$

In (5.23), the "exponential factor" becomes 1 when

$$\zeta_\text{n} = (\text{reference energy}), \tag{5.24}$$

which, in turn, makes

$$n = (\text{reference density}). \tag{5.25}$$

For example, in (5.9), $N_\text{c}$ is the "reference density" and $E_\text{c}$ is the "reference energy." In (5.10), $n_\text{i}$ is the "reference density" and $E_\text{i}$ is the "reference energy."

A finer point here is that the above "reference energy" is not "a certain fixed value on the $E$-axis" but "a relative energy, usually within the forbidden band, measured from $E_\text{c}$, $E_\text{v}$, or $E_\text{i}$." The distinction is irrelevant when $E_\text{c}$, $E_\text{v}$, and $E_\text{i}$ do not depend on the position $x$, but it becomes important when they are $x$-dependent (see Fig. 6.19 on p. 184, for example). If, for example, the "reference energy" is $E_\text{c}$, then the "reference energy" really is $E_\text{c}(x)$ at a given position $x$, even if there is band bending. This is because carrier densities are determined by the relative values of $\zeta_\text{n}$ and $\zeta_\text{p}$ in the energy gap, and not by their absolute (as opposed to relative) values on the $E$-axis. Recall that usually only relative values of Fermi level are considered for semiconductors (p. 91).

In general, the electron density of nondegenerate semiconductors can be variously expressed in the form of (5.23). "Reference density" and "reference energy" can be chosen for convenience. The energy band diagram also turns out useful in selecting or reading the "reference density" and "reference energy."

As an example, compare (5.9) and (5.10) on p. 123, or equivalently, (4.12) on p. 94 and (4.27) on p. 99, with Fig. 4.5 (p. 102). Equations (4.25) and (4.26), which appeared when we derived (4.27), are also expressions for electron density under certain conditions in the form of (5.23). To be more specific, in (4.25), $E_i$ is substituted for $\zeta_n$, and the "reference energy" is $E_c$. In (4.25) and (4.26), $N_c$ and $n_i$ are related to each other via the "vertical length" $E_c - E_i$ in Fig. 4.5. By reading the reference densities and the corresponding reference energy difference, converting a carrier density expression into another is easy (with some practice) (see Problem 5.3 on p. 164). This is the beauty and power of the theory of nondegenerate semiconductors. Basically the same derivation or manipulation of the carrier density is also done in §6.8 when deriving the current-voltage characteristics of the p-n junction (p. 195).

Now, of course, a parallel argument applies to the hole density. The hole density expression for nondegenerate semiconductors has the following form:

$$p = (\text{reference density}) \times \exp\left[\frac{(\text{reference energy}) - \zeta_p}{kT}\right]. \qquad (5.26)$$

Equations (5.11) and (5.12) on p. 123 have the same form as (5.26).

## 5.3 QUASI-FERMI LEVELS AND CURRENT DENSITY

### 5.3.1 Carrier Flux Density and Current Density

The current that flows due to the movement of carriers is called the *conduction current*. Another type of current that can flow in semiconductors is the *displacement current*. The current that flows through an insulator between the electrodes of a capacitor is usually the displacement current.[2] In this book, the term "current" refers to the conduction current unless otherwise stated.

To prepare for finding the conduction current density, let us consider the *carrier flux density*, which is the number of particles passing

[2] Another possible current that might flow through a very thin insulator is the *tunnel current* (p. 174). Of course, a breakdown current may flow even if the insulator is not so thin when one applies a very large voltage.

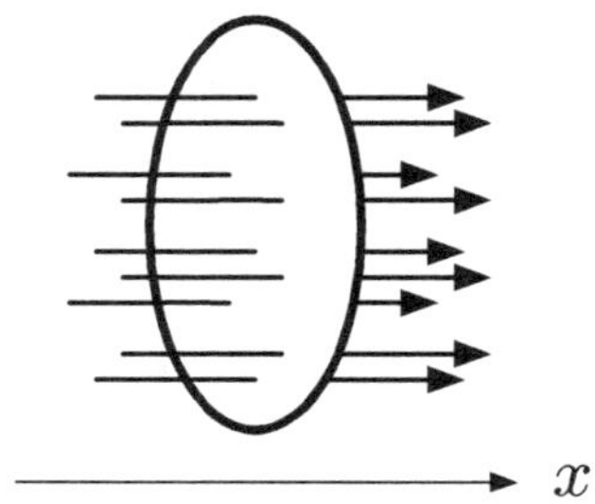


FIGURE 5.4 Carrier flux density equals the number of carriers crossing a unit area per unit time.

through a unit cross-sectional area (perpendicularly to the area) per unit time (Fig. 5.4).

If an ensemble of electrons with number density $n$ is moving collectively with velocity $v_n$ in the $x$ direction, then the electron flux density $\Phi_n$ is given by

$$\Phi_n \equiv n v_n. \qquad \text{(Electron flux density)} \tag{5.27}$$

Similarly, the hole flux density $\Phi_p$ is given by

$$\Phi_p \equiv p v_p, \qquad \text{(Hole flux density)} \tag{5.28}$$

where $p$ is the hole density, and $v_p$ is the hole velocity.

The current density $J_n$ due to electron conduction can be written using the electron flux density $\Phi_n$ as follows:

$$J_n = -q\Phi_n, \quad \text{(Electron current density in terms of flux density)} \tag{5.29}$$

where $-q$ is the electron charge. Electron current flows in the opposite direction to the flux density. Likewise, the hole current density can be written as

$$J_p = q\Phi_p, \qquad \text{(Hole current density in terms of flux density)} \tag{5.30}$$

where $q$ is the hole charge. Hole current flows in the same direction as the flux density. Equations (5.29) and (5.30) are highly general expressions that can be used regardless of the mechanism of carrier flux generation.

### 5.3.2 Quasi-Fermi Level Gradient and Current Density

Recalling the meaning of the quasi-Fermi level suggested by (5.2) on p. 118 and Fig. 5.3 on p. 122, we can infer that, if the gradient of the quasi-Fermi level is not too steep, the following proportionality relations should hold for the current density and the flux density:

$$\Phi_n \propto n \times \left(-\frac{d\zeta_n}{dx}\right), \tag{5.31}$$

$$\Phi_p \propto p \times \left[-\frac{d(-\zeta_p)}{dx}\right]. \tag{5.32}$$

The proportionality comes from the fact that an *I*-*V* curve that goes through the origin can be approximated by a straight line around the origin (§2.2.1). Note that $(-\zeta_p)$ in (5.32) is the electrochemical potential for holes (p. 123). Let us introduce proportionality coefficients, $\mu_n$ and $\mu_p$, to express the current densities (5.29) and (5.30) using the right-hand sides of (5.31) and (5.32), respectively.

$$J_n = \mu_n n \frac{d\zeta_n}{dx}, \tag{5.33}$$

$$J_p = \mu_p p \frac{d\zeta_p}{dx}. \tag{5.34}$$

$\mu_n$ and $\mu_p$ are called the *mobilities* of electrons and holes, respectively. Inserting the electron flux density (5.27) into the current density expression (5.29) and comparing it with (5.33), we find the following relationship for the electron velocity $v_n$:

$$v_n = -\frac{\mu_n}{q}\frac{d\zeta_n}{dx} = \mu_n \frac{d\psi_n}{dx}, \qquad \text{(Velocity of ensemble of electrons)} \tag{5.35}$$

where $\psi_n$ is the electron quasi-Fermi potential defined in (5.7) on p. 123. Likewise, from (5.28), (5.30), (5.34), and (5.8), we find

$$v_p = \frac{\mu_p}{q}\frac{d\zeta_p}{dx} = -\mu_p \frac{d\psi_p}{dx}. \qquad \text{(Velocity of ensemble of holes)} \tag{5.36}$$

From (5.35) and (5.36), it can be seen that mobility is a proportionality coefficient that links velocity and a quantity having the dimensions of the electric field.

With regard to (5.33) and (5.34), there are a few more important things worth pointing out. First, the existence (or not) and the direction of the electron current can be read from the gradient of $\zeta_n$ in an energy band diagram. Also from the gradient of $\zeta_p$, the existence (or not) and the direction of the hole current can be read. These were actually already shown in Fig. 5.3 (p. 122). Equations (5.33) and (5.34) are the reasons why carriers may move in the opposite direction to the electrostatic force, which was mentioned on p. 15. If the "gradient of $\zeta_n$" and the "gradient of $E_c$" have unlike signs, then electrons may move in the opposite direction from the electrostatic force (see the discussion on p. 119). In addition, the magnitude of current density can also be read from the energy band diagram. This is because the current density is given by the product of the carrier density and the quasi-Fermi level gradient as in (5.33) and (5.34), both of which can be read off from the energy band diagram. *The current density is determined mainly by the carrier density.* The reason is that $\mathrm{d}\zeta_n/\mathrm{d}x$ and $\mathrm{d}\zeta_p/\mathrm{d}x$ do not change that much in magnitude (as long as the slopes are clearly visible on a band diagram), whereas $n$ and $p$ can change by orders of magnitude (see the Example on p. 124).

### 5.3.3 Drift and Diffusion of Carriers

Let us look further at the gradients of the quasi-Fermi levels in (5.33) and (5.34). First, differentiate the quasi-Fermi level expression (5.17) on p. 125 by $x$ and insert the result into (5.33) (see Problem 5.4 on p. 164).

$$J_n = \mu_n n\left(\frac{\mathrm{d}E_c}{\mathrm{d}x} + kT\frac{\mathrm{d}}{\mathrm{d}x}\ln n\right) = \mu_n n\left(\frac{\mathrm{d}E_c}{\mathrm{d}x} + \frac{kT}{n}\frac{\mathrm{d}n}{\mathrm{d}x}\right)$$
$$= \mu_n n\frac{\mathrm{d}E_c}{\mathrm{d}x} + \mu_n kT\frac{\mathrm{d}n}{\mathrm{d}x}. \qquad \text{(Electron current density)} \qquad (5.37)$$

In the above, we used the fact that $N_c$ does not depend on $x$, assuming that we are dealing with a single semiconductor material. The first term of (5.37) comes from the gradient of the conduction band bottom $E_c$, and is called the *drift term*. Using (1.5) on p. 15,

$$\frac{\mathrm{d}E_c}{\mathrm{d}x} = \frac{\mathrm{d}}{\mathrm{d}x}(-q\psi) = q\mathcal{E}, \qquad (5.38)$$

and therefore the drift term arises from the electrostatic field $\mathcal{E}$. We will consider the drift term again in §5.4. The second term of (5.37) is proportional to the gradient of the electron density $n$ and is called the *diffusion term*. We will consider it further in §5.5.

Let us take a look at the hole current density. Differentiating the hole quasi-Fermi level (5.21) on p. 126 by $x$ and inserting the result into (5.34), we obtain

$$J_p = \mu_p p\left(\frac{dE_v}{dx} + kT\frac{d}{dx}\ln p\right) = \mu_p p\left(\frac{dE_v}{dx} - \frac{kT}{p}\frac{dp}{dx}\right)$$
$$= \mu_p p\frac{dE_v}{dx} - \mu_p kT\frac{dp}{dx}. \qquad \text{(Hole current density)} \qquad (5.39)$$

The first term of (5.39) is the drift term, and the second term is the diffusion term. From (1.6) on p. 15,

$$\frac{dE_v}{dx} = \frac{d}{dx}(-q\psi) = q\mathcal{E}. \qquad (5.40)$$

Considering also (4.23) on p. 98, the electrostatic field can be read from the gradient of $E_c$, $E_v$, or $E_i$ in an energy band diagram. Comparison of (5.37) and (5.39) shows that the signs of the diffusion terms are different (§5.5). This is because electrons and holes have opposite polarities. The carrier density gradient, which determines the sign of the diffusion term, can also be read from an energy band diagram (p. 123).

## 5.4 ELECTRIC CONDUCTION DUE TO ELECTRIC FIELD

This section describes the theory of electrical conduction based on the kinetic theory of gases. This theory was originally developed to explain electrical conduction in metals, but it is also applicable to semiconductors.

### 5.4.1 Drift of Carriers

As before, we consider a nondegenerate semiconductor and regard conduction electrons as an ideal gas (p. 92). Electrons are charged and should repel each other, but since there are usually cations (donor ions) around, and also the electron density of nondegenerate semiconductor is not that high, we will ignore the Coulomb interaction between electrons. Under these assumptions, we consider that electrons are in thermal motion, repeatedly colliding with the atoms and ions that constitute the crystal. In silicon, the *thermal velocity*$v_{\mathrm{th}}$, at which individual electrons move about, is about $10^7$ cm/s [31].

Let us consider the equation of motion for the $i$-th electron among many electrons in the presence of an external electric field $\mathcal{E}$. The force acting on the electron between its collision with an atom and the next collision with another atom can be considered to be only the electric force due to $\mathcal{E}$. The equation of motion is given by

$$m_{\mathrm{e}}\frac{\mathrm{d}v_i}{\mathrm{d}t} = -q\mathcal{E}, \quad \text{(Equation of motion in the absence of collisions)} \tag{5.41}$$

where $m_{\mathrm{e}}$ is the effective mass of the electron (p. 80), $v_i$ is the velocity of the $i$-th electron, and $-q$ is the electron charge. Let us assume for simplicity that the external field is uniform and does not depend on the position.

If there are $N$ electrons, the statistical mean of the velocities of individual electrons is given by

$$\langle v\rangle = \frac{1}{N}\sum_{i=1}^{N} v_i. \quad \text{(Average velocity of electrons)} \tag{5.42}$$

The electron velocity $v_{\mathrm{n}}$ in (5.27) on p. 130 was actually this average electron velocity, although no assumption was made there about the cause of the collective movement of electrons.

The effect of $\mathcal{E}$ on individual electrons is described by (5.41), but simply replacing $v_i$ by $\langle v\rangle$ would not give the correct equation of motion describing the time development of $\langle v\rangle$ because then electrons would be accelerated indefinitely. In reality, collisions between electrons and atoms are taking place all the time, so we need to incorporate the effect of these collisions into the equation of motion. It is not easy to consider the collision or scattering of individual electrons, but from a macroscopic point of view, the scattering of electrons by atoms can be thought of as a kind of friction acting on collectively flowing electrons. The higher the average electron velocity, the higher the frequency of scattering and hence the stronger the friction, so let us assume that the friction force proportional to $\langle v\rangle$ acts in the opposite direction to $\langle v\rangle$. Then, the average momentum, $m_{\mathrm{e}}\langle v\rangle$, of electrons is governed by the following equation of motion:

$$m_{\mathrm{e}}\frac{\mathrm{d}\langle v\rangle}{\mathrm{d}t} = -q\mathcal{E} - \frac{m_{\mathrm{e}}\langle v\rangle}{\tau_{\mathrm{e}}}, \quad \text{(Equation of collective motion)} \tag{5.43}$$

where $\tau_e$ is the *mean free time* between collisions for electrons, meaning that each electron is scattered, on average, every $\tau_e$ and its momentum is initialized randomly. $1/\tau_e$ is the average frequency of scattering. The distance, $l_e$, an electron travels during mean free time is called the *mean free path*.

$$l_e = v_{th}\tau_e. \qquad \text{(Electron mean free path)} \tag{5.44}$$

Solving the differential equation (5.43) with the initial condition $\langle v \rangle = 0$ at time $t = 0$ yields (see Problem 5.5 on p. 164)

$$\langle v \rangle (t) = -\frac{q\mathcal{E}\tau_e}{m_e}\left(1 - e^{-t/\tau_e}\right). \tag{5.45}$$

In (5.45), $\tau_e$ is the time constant of the exponential relaxation, so it is sometimes called the "relaxation time." But this kind of relaxation phenomenon can be seen in many other situations, too. To avoid confusion with the dielectric relaxation time (§5.2.2, §5.8) we refer to $\tau_e$ as the mean free time [27]. From (5.45), the terminal velocity is given by

$$\langle v \rangle (\infty) = \lim_{t\to\infty}\langle v \rangle (t) = -\frac{q\tau_e}{m_e}\mathcal{E}. \tag{5.46}$$

Equation (5.46) has a minus sign because the direction of motion of electrons is opposite to the direction of the electric field $\mathcal{E}$ due to the negative charge of electrons. The right-hand side of (5.46) actually does not depend on the initial condition. It can also be obtained by simply assuming that the time derivative on the left-hand side of (5.43) equals 0.

The phenomenon in which gas particles flow collectively due to force exerted by an external field is called *drift*. The left-hand side of (5.46) is the "average velocity," but the right-hand side was derived under the condition that electrons drifted due to an external force, so we call this velocity the electron *drift velocity* and rewrite it as $v_{n,drift}$:

$$v_{n,drift} = -\frac{q\tau_e}{m_e}\mathcal{E} = -\mu_n\mathcal{E}, \qquad \text{(Electron drift velocity)} \tag{5.47}$$

where the proportionality coefficient

$$\mu_n = \frac{q\tau_e}{m_e}, \qquad \text{(Electron mobility)} \tag{5.48}$$

introduced in (5.47), is the electron *mobility* that appeared earlier in (5.35) on p. 131. Mobility expresses how easily carriers respond to an electric field. The longer the mean free time $\tau_e$, the fewer scattering events there are, so electrons can move more easily. That is why $\mu_n$ is proportional to $\tau_e$. The smaller the electron effective mass $m_e$, the lighter and more mobile conduction electrons are. That is why $\mu_n$ is inversely proportional to $m_e$.

A parallel development applies to holes. Since holes are positively charged, they drift in the same direction as that of the electric field $\mathcal{E}$. The drift velocity of holes is given by

$$v_{p,drift} = \frac{q\tau_h}{m_h}\mathcal{E} = \mu_p \mathcal{E}, \qquad \text{(Hole drift velocity)} \qquad (5.49)$$

where the proportionality coefficient

$$\mu_p = \frac{q\tau_h}{m_h} \qquad \text{(Hole mobility)} \qquad (5.50)$$

is the hole mobility that appeared in (5.36). $\tau_h$ is the mean free time for holes, and $m_h$ is the hole effective mass.

The mean free time of carriers in metals and semiconductors is very short, for example, $10^{-13} \sim 10^{-12}$ s for silicon (see Table 5.2 on p. 122). This means that when an external electric field is applied, the average carrier velocity quickly reaches the drift velocity.

Incidentally, Table 1.3 (p. 159) shows that $\mu_n > \mu_p$ for both silicon (Si) and gallium arsenide (GaAs). In general, the mobility of electrons is larger than that of holes. This is because the electron effective mass $m_e$ is generally smaller than the hole effective mass $m_h$.

## INTUITIVE PICTURES OF MOTION OF HOLES

Consider a piece of semiconductor subjected to an external electric field $\mathcal{E}$ as shown in Fig. 1.8 (p. 14). The motion of holes in semiconductors is often explained as shown in the cartoon of Fig. 5.5(a). The valence band is almost completely filled with electrons, but there are some vacancies (or holes) here and there. One such hole is shown in Fig. 5.5(a). Then, the electron to the left of the hole moves to the right, and as a result, the hole moves to the left. If processes like this occur repeatedly, the hole will move more and more to the left. Since this is analogous to the motion of a bubble in a liquid, let us call it the "bubble model" of hole conduction.

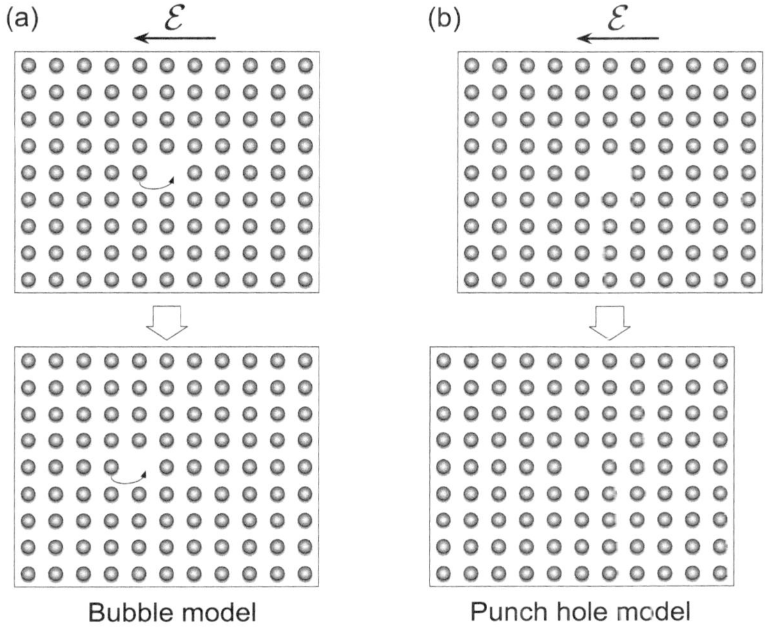


FIGURE 5.5 Two models of hole conduction.

Actually, the bubble model has a serious flaw. In spite of the presence of the external field $\mathcal{E}$, almost all electrons stay still, in contradiction to the discussion in §5.4. At least, it seems necessary to explain why the electrons do not drift despite the electric force due to $\mathcal{E}$.

Next, let us look at the "punch hole model" shown in Fig. 5.5(b). In this model, a hole is considered to be like a hole punched in a piece of paper. Then, as you move the piece of paper to the left, the hole also moves leftward. In this model, all electrons drift as they should, but an obvious flaw is that they drift in the wrong direction, in the same direction as the electric field. Since electrons are negatively charged, they ought to move in the opposite direction from $\mathcal{E}$. Now what can we do?

**INTUITIVE PICTURES OF MOTION OF HOLES (CONT.)**

Recalling Fig. 3.21 (p. 80) and (3.77) for the effective mass, the electrons around the hole (i.e., electrons in the valence band) have *negative effective mass*. The effective mass being negative ($m_e < 0$) in the equation of motion $m_e a = F$ means that the acceleration $a$ is in the opposite direction to the exerted force $F = -q\mathcal{E}$. That is, electrons in the valence band are subjected to a force ($-q\mathcal{E}$) in the opposite direction to the electric field $\mathcal{E}$, and nevertheless, accelerated in the same direction as $\mathcal{E}$! Thus, from the viewpoint of carrier drift, the punch hole model is superior.

Just to be clear, the punch hole model, improved with negative effective mass, is yet another poor man's analogy. It fails to explain some other things (see Problem 5.6 on p. 164). Actually, the two models share the same traits: they are flat-out inconsistent with the accepted theoretical description of holes. Interested readers are encouraged to study solid-state physics [3, 14] and see for themselves what the theory has to say about the motion of holes [14]. Can you picture it in words or drawings?

### 5.4.2 Relationship between Mobility and Conductivity

The conduction current that flows due to the drift of carriers is called *drift current*, and the corresponding current density is called drift current density. The drift terms of (5.37) and (5.39) on p. 133 give the drift current densities. The electron drift current density can be written as

$$J_{\mathrm{n,drift}} = qn\mu_{\mathrm{n}}\mathcal{E} = \sigma_{\mathrm{n}}\mathcal{E}, \qquad \text{(Electron drift current density)} \qquad (5.51)$$

where the proportionality coefficient

$$\sigma_{\mathrm{n}} \equiv qn\mu_{\mathrm{n}} \qquad \text{(Conductivity due to electron conduction)} \qquad (5.52)$$

is the conductivity associated with the conduction of electrons. Its inverse, $\rho_{\mathrm{n}} = 1/\sigma_{\mathrm{n}}$, is the resistivity.

The hole drift current density is given by the first term of (5.39) on p. 133 and can be written as

$$J_{\mathrm{p,drift}} = qp\mu_{\mathrm{p}}\mathcal{E} = \sigma_{\mathrm{p}}\mathcal{E}, \qquad \text{(Hole drift current density)} \qquad (5.53)$$

where the proportionality coefficient

$$\sigma_{\mathrm{p}} \equiv q p \mu_{\mathrm{p}} \qquad \text{(Conductivity due to hole conduction)} \qquad (5.54)$$

is the conductivity associated with the conduction of holes. Its inverse, $\rho_{\mathrm{p}} = 1/\sigma_{\mathrm{p}}$, is the resistivity.

Naturally, both the electron drift current (5.51) and the hole drift current (5.53) flow in the same direction as the electric field $\mathcal{E}$. If electrons and holes coexist, the total drift current density is given by

$$J_{\mathrm{drift}} = J_{\mathrm{n,drift}} + J_{\mathrm{p,drift}}. \qquad \text{(Total drift current density)} \qquad (5.55)$$

The associated conductivity, therefore, is given by

$$\sigma \equiv \sigma_{\mathrm{n}} + \sigma_{\mathrm{p}} = q(n\mu_{\mathrm{n}} + p\mu_{\mathrm{p}}). \qquad \text{(Conductivity)} \qquad (5.56)$$

Since the resistivity is the inverse of conductivity,

$$\rho \equiv \frac{1}{\sigma} = \frac{1}{q\left(n\mu_{\mathrm{n}} + p\mu_{\mathrm{p}}\right)} = \frac{1}{\rho_{\mathrm{n}}^{-1} + \rho_{\mathrm{p}}^{-1}}. \qquad \text{(Resistivity)} \qquad (5.57)$$

Note that both electrons and holes coexist in semiconductors, but usually, the majority carrier density is orders of magnitude greater than the minority carrier density due to (5.13) on p. 124. Therefore, only one of the terms of the right-hand side of (5.55) is dominant.

## 5.5 ELECTRIC CONDUCTION DUE TO CARRIER DIFFUSION

External forces are not the only mechanism by which gas particles flow. When the density of gas particles is not spatially uniform, they *diffuse* to eliminate the density gradient (see (vi) and (viii) on p. 107). *Diffusion* is a phenomenon that occurs spontaneously even when there is no force acting on the particles.

**LIMITATIONS OF THE MOBILITY-BASED DESCRIPTION OF CARRIER VELOCITY**

You might be under the impression that the proportionality relation between drift velocity and electric field as in (5.47) and (5.49) on p. 135 (i.e., equations for a straight line passing through the origin of the $\mathcal{E}$-$v$ plane) is "correct" because it is derived from the differential equation (5.43). But (5.43) is derived by assuming that the friction force is proportional to $\langle v \rangle$ and that $\tau_e$ is constant, and we do not know (or this book does not discuss) whether or to what extent these assumptions are correct. As a matter of fact, it is known that the proportionality relations (5.47) and (5.49) hold when the electric field is sufficiently weak, but do not hold when the field becomes intense (p. 261).

We already saw similar cases in Chapter 2. For example, $R$ in (2.1) on p. 29 represents the resistance of a linear resistor, but as mentioned in §2.2.1, "resistance" has only a qualitative meaning in nonlinear resistors, and for quantitative discussions, incremental resistance and/or chord resistance have to be used. This implies that if we want to discuss carrier velocity using a quantity with the dimensions of mobility when (5.47) and (5.49) do not hold, we have to consider "incremental mobility" and "chord mobility." Alternatively, we can forget about mobility and use a nonlinear function $v_{\mathrm{drift}} = v_{\mathrm{drift}}(\mathcal{E})$, just as we considered a nonlinear function $V(I)$ in §2.2.1. The same can be said for conductivity and resistivity.

In this book, we assume that (5.47) and (5.49) hold unless otherwise stated. A response of a system as in (5.47) and (5.49), proportional to the exerted external force, is called a *linear response*. Most systems respond linearly as long as the exerted force is weak. The mobilities that appear in (5.48) and (5.50) are also called *low-field mobilities*.

### 5.5.1 Diffusion Current

A mathematical description of particle diffusion is known as Fick's law. Let us consider again electrons as a gas and neglect interaction between electrons. The electron flux density, $\Phi_{\mathrm{n}}$, due to diffusion is expressed by using a proportionality factor $D_{\mathrm{n}}$ as follows:

$$\Phi_{\mathrm{n}} = -D_{\mathrm{n}}\frac{\mathrm{d}n}{\mathrm{d}x}, \qquad \text{(Fick's law for electrons)} \qquad (5.58)$$

where $D_{\mathrm{n}}\,(> 0)$ is called the *diffusion coefficient* for electrons. The minus sign in (5.58) indicates that the electrons diffuse in the direction opposite to the increase in electron density. According to the current density equation (5.29) on p. 130, the electron diffusion current density is given by

$$J_{n,\text{diff}} = -q\Phi_n = qD_n\frac{dn}{dx}. \text{ (Electron diffusion current density)} \quad (5.59)$$

The electron diffusion current flows in the direction of increasing electron density.

The diffusion coefficient $D_p$ is likewise defined for holes, and Fick's law can be written as follows:

$$\Phi_p = -D_p\frac{dp}{dx}. \qquad \text{(Fick's law for holes)} \qquad (5.60)$$

Using (5.30) on p. 130, the hole diffusion current density is given by

$$J_{p,\text{diff}} = q\Phi_p = -qD_p\frac{dp}{dx}. \qquad \text{(Hole diffusion current density)} \quad (5.61)$$

The hole diffusion current flows in the direction of decreasing hole density.

### 5.5.2 Einstein's Relation

The diffusion term of the electron current density expression (5.37) on p. 132 is the electron diffusion current density, given also by (5.59). Equating the two expressions for the diffusion current density, we find the following relationship between the mobility and the diffusion coefficient.

$$\frac{D_n}{\mu_n} = \frac{kT}{q}. \qquad \text{(Einstein's relation for electrons)} \qquad (5.62)$$

This is known as *Einstein's relation*. If the mobility is known from measurement, for example, the diffusion coefficient can be determined, and vice versa.

Similarly, the diffusion term of the hole current density expression (5.39) on p. 133 is the hole diffusion current density, also given by (5.61). Einstein's relation for holes is given by

$$\frac{D_p}{\mu_p} = \frac{kT}{q}. \qquad \text{(Einstein's relation for holes)} \qquad (5.63)$$

Note that (5.62) and (5.63) hold for nondegenerate semiconductors but not as-is for degenerate semiconductors, as can be seen by going through the derivation.

Now, what is the physical significance of Einstein's relation? Mobility is a coefficient that expresses how carriers respond to an electric field. The diffusion coefficient, on the other hand, describes a spontaneous phenomenon that eliminates carrier density gradient. Einstein's relation shows that these different physical phenomena are related to each other. Both processes involve interactions between carriers and the crystal lattice, and it is natural that they are related. Einstein's relation is an example of a more general result known as the *fluctuation-dissipation theorem* in statistical mechanics.

### EINSTEIN'S RELATION FOR THE ATMOSPHERE NEAR EARTH'S SURFACE

Is Einstein's relation only for charged particles such as electrons and holes? Gravity is a force that also acts on neutral gas particles and is quite similar to the electrostatic force acting on electrons. We can expect a relationship similar to (5.62) and (5.63) on p. 141, to exist between the diffusion coefficient for gas particles and the proportionality coefficient corresponding to mobility.

Although Earth's gravity depends on the altitude, for simplicity, let us use the acceleration of gravity $g$ near the surface, as shown in Fig. 4.12 (p. 112). A dimensional analysis shows that the coefficient of proportionality (also known as transport coefficient) linking the drift velocity, $v_{\mathrm{drift}}$, of gas particles and the external field $g$ has a dimension of time. Let this time constant be $\tau_{\mathrm{g}}$, that is, $v_{\mathrm{drift}} = \tau_{\mathrm{g}} g$. Recalling the discussion in §5.4.1, it seems reasonable to consider $\tau_{\mathrm{g}}$ as the mean free time of the atmospheric gas, given that $\mu_{\mathrm{n}} \propto \tau_{\mathrm{e}}$ in the mobility expression (5.48) on p. 135. However, the collisions here must be against surrounding gas particles.

Although we skip the derivation, the density of atmospheric gases at the surface can also be written in the same form as (5.23) on p. 128 [15]. Also using the external chemical potential (4.60) on p. 113, the total chemical potential is given by a similar expression to the electron quasi-Fermi level expression (5.19) on p. 126:

$$\zeta_{\mathrm{tot}} = mgh + kT \ln\left(\frac{n_{\mathrm{g}}}{N_{\mathrm{g}}}\right), \tag{5.64}$$

where $n_g$ is the gas density, and $N_g$ is the reference density (§5.2.5).

Einstein's relation for electrons and holes states that the ratio of the diffusion coefficient to the mobility, which represents a linear response to an external field, is a constant. A similar derivation for the surface atmosphere yields $D_g/\tau_g = kT/m$. This is Einstein's relation for the atmosphere near the Earth's surface. The results are summarized in Table 5.1 on p. 121.

## 5.6 CARRIER GENERATION AND RECOMBINATION

The electron density $n$ and hole density $p$ can change at a certain location in a semiconductor without carrier transfer (i.e., conduction current). The process of the creation of electrons and holes is called *carrier generation* or simply *generation*. The reverse process, namely the annihilation of electrons and holes is called *carrier recombination* or simply *recombination*. For example, all electrons and holes in an intrinsic semiconductor exist as a result of pairwise generation of electrons and holes (see Fig. 1.7 on p. 13). Note that the conservation of charge always holds even when carriers are created or annihilated. No electron or hole is created or annihilated without the accompanying creation or annihilation of an opposite charge. Energy and momentum are also conserved before and after generation or recombination. The conservation of energy and momentum are also relevant to light emission and reception by semiconductors, although these are not discussed in this book.

### 5.6.1 Direct Generation and Recombination

As shown in Fig. 5.6(a) (p. 144), the process in which an electron-hole pair is formed by the direct transition of an electron in the valence band to the conduction band is called *direct generation*. The process in which an electron-hole pair is annihilated by a direct transition of an electron in the conduction band to the valence band, as shown in Fig. 5.6(b), is called *direct recombination*. These processes are collectively called direct generation-recombination.

The direct generation-recombination was written in the form of a chemical reaction formula in (5.3) on p. 119 (reproduced below).

(covalent bond) $\rightleftharpoons$ (electron) + (hole) (Direct generation-recombination)

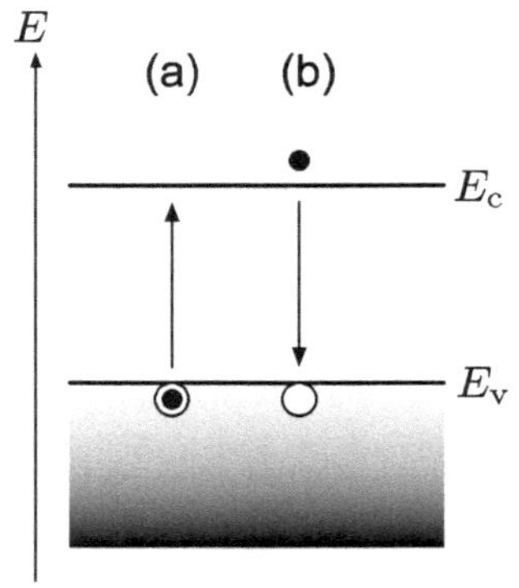


FIGURE 5.6 (a) Direct generation. (b) Direct recombination. A black-filled circle (•) represents the electron, and an open circle (∘) represents the hole. The states before the transition occurs are shown.

The energy difference between the two sides of this chemical reaction formula is given by the energy gap $E_g$. At room temperature $T = 300\,\text{K}$, the thermal energy $kT$ is $26\,\text{meV}$ (see Problem 1.3 on p. 26). For silicon, $E_g \simeq 1.1\,\text{eV}$ (see Table 1.3 on p. 5), and is much greater than the thermal energy. Therefore, direct generation-recombination is unlikely to occur very much at room temperature. The dominant generation-recombination process in silicon is indirect generation-recombination, which will be explained in §5.6.2.

### 5.6.2 Indirect Generation and Recombination

Often, there are what are called *generation-recombination centers* in the forbidden band that mediate generation-recombination, as shown in Fig. 5.7. They are also known as *carrier traps* or more simply as *traps*. Physically, these are impurities or crystal defects. Carrier generation and recombination mediated by traps are called indirect generation-recombination or Shockley–Read–Hall processes. Indirect generation-recombination is much more likely to occur than direct generation-recombination. At times, traps are deliberately introduced to semiconductors to control device characteristics (more specifically, minority carrier lifetime, p. 149).

Traps can be classified into *acceptor-type traps* and *donor-type traps*. Acceptor-type traps can, just like acceptors, assume one of the two states: *neutral* or *negatively charged*. Donor-type traps can, just like donors, assume one of the two states: *neutral* or *positively*

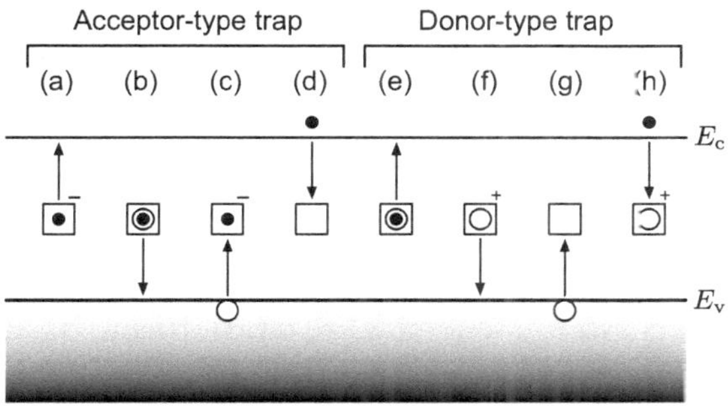


FIGURE 5.7 Eight elementary processes of indirect generation-recombination. A square (□) represents a trap. A black filled circle (•) represents an electron. An open circle (○) represents a hole. The states before the transition occurs are shown.

*charged*. Acceptor-type traps are usually located around the middle of the forbidden band or above. Donor-type traps are usually located around the middle of the forbidden band or below. In this connection, an acceptor can be understood as an "acceptor-type trap" located just above the valence band top, $E_V$ (see Fig. 1.18 on p. 22). Similarly, a donor can be understood as a "donor-type trap" located just below the conduction band bottom, $E_C$ (see Fig. 1.13 on p. 20). Since charged traps are spatially fixed charges, they appear in the equation of the charge neutrality condition and the charge term of the Poisson equation (p. 153). There are eight elementary processes of indirect generation-recombination, as shown in Fig. 5.7. Fig. 5.7(a) to (d) are the elementary processes involving an acceptor-type trap, and Fig. 5.7(e) to (h) are the elementary processes involving a donor-type trap.

(a) A negatively charged acceptor-type trap emits an electron into the conduction band and becomes neutral.

(b) A neutral acceptor-type trap emits a hole into the valence band and becomes negatively charged.

(c) A negatively charged acceptor-type trap captures a hole from the valence band and becomes neutral.

(d) A neutral acceptor-type trap captures an electron from the conduction band and becomes negatively charged.

(e) A neutral donor-type trap emits an electron into the conduction band and becomes positively charged.

(f) A positively charged donor-type trap emits a hole into the valence band and becomes neutral.

(g) A neutral donor-type trap captures a hole from the valence band and becomes positively charged.

(h) A positively charged donor-type trap captures an electron from the conduction band and becomes neutral.

These can be expressed in chemical reaction formulas as follows:

$$\text{(neutral acceptor-type trap)} + \text{(electron)} \underset{\text{(a)}}{\overset{\text{(d)}}{\rightleftharpoons}} \text{(negative acceptor-type trap)}, \tag{5.65}$$

$$\text{(neutral acceptor-type trap)} \underset{\text{(c)}}{\overset{\text{(b)}}{\rightleftharpoons}} \text{(negative acceptor-type trap)} + \text{(hole)}, \tag{5.66}$$

$$\text{(neutral donor-type trap)} \underset{\text{(h)}}{\overset{\text{(e)}}{\rightleftharpoons}} \text{(positive donor-type trap)} + \text{(electron)}, \tag{5.67}$$

$$\text{(neutral donor-type trap)} + \text{(hole)} \underset{\text{(f)}}{\overset{\text{(g)}}{\rightleftharpoons}} \text{(positive donor-type trap)}. \tag{5.68}$$

Generation of an electron-hole pair can occur via the following combinations of the elementary processes.

$$\text{(a)} \rightarrow \text{(b)} \quad \text{or} \quad \text{(b)} \rightarrow \text{(a)} \tag{5.69}$$

$$\text{(e)} \rightarrow \text{(f)} \quad \text{or} \quad \text{(f)} \rightarrow \text{(e)} \tag{5.70}$$

Annihilation of an electron-hole pair or equivalently, electron-hole recombination, can occur via the following combinations of the elementary processes.

$$\text{(c)} \rightarrow \text{(d)} \quad \text{or} \quad \text{(d)} \rightarrow \text{(c)} \tag{5.71}$$

$$(g) \rightarrow (h) \quad \text{or} \quad (h) \rightarrow (g) \tag{5.72}$$

Note that in all cases the total amount of charge is conserved. In order for charge conservation to hold, it is essential to consider the charge of the traps, too. The following processes, known as electron *trapping and detrapping* can also occur via traps.

$$(d) \rightarrow (a) \rightarrow (d) \rightarrow (a) \rightarrow \cdots \tag{5.73}$$

$$(h) \rightarrow (e) \rightarrow (h) \rightarrow (e) \rightarrow \cdots \tag{5.74}$$

One might think that a trap that has captured an electron would become negatively charged, but that is not necessarily the case. A donor-type trap captures an electron and becomes neutral, as shown in Fig. 5.7(h). Likewise, the following are hole trapping and detrapping processes.

$$(c) \rightarrow (b) \rightarrow (c) \rightarrow (b) \rightarrow \cdots \tag{5.75}$$

$$(g) \rightarrow (f) \rightarrow (g) \rightarrow (f) \rightarrow \cdots \tag{5.76}$$

Only donor-type traps become positively charged by capturing a hole (Fig. 5.7(g)). Acceptor-type traps become neutral when they capture a hole (Fig. 5.7(c)). In general, whether a trap is positively or negatively charged on average is determined solely by the trap type (donor-type or acceptor-type). In Fig. 5.7, the trap levels are located at the midgap. The amount of energy change before and after any transition involving a trap depends on where the trap level is actually located in the forbidden band. What is more likely to happen (generation-recombination or trapping-detrapping) also depends on the location of the trap level. If the trap level is located near the midgap, carrier generation-recombination is more likely to occur. Conversely, if the trap level is located near an edge of the forbidden band, $E_c$ or $E_v$, trapping-detrapping is more likely to occur. The chemical reaction formula (5.5) for donors on p. 120 corresponds to the electron trapping-detrapping, (5.74). The reason why donors are not involved in generation-recombination so much is that the donor level is located near the bottom of the conduction band $E_c$. In the case of donors, as long as the donor density is low and the Fermi level is not too close to $E_c$, the rightward reaction in (5.5) is dominant, and the ionization rate is high. Similarly, the chemical reaction formula (5.4) involving acceptors corresponds to the hole trapping-detrapping, (5.75). Apart from the location of the trap level within the energy gap, the presence or absence of a *Coulomb interaction* between the trap and carriers deserves attention. For example, both Figs. 5.7(d) and (h) are processes in which an electron in the conduction band is captured. But in Fig. 5.7(h), a positively charged donor-type trap captures an electron, and thus a Coulomb attraction between the positive and negative charges takes place. In contrast, in Fig. 5.7(d), there is no Coulomb interaction because a neutral acceptor-type trap captures an electron. In the case of Fig. 5.7(h), there is a Coulomb interaction between the positively charged trap and nearby electrons, even if none of them are captured. That is, electrons undergo Coulomb scattering (see Problem 5.7 on p. 164). A parallel argument applies to holes.

### 5.6.3 Carrier Generation-Recombination Rates

Suppose that $g_n$ electrons are generated per unit time in a unit volume, directly or indirectly. Then $g_n$ is called the electron *generation rate*. Similarly, suppose that $r_n$ electrons are annihilated per unit time in a unit volume by recombination. Then $r_n$ is called the electron *recombination rate* or *annihilation rate*. The generation and recombination of electrons always occur simultaneously, albeit not necessarily at the same rate. The *net generation-recombination rate* of electrons is given by

$$U_n \equiv g_n - r_n. \qquad \text{(Net electron generation-recombination rate)} \tag{5.77}$$

$U_n$ can be positive or negative. If $U_n > 0$, net electron generation is taking place, and if $U_n < 0$, net electron annihilation is taking place. In equilibrium, generation and recombination are balanced and $U_n = 0$. Note, however, that the two terms of (5.77) individually satisfy $g_n > 0$ and $r_n > 0$. Similarly for holes, we can consider the generation rate $g_p$ and the recombination rate $r_p$. The net generation-recombination rate of holes is given by

$$U_p \equiv g_p - r_p. \qquad \text{(Net hole generation-recombination rate)} \tag{5.78}$$

Carrier generation and recombination usually occur as the generation and annihilation of electron-hole pairs (p. 146), in which case

$g_n = g_p$ and $r_n = r_p$. It is, however, also possible for electrons, for example, to be generated without the generation of the same number of holes. For example, when an n-type semiconductor is formed by adding donors to an intrinsic semiconductor, electrons are generated by the rightward process in (5.4) on p. 120. However, this only occurs for a very short period of time. Processes like this do not usually continue indefinitely. Steady generation and recombination of carriers occur as the generation and annihilation of electron-hole pairs.

### 5.6.4 Minority Carrier Lifetime

In semiconductor devices, minority carriers often play an important role. Using a p-type semiconductor as an example, let us consider how electrons, which are minority carriers in this case, behave in connection with generation-recombination. Let $p_P$ and $n_P$ denote hole and electron densities in a p-type semiconductor, respectively. Suppose that the carrier densities, $p_P$ and $n_P$, deviate from $p_{P0}$ and $n_{P0}$, which are the carrier densities at equilibrium when there is no band bending. That is, $p_{P0}$ is the equilibrium hole density in a uniform p-type semiconductor, and $n_{P0}$ is the corresponding electron density. Suppose further that the deviation occurs uniformly in the entire p-type semiconductor. No current flows because of the uniformity. From (4.19) on p. 96, $p_{P0}n_{P0} = n_i^2$, and since we are considering a p-type semiconductor, $p_{P0} \gg n_{P0}$. The nonequilibrium electron and hole densities can be written as

$$n_P = n_{P0} + \Delta n, \qquad \text{(Electron density in nonequilibrium p-type)} \tag{5.79}$$

$$p_P = p_{P0} + \Delta p \simeq p_{P0}, \qquad \text{(Hole density in nonequilibrium p-type)} \tag{5.80}$$

where $\Delta n$ is the excess electron density and $\Delta p$ is the excess hole density. $\Delta n$ and $\Delta p$ may assume positive and negative values. In order to make it easier to write down equations, let us assume that $\Delta n > 0$. The differential equation describing the time variation of the density of minority carriers (i.e., electrons) can be written using the net generation-recombination rate $U_n$, given by (5.77), as

$$\frac{dn_P}{dt} = \frac{d\Delta n}{dt} = U_n = g_n - \frac{n_P}{\tau_n}, \tag{5.81}$$

where we used the fact that $n_{P0}$ does not depend on time. The second term of (5.81) is the electron recombination rate.

$$r_n \equiv \frac{n_P}{\tau_n}. \qquad \text{(Electron recombination rate)} \qquad (5.82)$$

It represents the effect of trying to pull the electron density back to its equilibrium value, $n_{P0}$. This term is similar to the second term of (5.43) on p. 134. The constant $\tau_n$ has the dimensions of time and is called the *electron lifetime*. Since the time derivative of the left-hand side of (5.81) equals zero at equilibrium, the right-hand side shows that the equilibrium electron generation rate $g_{n0}$ and the equilibrium electron density $n_{P0}$ satisfy the following relationship.

$$g_{n0} = r_{n0} \equiv \frac{n_{P0}}{\tau_n}. \qquad \text{(Electron generation rate at equilibrium)} \qquad (5.83)$$

Let us return to a nonequilibrium state with $\Delta n > 0$. If the deviation from the equilibrium state is not very significant, then we can consider the following equation to hold.

$$g_n \simeq g_{n0} = \frac{n_{P0}}{\tau_n}. \qquad \text{(Electron generation rate)} \qquad (5.84)$$

Inserting this into (5.81) and eliminating $g_n$, we obtain

$$\frac{d\Delta n}{dt} = U_n = \frac{n_{P0} - n_P}{\tau_n} = -\frac{\Delta n}{\tau_n}. \qquad (5.85)$$

This differential equation can be solved in the same way as (5.43) on p. 134, and we see that the excess electron density $\Delta n$ decreases exponentially with the time constant $\tau_n$ toward 0. In the above, we assumed that $\Delta n > 0$. If $\Delta n < 0$, the roles of generation and recombination are interchanged, and generation takes the role of pulling the state back to equilibrium. Basically the same is true for holes in n-type semiconductors. Let $p_N$ and $n_N$, respectively, denote hole and electron densities in the n-type semiconductor. The hole density can be written as

$$p_N = p_{N0} + \Delta p, \qquad \text{(Hole density in nonequilibrium n-type)} \quad (5.86)$$

where $p_{N0}$ is the equilibrium hole density in a uniform n-type semiconductor. The electron density can be written as

$$n_N = n_{N0} + \Delta n \simeq n_{N0}, \quad \text{(Electron density in nonequilibrium n-type)} \qquad (5.87)$$

where $n_{N0}$ is the equilibrium electron density in a uniform n-type semiconductor. The governing differential equation for the excess hole density $\Delta p$ is given by

$$\frac{d\Delta p}{dt} = U_p = \frac{p_{N0} - p_N}{\tau_p} = -\frac{\Delta p}{\tau_p}. \tag{5.88}$$

The lifetime, $\tau_p$, of holes as minority carriers satisfies the following relation:

$$g_p \simeq g_{p0} = r_{p0} \equiv \frac{p_{N0}}{\tau_p}. \tag{5.89}$$

To put it loosely, *the minority carrier lifetime is the "waiting time" before carriers move vertically in an energy band diagram*, as shown in Fig. 5.7 (p. 145). As noted on p. 120, generation and recombination try to pull the nonequilibrium state back to equilibrium. Whether generation is dominant or recombination is dominant depends on whether the effective intrinsic carrier density $n_i'$, given by (5.14) on p. 124, is greater than the intrinsic carrier density $n_i$ ($n_i' > n_i$) or not ($n_i' < n_i$). If $n_i' > n_i$, the $pn$ product, (5.13), is greater than the equilibrium value $n_i^2$, and therefore recombination becomes dominant ($U_n < 0$, $U_p < 0$). Conversely, if $n_i' < n_i$, generation becomes dominant ($U_n > 0$, $U_p > 0$). $n_i' \gtrless n_i$ corresponds to $\zeta_n - \zeta_p \gtrless 0$. So *whether generation is dominant* ($\zeta_n < \zeta_p$) *or recombination is dominant* ($\zeta_n > \zeta_p$) *can be read from an energy band diagram*, provided quasi-Fermi levels are drawn. The minority carrier lifetime is highly dependent on the quality of the crystal. The more crystal defects and impurities, including dopants, there are, or to put it another way, the more traps there are, the shorter the minority carrier lifetime. Typical values of minority carrier lifetime in silicon are given in Table 5.2 on p. 122.

## 5.7 BASIC EQUATIONS FOR SEMICONDUCTOR DEVICES

In §5.6.4, we made the somewhat bizarre assumption that the minority carrier density uniformly deviates from the equilibrium value ($\Delta n \gtrless 0$ or $\Delta p \gtrless 0$) throughout a given piece of semiconductor. In practice, the occurrence of excess minority carriers is usually localized to a small region, and the localized excess minority carriers are accompanied by a current. In some cases, current flow leads to localized excess minority carriers in a certain region, while in other cases, the excess minority carriers are generated first (due, for instance, to a high electric field or light irradiation), and as a result, a current flows. The differential equations that describe carrier generation-recombination and electron and hole currents are the *continuity equations* for charge and current. The continuity equations can be considered to be expressions of charge conservation, taking into account both the conduction current and the generation-recombination of carriers. Up to now, we have not made the position- and time-dependence of physical quantities such as carrier densities explicit; that is, we wrote the electron density as $n$ instead of $n(x, t)$, for example. But in this section, we will make the dependence on position $x$ and time $t$ explicit. The time variation of the electron density $n(x, t)$ and the hole density $p(x, t)$at a point in a semiconductor is described as follows:

$$\frac{\partial n(x,t)}{\partial t} = \frac{1}{q}\frac{\partial J_{\mathrm{n}}(x,t)}{\partial x} + U_{\mathrm{n}}(x,t), \quad \text{(Continuity equation for electrons)} \tag{5.90}$$

$$\frac{\partial p(x,t)}{\partial t} = -\frac{1}{q}\frac{\partial J_{\mathrm{p}}(x,t)}{\partial x} + U_{\mathrm{p}}(x,t). \quad \text{(Continuity equation for holes)} \tag{5.91}$$

The first terms of (5.90) and (5.91) represent the increase or decrease in carrier density due to conduction current, $J_{\mathrm{n}}(x, t)$ or $J_{\mathrm{p}}(x, t)$. The signs of the first terms are different because electrons and holes have opposite charges. The generation-recombination rates, $U_{\mathrm{n}}(x, t)$ and $U_{\mathrm{p}}(x, t)$, are given by (5.85) and (5.88), respectively. The current densities, $J_{\mathrm{n}}(x, t)$ and $J_{\mathrm{p}}(x, t)$, are given by (5.37) and (5.39) on p. 132, respectively. By using (5.38) and (5.40), we can rewrite the expressions for current densities using the electrostatic potential $\psi(x, t)$ as follows:

$$J_{\mathrm{n}}(x,t) = -q\mu_{\mathrm{n}}(x)\, n(x,t)\frac{\partial \psi(x,t)}{\partial x} + \mu_{\mathrm{n}}(x)\, kT\frac{\partial n(x,t)}{\partial x} \tag{5.92}$$

$$J_{\mathrm{p}}(x,t) = -q\mu_{\mathrm{p}}(x)\, p(x,t)\frac{\partial \psi(x,t)}{\partial x} - \mu_{\mathrm{p}}(x)\, kT\frac{\partial p(x,t)}{\partial x} \tag{5.93}$$

Since (5.92) and (5.93) involve $\psi(x)$, the *Poisson equation* must also be solved simultaneously to solve the continuity equations. The Poisson equation can be derived from Gauss' law. The Poisson equation written as a differential equation for $\psi(x)$ is given by

$$\frac{\mathrm{d}^2\psi(x,t)}{\mathrm{d}x^2} = -\frac{q\left[p(x,t) - n(x,t) + N_\mathrm{D}^+(x,t) - N_\mathrm{A}^-(x,t)\right] + \rho_\mathrm{t}(x,t)}{\epsilon(x)}, \tag{5.94}$$

where $\epsilon(x)$ is the permittivity, and $\rho_\mathrm{t}(x,t)$ is the trap charge density, which may be positive or negative. Since the Poisson equation is an equation for an electrostatic field, strictly speaking, it cannot be used when the electromagnetic field changes with time. Magnetic fields are not taken into consideration, either. To take these into account, one must solve the Maxwell equations, which consist of several differential equations. Thus, both time-varying electromagnetic fields and currents are correctly taken into account. Actually, the Poisson equation and the continuity equations are both subsets of Maxwell equations. However, for many devices, it has been found that a good approximation is to neglect the magnetic field and use the Poisson equation to find $\psi(x)$, including cases where voltages and currents vary with time. The simultaneous partial differential equations consisting of the continuity equations, (5.90) and (5.91), and the Poisson equation (5.94) are the *basic equations for semiconductor devices* when the magnetic field can be neglected. These equations are also known as the *Shockley equations* [22]. Let us assume for simplicity that the dopant and trap ionization rates are independent of $t$ and bias conditions. Then the three unknowns of the Shockley equations are the electron density $n(x,t)$, the hole density $p(x,t)$, and the electrostatic potential $\psi(x,t)$ (see the box on p. 105). In the above, we considered only one spatial dimension ($x$-axis) explicitly. But this does not mean that the $y$- and $z$-axes do not exist. We implicitly assumed that the carrier densities and the electrostatic potential do not change in the $y$- and $z$-directions. Systems that are truly one- or two-dimensional have different density-of-state functions [24] (p. 88) and must be treated quite differently. As a reminder, the continuity equations (5.90) and (5.91) and the Poisson equation (5.94) can be used for both nondegenerate and degenerate semiconductors. However, the current density formulas (5.92) and (5.93) can only be used for nondegenerate semiconductors that exhibit a linear response (see the Box on p. 139). In order to correctly handle nonlinear responses and degenerate semiconductors, the current density equations need to be modified.

### DEVICE SIMULATOR

Device simulator or device simulation is also often referred to as *TCAD*, which stands for *technology computer-aided design*. "Technology" here refers specifically to technology related to semiconductor devices or integrated circuit fabrication or manufacturing.

Device simulators are indispensable tools for the design and analysis of semiconductor devices. Of course, a good understanding of device physics is essential to do anything meaningful with a device simulator.

Writing a one-dimensional device simulator program, which assumes $y$- and $z$-directions are spatially uniform (p. 153), may be relatively easy. Full-featured device simulators are commercially available from specialized software vendors.

One of the authors (K.M.) once had a student write a device simulator to analyze the low-temperature operation of semiconductor devices. But calculations did not converge and we had a hard time trying to fix it. It turned out to be due to a lack of the number of significant digits in the calculation. Although this problem can be alleviated somewhat by writing the program carefully, the only fundamental fix is to increase the number of significant digits used in the computation. However, it is not so trivial to increase the number of significant digits beyond what is offered by the typical computer hardware. Today's number-crunching processors typically offer 15 significant digits when converted to the decimal number system. This is often insufficient for handling problems in electrical and electronic engineering (see the example on p. 124). The use of numbers with a larger number of significant digits is possible by software but is much much slower than processing by hardware.

If the time derivatives of (5.90) and (5.91) are set to 0, the equations become steady-state equations, which are much simpler. However, they are still simultaneous differential equations, and, in general, they can be solved only numerically. A computer program that numerically solves the Shockley equations is called a *device simulator*. It can, of course, also be used to draw energy band diagrams.

## 5.8 DIELECTRIC RELAXATION

As an example of applying the continuity equations from §5.7, we consider in this section the response of majority carriers. When we discussed the diffusion of carriers that occurs due to the density gradient of carriers in §5.5, we forgot that the carriers are charged particles, except in the calculation of the current density. However, if carriers are distributed spatially nonuniformly, an electric field should also be generated (unless the nonuniform carrier charges are exactly neutralized by ions or carriers of opposite polarity). Then, a drift current should also flow due to the electric field. Since the drift terms of the current density equations (5.92) and (5.93) are proportional to the carrier density, the drift current that flows in this case should be dominated by the majority carrier current. This drift current flows in the direction of reducing the carrier density gradient and the potential gradient due to the nonuniform carrier distribution, and thus the system evolves toward the charge neutrality condition. This phenomenon is called *dielectric relaxation* [29]. In dielectric relaxation, an electric force acts on carriers, so the relaxation process should proceed faster than with purely spontaneous diffusion. The time constant of dielectric relaxation is called the *dielectric relaxation time* or *majority-carrier response time*. In the case of n-type semiconductors, it is given by

$$\tau_{\mathrm{drn}} \equiv \rho_{\mathrm{n}}\epsilon_{\mathrm{s}} = \frac{\epsilon_{\mathrm{s}}}{\sigma_{\mathrm{n}}}, \qquad \text{(Dielectric relaxation time in n-type)} \qquad (5.95)$$

where $\epsilon_{\mathrm{s}}$ is the permittivity of the given n-type semiconductor, and $\rho_{\mathrm{n}} = 1/\sigma_{\mathrm{n}}$ is its resistivity (see (5.52) on p. 138). The higher the resistivity, the harder it is for the current to flow, and thus the longer it takes to relax into the final state.

### *Example: Derivation of Dielectric Relaxation Time*

Let us derive (5.95). Suppose that in n-type silicon, the density of electrons, which are the majority carriers, becomes locally excessive at a certain position $x$ (it is, admittedly, unnatural to consider this in one spatial dimension (p. 153)). Assuming that only the excess electron density, $\Delta n$ (p. 153), breaks the charge neutrality condition, inserting $\Delta n$ into the Poisson equation (5.94) on p. 153 yields

$$\frac{\mathrm{d}^2\psi}{\mathrm{d}x^2} = -\frac{\mathrm{d}\mathcal{E}}{\mathrm{d}x} = \frac{q\Delta n}{\epsilon_{\mathrm{Si}}}, \qquad \text{(Poisson equation)} \qquad (5.96)$$

where $\epsilon_{\text{Si}}$ is the permittivity of silicon. Neglecting the diffusion term in the electron current density equation (5.92) and using the conductivity $\sigma_{\text{n}}$ in (5.52) on p. 138, we obtain (see Problem 5.8 on p. 164)

$$J_{\text{n}} = -\sigma_{\text{n}} \frac{\text{d}\psi}{\text{d}x}. \qquad \text{(Electron drift current density)} \qquad (5.97)$$

Inserting this into the continuity equation (5.90) on p. 152 and neglecting the generation-recombination term $U_{\text{n}}$, we get

$$\frac{\partial n}{\partial t} = \frac{\partial \Delta n}{\partial t} = -\frac{\sigma_{\text{n}}}{q} \frac{\partial^2 \psi}{\partial x^2} = -\frac{\Delta n}{\rho_{\text{n}} \epsilon_{\text{Si}}}, \qquad \text{(Continuity equation)} \quad (5.98)$$

where we used (5.96). We saw differential equations of this form in (5.43) on p. 134 and (5.85) on p. 150. From the above, the time constant of the relaxation process that makes $\Delta n \to 0$ is $\tau_{\text{drn}} = \rho_{\text{n}} \epsilon_{\text{Si}}$ (Problem 5.9 on p. 165). ■

Loosely put, *the dielectric relaxation time is the "waiting time" before carriers move sideways in an energy band diagram by drift.* Using the conductivity formula (5.52) on p. 138 and the mobility formula (5.48) on p. 135 in (5.95), we obtain

$$\tau_{\text{drn}} = \frac{\epsilon_{\text{s}}}{q n \mu_{\text{n}}} = \frac{\epsilon_{\text{s}} m_{\text{e}}}{q^2 n \tau_{\text{e}}}. \qquad \text{(Dielectric relaxation time in n-type)} \qquad (5.99)$$

Equation (5.99) indicates that the dielectric relaxation time is inversely proportional to the majority carrier density $n$ and the mean free time $\tau_{\text{e}}$. Table 5.2 summarizes typical values of the time constants related to carriers in silicon. Typically, the dielectric relaxation time is orders of magnitude shorter than the lifetime of minority carriers:

$$\text{(dielectric relaxation time)} \ll \text{(minority carrier lifetime)}. \qquad (5.100)$$

Charge neutralization by dielectric relaxation (sideways movement of carriers in band diagram) precedes the disappearance of excess minority carriers due to carrier generation-recombination (vertical movement of carriers in band diagram (p. 151)). As mentioned in §5.2.2 (p. 119), this big difference in the time scales is the reason for having to define separate quasi-Fermi levels for electrons and holes in nonequilibrium states. Also, the long lifetime, $\tau_{\text{n}}$, justifies the neglect of the generation-recombination term $U_{\text{n}}$ in (5.98) (see (5.85) on p. 150). Equation (5.99) suggests that the dielectric relaxation time varies considerably depending on the majority carrier density, but since the authors do not know the specific values, Table 5.2 only quotes a value from [31] (see Problem 5.10 on p. 165).

## 5.9 DEBYE LENGTH

In §5.8, we derived an important time scale, the dielectric relaxation time. In the process, we ignored the diffusion term when we obtained (5.97). Ignoring diffusion is normal in metals, but in semiconductors, carrier densities can change by many orders of magnitude, so diffusion must usually be taken into account as well. An important length scale appears when the diffusion term is also considered in the dielectric relaxation. Assuming n-type silicon again, and this time without neglecting the diffusion term, we obtain the following equation instead of (5.97):

$$J_{\mathrm{n}} = -\sigma_{\mathrm{n}}\frac{\mathrm{d}\psi}{\mathrm{d}x} + \mu_{\mathrm{n}}kT\frac{\mathrm{d}\Delta n}{\mathrm{d}x}. \qquad \text{(Electron current density)} \qquad (5.101)$$

Inserting this into the continuity equation (5.90) on p. 152 with ignoring the generation-recombination term $U_{\mathrm{n}}$ yields

$$\frac{\partial \Delta n}{\partial t} = -\frac{\Delta n}{\rho_{\mathrm{n}}\epsilon_{\mathrm{Si}}} + \frac{\mu_{\mathrm{n}}kT}{q}\frac{\partial^2 \Delta n}{\partial x^2}. \qquad (5.102)$$

This is a spatially second order and temporally first order partial differential equation (see Table 3.2 on p. 51). In a steady state, the time derivative on the left-hand side equals zero. In this case, (5.102) reduces to

$$\frac{\mathrm{d}^2 \Delta n}{\mathrm{d}x^2} = \frac{q\Delta n}{\mu_{\mathrm{n}}kT\rho_{\mathrm{n}}\epsilon_{\mathrm{Si}}}. \qquad (5.103)$$

Using (5.52) on p. 138 for the relationship between mobility and conductivity, we have

$$\mu_{\mathrm{n}} = \frac{1}{qn\rho_{\mathrm{n}}}. \qquad (5.104)$$

Putting this in (5.103), we obtain

$$\frac{\mathrm{d}^2 \Delta n}{\mathrm{d}x^2} = \frac{q^2 n}{kT\epsilon_{\mathrm{Si}}}\Delta n. \qquad (5.105)$$

This differential equation has the same form as the wave equation for transmission line (p. 62) and the time-independent Schrödinger equation shown in Fig. 3.15 on p. 71. Equation (5.105) says that the second derivative of $\Delta n$ equals $\Delta n$ multiplied by $q^2 n/(kT\epsilon_{\mathrm{Si}})$. So the solution to (5.105) should have the following form:

$$\Delta n(x) \propto \exp\left(-\frac{x}{L_{\mathrm{D}}}\right), \tag{5.106}$$

where

$$L_{\mathrm{D}} \equiv \sqrt{\frac{kT\epsilon_{\mathrm{Si}}}{q^2 n_{\mathrm{N0}}}} = \sqrt{\frac{kT\epsilon_{\mathrm{Si}}}{q^2 N_{\mathrm{D}}^{+}}}, \qquad \text{(Debye length in n-type silicon)} \tag{5.107}$$

is called the *Debye length*. Note that in (5.107) we used (5.113) on p. 161 for n-type semiconductors, and approximated $n$ to $n_{\mathrm{N0}}$ ($n \simeq n_{\mathrm{N0}}$). Equation (5.106) shows that the electron density cannot change spatially abruptly due to diffusion and that some distance is required for the change. The Debye length gives an estimate of this. To be more specific, $L_{\mathrm{D}}$ is the length required for the carrier density to change by a factor of e $\simeq$ 2.7 or 1/e $\simeq$ 0.37. If the ratio of the carrier densities of two regions is larger than e (or smaller than 1/e), the length required for the transition[3] is longer than the Debye length. In semiconductor devices, impurity doping is finely controlled, but no matter how steeply the impurity density is changed, it is impossible to change the carrier densities, $E_{\mathrm{c}}$, and $E_{\mathrm{v}}$ on a scale shorter than the Debye length. Therefore, the Debye length is also related to the limit of device miniaturization. The Debye length in p-type silicon is also given by the same form as in (5.107).

$$L_{\mathrm{D}} \equiv \sqrt{\frac{kT\epsilon_{\mathrm{Si}}}{q^2 p_{\mathrm{P0}}}} = \sqrt{\frac{kT\epsilon_{\mathrm{Si}}}{q^2 N_{\mathrm{A}}^{-}}} \qquad \text{(Debye length in p-type silicon)} \tag{5.108}$$

Table 5.3 shows numerical examples of the Debye length of silicon obtained using (5.107).

[3] The region of transition may be the depletion layer (p. 186).

TABLE 5.3 Debye Length in Silicon

| Ionized dopant density ($cm^{-3}$) | Debye length (nm) |
|---|---|
| $10^{15}$ | 129 |
| $10^{16}$ | 41 |
| $10^{17}$ | 13 |
| $10^{18}$ | 4 |
| $10^{19}$ | 1 |

## 5.10 HALL EFFECT

In §1.3, we explained that there are positively charged carriers called holes in semiconductors, but we did not explain how this was found out. In §4.1, we explained that holes are unoccupied states near the top of the valence band, whereas elsewhere in this book we have basically treated holes as positively charged particles. However, there seems to be a considerable gap between the pictures of "holes," as considered in the Box on p. 136, and the picture of "positively charged particles." That holes are indeed positively charged particles has been experimentally confirmed by a phenomenon known as the *Hall effect*, named after Edwin Herbert Hall. As shown in Fig. 5.8, an electric field **E** is applied to the semiconductor strip, which is connected to a DC power supply (not shown), and a DC current **J** flows through it. Note that **E** and **J** are vectors. A uniform magnetic field **B** is applied from bottom to top. Then, the Lorentz force acts on carriers moving in the magnetic field according to Fleming's left-hand rule. This causes carriers to move in a circular motion (i.e., cyclotron motion). However, for the sake of simplicity, let us forget about this and consider only in which direction the Lorentz force $\mathbf{F}_L$ is exerted by the magnetic flux density vector **B** when the charged particle starts to move in the direction of the current density vector **J**.

When the majority carriers are electrons, they flow in the opposite direction to **J** as shown in Fig. 5.8(a). If a magnetic field **B** is applied, the Lorentz force $\mathbf{F}_L$ acts on electrons in the direction as shown. As a result, a potential gradient is generated in the direction orthogonal to **J** so that the left front side has a lower potential. On the other hand, when the majority carriers are holes as shown in Fig. 5.8(b), they flow in the same direction as **J**. If a magnetic field **B** is applied, the Lorentz force $\mathbf{F}_L$ acts on holes in the same direction as in Fig. 5.8(a). However, since the polarity of the carriers is opposite, the resulting

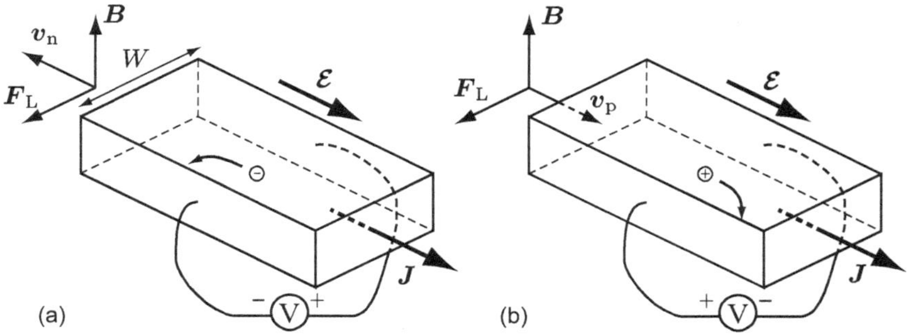


FIGURE 5.8 (a) Hall effect in n-type semiconductor. (b) Hall effect in p-type semiconductor.

potential gradient orthogonal to $\mathbf{J}$ is opposite (the potential on the left front side is higher). Thus, by examining the potential gradient in the direction orthogonal to $\mathbf{J}$, the polarity of the majority carriers can be determined. The Hall effect is an example that cannot be properly described by the Shockley equations in §5.7 because a magnetic field is involved. Moreover, three-dimensional space and cyclotron motion must also be considered, so its in-depth treatment is quite advanced. Hall effect experiments are also conducted to measure carrier mobility. The voltage $V_H$ measured by the voltmeter in Fig. 5.8 is called the Hall voltage. In a steady state, the Lorentz force $\mathbf{F}_L = \mp q\mathbf{v} \times \mathbf{B}$, originating from the external magnetic field acting on individual carriers, is balanced by the electric force $\mp q\mathbf{E}$, originating from the electric field resulting from the carriers being pressed against the side wall. Here, $\mathbf{v}$ is the velocity vector of the carrier. Specifically, $\mathbf{v} = \mathbf{v}_n$ for electrons and $\mathbf{v} = \mathbf{v}_p$ for holes. The equation for the balance of these forces can be written as

$$qvB = q\mathcal{E}, \tag{5.109}$$

where $\mathcal{E} = |\mathbf{E}|$, $v = |\mathbf{v}|$, and $B = |\mathbf{B}|$. If the width of the semiconductor strip is $W$, the Hall voltage can be written as

$$V_H = W\mathcal{E} = WvB. \tag{5.110}$$

$v$ on the right-hand side can be expressed in terms of the current density $J = |\mathbf{J}|$ using (5.27) and (5.29) on p. 130 and the like.

$$V_{\mathrm{H}} = \begin{cases} \dfrac{WBJ}{-qn} & \text{(Hall voltage for electron conduction)} \\ \dfrac{WBJ}{qp} & \text{(Hall voltage for hole conduction)} \end{cases} \tag{5.111}$$

Equation (5.111) can be rewritten as

$$V_{\mathrm{H}} = R_{\mathrm{H}} WBJ, \qquad \text{(Hall voltage)} \tag{5.112}$$

where $R_{\mathrm{H}}$ is called the *Hall coefficient*. The polarity of majority carriers can be found from the sign of $R_{\mathrm{H}}$. Note, however, that (5.111) is known not to hold as is in general. To take this into account, the Hall factor in (5.112) is written as

$$R_{\mathrm{H}} = \begin{cases} -\frac{r_{\mathrm{H}}}{qn} \, (< 0) & \text{(Hall coefficient for electron conduction)} \\ \frac{r_{\mathrm{H}}}{qp} \, (> 0) & \text{(Hall coefficient for hole conduction)} \end{cases} \tag{5.113}$$

where $r_{\mathrm{H}} \, (\neq 1)$ is a positive coefficient known as the *Hall factor*.

## POLARITY OF CARRIERS IN METALS

We learn at school that electrons are responsible for electrical conduction in metals. The majority carriers in semiconductors can be electrons or holes, depending on the doping. But the possibility of carrier polarity being positive or negative is not unique to semiconductors. In the case of metals, too, depending on the material and conditions, majority carriers can be holes. One of the authors (K.M.) once performed Hall measurements on a few metals with his student and confirmed that majority carriers can be holes. Fig. 5.9 shows that the Hall coefficient of molybdenum (Mo) is positive, meaning that the majority carriers are holes. Note that the unit of the vertical axis is wrong. The correct unit should be $m^3/C$.

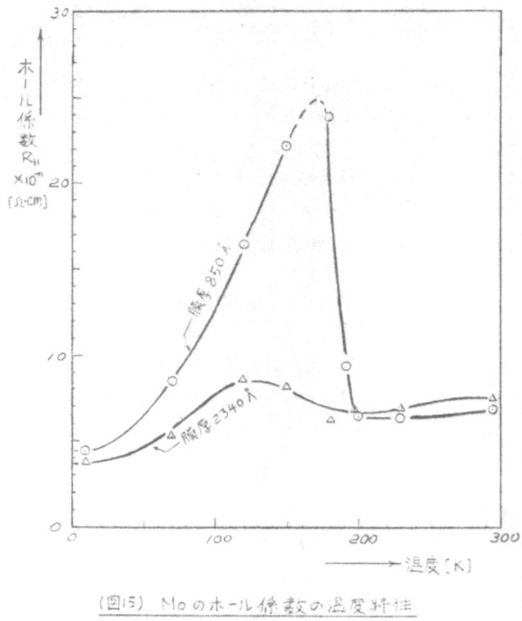


FIGURE 5.9 Temperature dependence of the Hall coefficient of molybdenum films of different thicknesses [38].

## 5.11 SUMMARY

In this chapter, we discussed electrical conduction and related phenomena in semiconductors.

- An equilibrium state is a state in which the temperature is constant throughout the system and there is no net particle flow or chemical reactions. A steady state is a nonequilibrium state that does not change with time.
- Sinusoidal steady states can be treated in a similar manner to steady states by the replacement $d/dt \rightarrow j\omega$ in the governing differential equation.
- In equilibrium, the electron and hole densities can be expressed in terms of the Fermi level, but in nonequilibrium, we need to consider separate quasi-Fermi levels for electrons and holes.

- The electron quasi-Fermi level, $\zeta_n$, is the nonequilibrium electron density converted into a quantity that can be drawn on an energy band diagram. The hole quasi-Fermi level, $\zeta_p$, is the nonequilibrium hole density converted into a quantity that can be drawn on an energy band diagram.
- The conduction current density is proportional to the carrier density and the gradient of the quasi-Fermi level, as long as the gradient is not too steep. The constant of proportionality is the mobility.
- The conduction current consists of two components. One is the drift current that flows due to external forces acting on carriers and the other is the diffusion current that flows spontaneously due to the density gradient of carriers.
- Drift and diffusion are related to each other by Einstein's relation.
- There are two types of carrier generation-recombination: direct generation-recombination and indirect generation-recombination. The latter is dominant in most semiconductors.
- The minority carrier lifetime is the time constant associated with the generation-recombination processes.
- From an energy band diagram with quasi-Fermi levels for electrons and holes, one can read the gradient of the electrostatic potential, carrier densities and their gradients, the direction and the magnitude of the current, which of drift and diffusion is the dominant current component, and which of generation and recombination is dominant.
- The basic equations for semiconductor devices, also known as the Shockley equations, consist of the continuity equations for charge and current and the Poisson equation for the electrostatic field.
- The dielectric relaxation time is the time constant associated with charge neutralization by the drift of majority carriers. It is usually much shorter than the lifetime.
- The Debye length is an important length scale that characterizes the limit to spatial carrier density manipulation by doping.

## 5.12 PROBLEMS

5.1 The difference between the electron and hole quasi-Fermi levels, $|\zeta_n - \zeta_p|$, represents the degree of deviation from equilibrium (p. 124). How would the value of $|\zeta_n - \zeta_p|$ change if you somehow shortened the lifetime of minority carriers?

5.2 Although we stated on p. 106 that there are no states in the forbidden band, there are, in fact, spatially localized energy levels such as dopant levels and trap levels in the forbidden band. In equilibrium, the occupancy of these states is described by a distribution function that looks similar to the Fermi–Dirac distribution function (4.4) on p. 89. Now, how should we consider the occupancy of these states at nonequilibrium?

5.3 Let us try to understand the expression (4.26) on p. 98 of the effective density of states $N_c$ for the conduction band as an example of the general form of the carrier density expression, (5.23) on p. 128. Answer what corresponds to $\zeta_n$ and the "reference energy" in (4.26).

5.4 Just as in (4.59) on p. 111 for the total chemical potential, suppose we have split the electron quasi-Fermi level into two terms as $\zeta_n = \zeta_{n,ext} + \zeta_{n,int}$ (p. 126). From the electron current density expression (5.37) on p. 132, find the terms corresponding to $d\zeta_{n,ext}/dx$ and $d\zeta_{n,int}/dx$.

5.5 Solve the differential equation (5.43) on p. 134 for the average velocity $\langle v \rangle$ of electrons with the initial condition $\langle v \rangle(0) = 0$ and derive the expression (5.45) for the time variation of the average velocity.

5.6 Give an example of something that cannot be well explained by the "punch hole model" of hole conduction in Fig. 5.5(b) (p. 137).

5.7 Classify the eight elementary processes of indirect generation and recombination shown in Fig. 5.7 (p. 145) into those involving Coulomb interaction between the trap and the carrier and those not involving Coulomb interaction. Coulomb interaction involving a nonuniform electric field originating from a charged trap can be considered as a form of carrier scattering, discussed in §5.4.1, and shortens the mean free time.

5.8 When we derived (5.97) on p. 156, we treated the conductivity $\sigma_n$ as a constant. However, since $\sigma_n = qn\mu_n$, a diffusion term

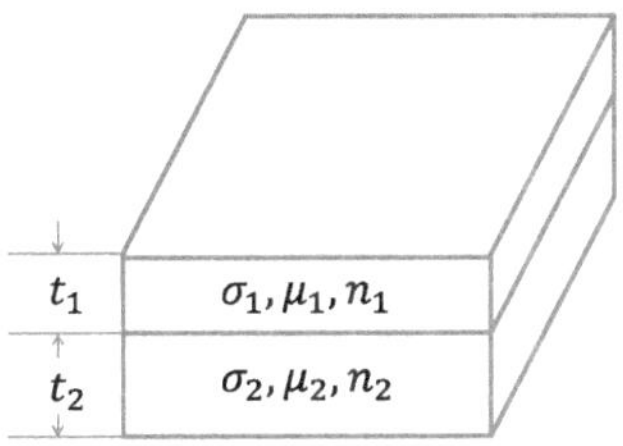


FIGURE 5.10 Two-layer structure.

proportional to $dn/dx$ should also appear. Work out the ensuing derivation considering the diffusion term.

5.9 In the derivation of the dielectric relaxation time in n-type semiconductors on p. 156, we considered the excess or deficiency, $\Delta n$, of electrons, which are the majority carriers. However, the discussion on p. 121 and p. 151 focused rather on the excess or deficiency of minority carriers. Assume that the excess density of holes (minority carriers) is given by $\Delta p$ and modify the discussion of dielectric relaxation.

5.10 Table 5.2 (p. 122) gives only a single value for the dielectric relaxation time, but its possible range can easily be estimated using (5.95) on p. 155. When the dopant density in silicon goes from $10^{16}\,\text{cm}^{-3}$ to $10^{19}\,\text{cm}^{-3}$, the resistivity goes from about $10^{0}\,\Omega\cdot\text{cm}$ to $10^{-2}\,\Omega\cdot\text{cm}$ [30]. Use $\epsilon_0 \simeq 8.85\times10^{-12}$ F/m for the permittivity of vacuum, and refer to Table 1.3 (p. 5) for the relative permittivity of silicon, and estimate the corresponding range of the dielectric relaxation time.

5.11 Consider the structure shown in Fig. 5.10 on p. 165, composed of two layers of thin films with thicknesses $t_1$ and $t_2$. The conductivity, mobility, and carrier density of layer $i$ are $\sigma_i$, $\mu_i$, and $n_i$ $(i = 1, 2)$, respectively. Express the overall apparent conductivity $\sigma_{12}$, mobility $\mu_{12}$, and carrier density $n_{12}$ of the two-layer structure in terms of $\sigma_i$, $\mu_i$, and $n_i$, assuming that a current uniformly flows in the vertical direction.

5.12 Another setup of more practical interest than the above problem. The conductivity of a thin film can be found from current-voltage characteristics and the Hall coefficient from Hall measurement. Suppose that the apparent conductivity $\sigma_{12}$ and the apparent Hall coefficient $R_{\text{H}12}$ of a two-layer structure, shown in Fig. 5.11(a),

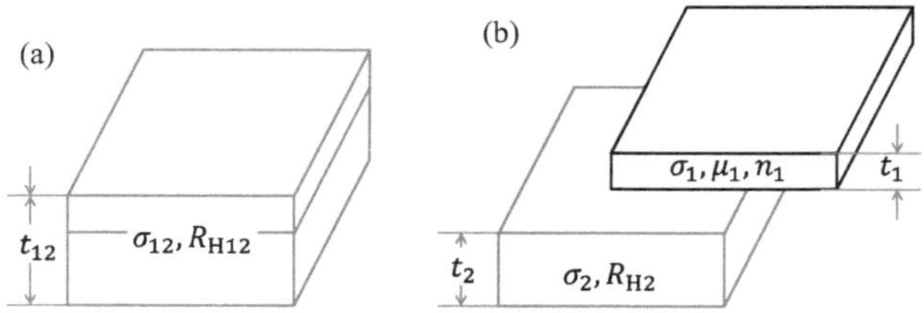


FIGURE 5.11 (a) Two-layer structure. (b) Upper layer is etched out.

have been measured. Now, suppose that the upper layer with the thickness $t_1$ is etched out, and that the conductivity $\sigma_2$ and the Hall coefficient $R_{H2}$ of the lower layer is measured (Fig. 5.11(b)). Express $\sigma_1$, $\mu_1$, and $n_1$ in terms of $\sigma_{12}$, $R_{H12}$, $\sigma_2$, and $R_{H2}$.

CHAPTER 6

# p-n Junctions

In this chapter, we will look at *p-n junctions*, which are extremely important building blocks of semiconductor devices. A p-n junction diode is a device that consists of a p-n junction itself. First, a qualitative understanding of the physics of p-n junctions will be attempted. Then the depletion layer and DC current-voltage characteristics will be discussed using mathematical equations.

## 6.1 WHAT IS A P-N JUNCTION?

On p. 35, we mentioned p-n junction diodes as an example of nonlinear resistors made of semiconductors. The basic structure of a p-n junction diode is shown in Fig. 6.1. It consists of a p-type region and an n-type region. The electrode attached to the p-type region is sometimes called the *anode*, and the electrode attached to the n-type region is called the *cathode*. These terms come from vacuum tubes.

The structure formed by contacting a p-type semiconductor and an n-type semiconductor is called a *p-n junction*. The p-n junction is a basic and extremely important structure found in almost all semiconductor devices.

For an applied bias voltage $V$ shown in Fig. 6.1, $V > 0$ is called a *forward bias*, and $V < 0$ is called a *reverse bias*. As shown in Fig. 6.2, the p-n junction has a rectifying action, and the triangle in the schematic symbol in Fig. 6.1 indicates the direction of the forward current.

The goals of this chapter are:

- to gain a qualitative understanding of the reason for the nonlinear current-voltage characteristic as in Fig. 6.2,
- to understand the basic physics of p-n junctions,
- and to derive the equation for the current-voltage characteristic under certain simplifying assumptions.

DOI: 10.1201/ 9781003439417-6

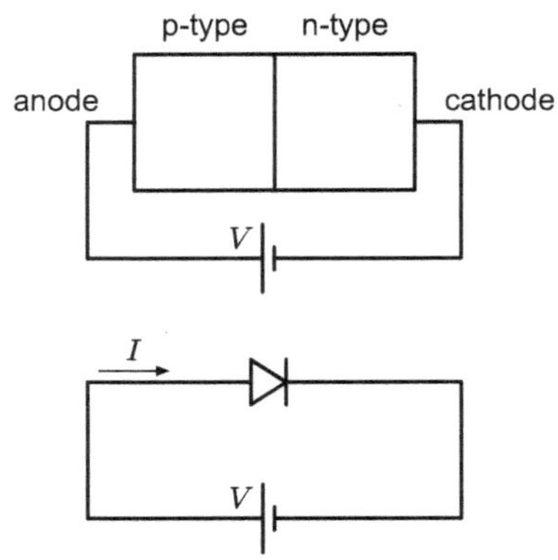


FIGURE 6.1 A schematic drawing and a circuit diagram of a biased p-n junction diode.

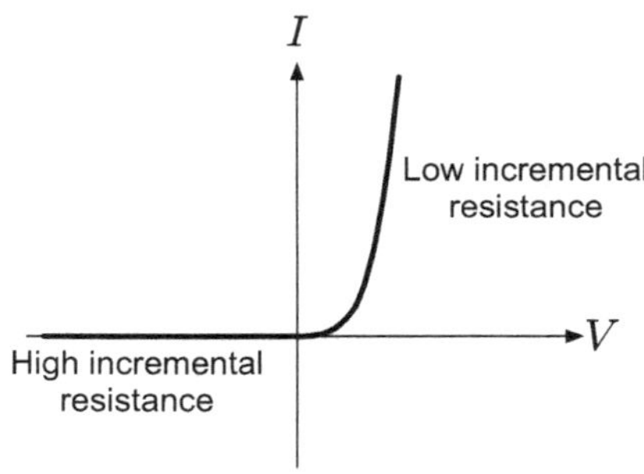


FIGURE 6.2 Current-voltage characteristics of a p-n junction.

From physical considerations, it turns out that the p-n junction is not just a nonlinear resistor, but also has a capacitive component.

## 6.2 CONTACT POTENTIAL

### 6.2.1 What Is Contact Potential?

Let us start our study of the physics of p-n junctions with the concept of contact potential. In general, when two solid substances with different properties are brought into contact with each other and reach an equilibrium state, a potential difference (i.e., difference in electrostatic potential $\psi$) is generated between them. This potential difference is called the *contact potential.*

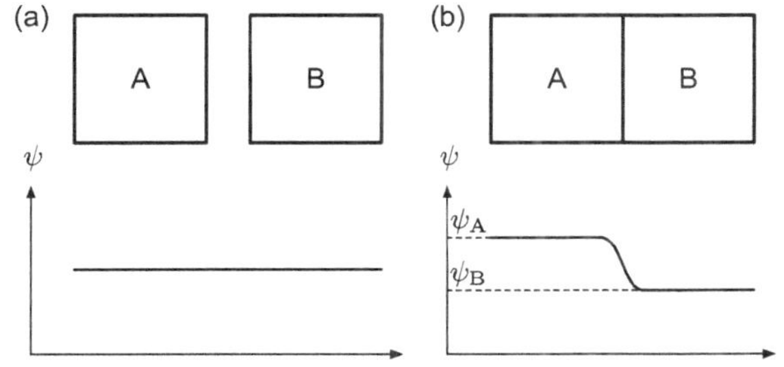


FIGURE 6.3 (a) Two conductive solid substances, A and B, not in contact. (b) After being brought into contact. $\psi$ is the electrostatic potential.

Let us consider bringing two different conductive solid substances, A and B, into contact. Before bringing them into contact, as shown in Fig. 6.3(a), both A and B are electrically neutral. Since the electron densities in A and B are different before contact, their respective Fermi levels do not coincide ($\zeta_A \neq \zeta_B$). Since the electron densities are different, when A and B are brought into contact, electrons (and holes) diffuse from one to the other. The one that accepts electrons becomes negatively charged. The other one that emitted the electrons is left with cations, so it becomes positively charged. As a result, an electric field is generated between A and B, and an electric force acts to prevent carrier diffusion. In theory, a drift current component appears in the opposite direction to the diffusion current, albeit smaller in magnitude than the latter. When the (quasi)-Fermi level becomes flat throughout, the diffusion and drift currents are balanced, and there is no net carrier transfer. In the example shown in Fig. 6.3(b), A is positively charged and B is negatively charged, and there is an electrostatic potential difference between them.

The contact potential between A and B is given by

$$\varphi_{AB} = \psi_A - \psi_B \quad \text{(B-referenced contact potential between A and B)} \tag{6.1}$$

or

$$\varphi_{BA} = \psi_B - \psi_A. \quad \text{(A-referenced contact potential between A and B)} \tag{6.2}$$

Incidentally, in (6.1) and (6.2), the Greek letters $\psi$ and $\varphi$ are used. In this book, electrostatic potentials measured from an absolute

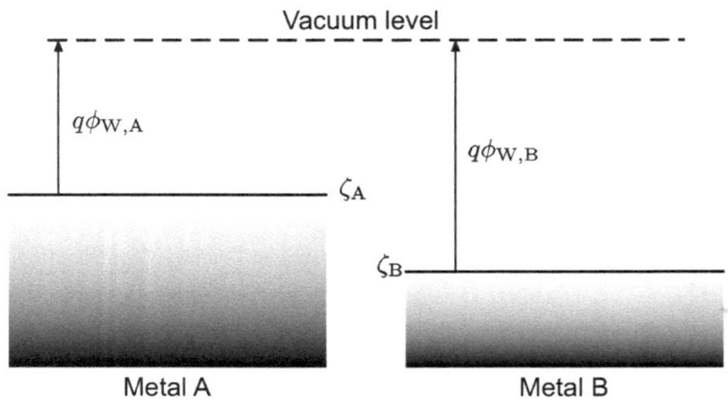


FIGURE 6.4 Work functions of metals A and B.

reference (not always explicitly specified) are denoted by $\psi$ (with subscripts), and quantities defined as the difference between electrostatic potentials are denoted by $\varphi$ (with subscripts).

## 6.2.2 Work Function and Electron Affinity

The contact potential between two solid substances is related to the physical properties of the given solids.

### *6.2.2.1 Work Function*

The *work function* (mainly of metals) is the energy required to emit an electron at the Fermi level to the free space (assumed to be a vacuum) outside the solid. Fig. 6.4 shows the work functions of two metals, A and B, on an energy band diagram. The lowest energy level of an electron that has been freed from binding by a solid is called the *vacuum level* or *free-electron level* [32]. The energy difference between the vacuum level and the Fermi level is the work function. The work function of a metal is a material-specific constant. If metals A and B in Fig. 6.4 are brought into contact, the contact potential is given by (see Problem 6.1 on p. 212)

$$q\varphi_{AB} = -q\left(\varphi_{W,A} - \varphi_{W,B}\right) \tag{6.3}$$

Thus, the contact potential is determined by the work function difference. Equation (6.3) holds even when A and B are not metals.

### *6.2.2.2 Electron Affinity*

For semiconductors, too, the work function can be considered the energy difference between the vacuum level and the Fermi level.

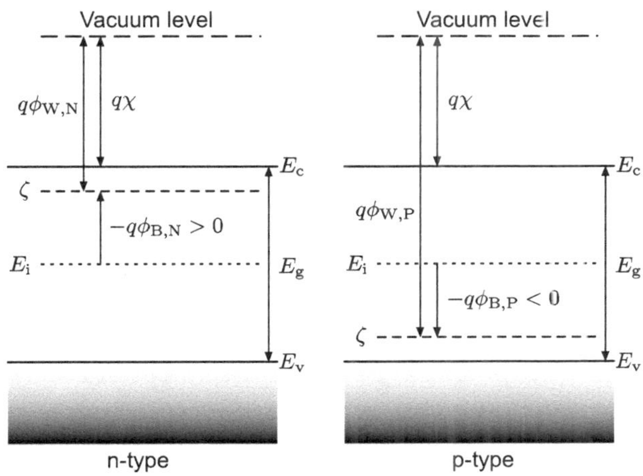


FIGURE 6.5 The work function and electron affinity of a semiconductor.

However, since the Fermi level of a semiconductor depends on impurity doping (§4.3), the work function of a semiconductor is not a material-specific constant. In addition, since the Fermi level of a semiconductor is usually located in the forbidden band (see (4.5) on p. 89), it is not the electrons at the Fermi level but those in the conduction band that will be emitted when the semiconductor is irradiated with electromagnetic rays. Therefore, the work function and Fermi level of semiconductors are not as important as they are for metals (see p. 91).

In semiconductors, the *electron affinity*, $q\chi$—the difference between the vacuum level and the bottom of the conduction band, $E_c$—is a material-specific constant (Fig. 6.5).

### 6.2.2.3 Semiconductor Fermi Level and Contact Potential

Let us denote the difference between the Fermi level $\zeta$ and the intrinsic Fermi level $E_i$ by the symbol $\varphi_B$, which has the dimensions of voltage.

$$\varphi_B \equiv \frac{\zeta - E_i}{-q} \qquad \text{(Bulk potential)} \qquad (6.4)$$

We refer to $\varphi_B$ as the *bulk potential*. "Bulk" here implies that $\varphi_B$ is to be defined at a position far enough from any interface or band bending. As such, $\varphi_B$ correctly represents the amount of separation between the Fermi level and the intrinsic Fermi level only at equilibrium and where the band is not bent. We will encounter cases where the band may be bent on p. 175 onward. The bulk potential $\varphi_B$ is essentially the same as the Fermi potential $\psi_F$ defined in (4.24) on p. 98, but the value of the constant term is explicitly specified to eliminate arbitrariness in (4.24).

For an n-type semiconductor, we see from (4.47) on p. 103 that $\varphi_B = \varphi_{B,N} < 0$, and for a p-type semiconductor, $\varphi_B = \varphi_{B,P} > 0$ from (4.52). The subscript "N" indicates that the quantity is that in the n-type semiconductor, "P" indicates that the quantity is that in the p-type semiconductor. $-\varphi_{B,N}$ is the contact potential of an n-type semiconductor with respect to an intrinsic semiconductor. Similarly, $-\varphi_{B,P}$ is the contact potential of a p-type semiconductor with respect to an intrinsic semiconductor.

The electron density, $n_{N0}$, of an n-type semiconductor in equilibrium without band bending can be written as follows using (4.27) on p. 99:

$$n_{N0} = n_i \exp\left(\frac{\zeta - E_i}{kT}\right) = n_i \exp\left(-\frac{q\varphi_{B,N}}{kT}\right). \tag{6.5}$$

Solving this equation for $\varphi_{B,N}$ leads to

$$\varphi_{B,N} = -\frac{kT}{q}\ln\left(\frac{n_{N0}}{n_i}\right) \approx -\frac{kT}{q}\ln\left(\frac{N_D^+}{n_i}\right)(< 0). \tag{6.6}$$

As can be seen from (6.6), the n-type bulk potential $\varphi_{B,N}$ is determined by the ionized donor density.

Similarly, the hole density $p_{P0}$ of a p-type semiconductor in equilibrium without band bending can be written as follows using the hole density equation (4.28) on p. 99:

$$p_{P0} = n_i \exp\left(\frac{E_i - \zeta}{kT}\right) = n_i \exp\left(\frac{q\varphi_{B,P}}{kT}\right). \tag{6.7}$$

Solving this equation for $\varphi_{B,P}$ yields

$$\varphi_{B,P} = \frac{kT}{q}\ln\left(\frac{p_{P0}}{n_i}\right) \approx \frac{kT}{q}\ln\left(\frac{N_A^-}{n_i}\right) \quad (> 0). \tag{6.8}$$

Clearly, the p-type bulk potential $\varphi_{B,P}$ is determined by the ionized acceptor density.

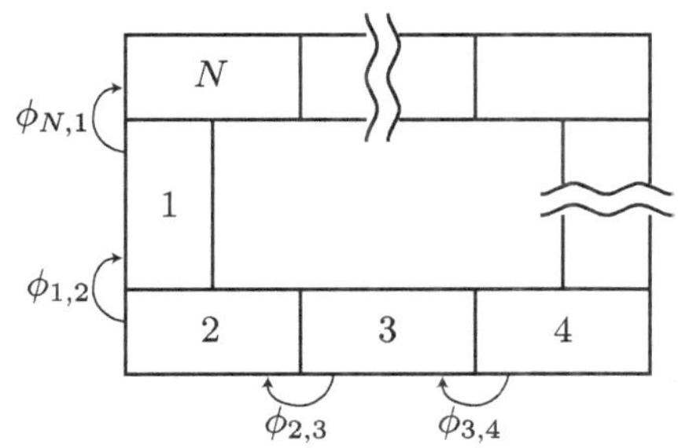


FIGURE 6.6 A loop structure composed of several materials.

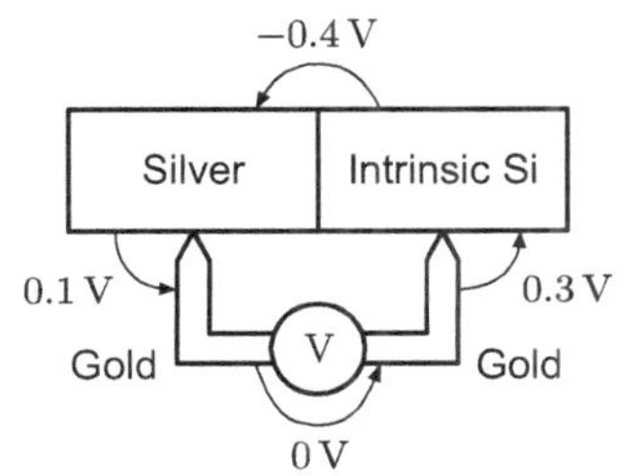


FIGURE 6.7 Loop consisting of silver, intrinsic silicon, and a voltmeter.

### 6.2.3 Properties of Contact Potential

Suppose that several conductive materials are combined to make a loop structure, as shown in Fig. 6.6. Since the sum of the contact potentials around the loop equals zero, we can express it as follows:

$$\varphi_{1,2} + \varphi_{2,3} + \varphi_{3,4} + \cdots + \varphi_{N,1} = 0. \tag{6.9}$$

This could be thought of as Kirchhoff's voltage law (KVL).

Let us now incorporate an ideal voltmeter in the loop. As a concrete example, let us suppose that we measure the potential difference between silver (Ag) and intrinsic silicon (Si) using a voltmeter with its electrodes made of gold (Au), as shown in Fig. 6.7. The contact potentials between the materials are shown.

Fig. 6.7 is only an example, but according to (6.9), the voltmeter shows 0 V regardless of the combination of the two materials under test. In other words, the contact potential between materials cannot be measured with an ordinary voltmeter. A different method is needed to measure the contact potential.

The question then arises as to what exactly is the quantity that the voltmeter measures. As is clear from the above discussion, it is

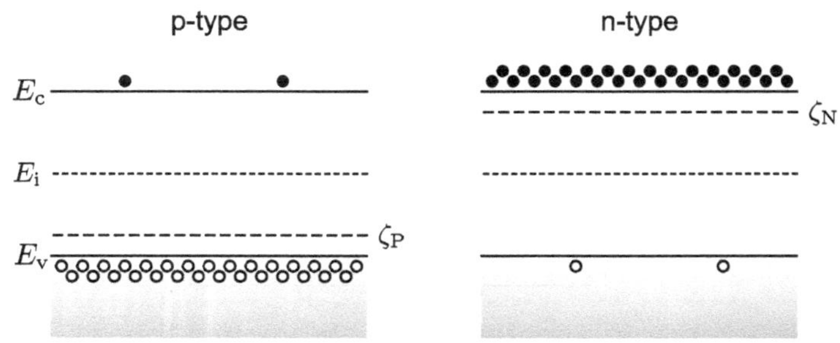


FIGURE 6.8 Energy band diagrams of p-type and n-type semiconductors not in contact.

*not* the electrostatic potential difference that the voltmeter measures. The voltmeter measures the (quasi-)Fermi level difference, that is, the motive force for the conduction current to flow [27] (see (5.33) and (5.34) on p. 131). If a loop structure as in Fig. 6.6 or Fig. 6.7 is made without a power supply, the Fermi level will be constant everywhere, so it is natural that the voltmeter shows 0 V.

## 6.3 FORMATION OF A P-N JUNCTION

### 6.3.1 Contact between p-Type and n-Type Semiconductors

This is basically the same as in Fig. 6.3 on p. 169, but let us consider a thought experiment in which p-type and n-type semiconductors are brought into contact. Fig. 6.8 shows the energy band diagrams of two semiconductors before being brought into contact.

When the two are brought close to each other, holes start to flow from the p-type region to the n-type region, and electrons start to flow from the n-type region to the p-type region by diffusion just before the actual contact, as shown in Fig. 6.9, due to the carrier density gradient.[1] The current that flows through such a very thin insulating layer (or a vacuum) is called the *tunnel current*. When the two actually make contact, a larger diffusion current flows. As a result, the p-type region is negatively charged due to electrons and acceptor ions, whereas the

[1] Note that there is no electrostatic potential difference between the two neutral semiconductors.

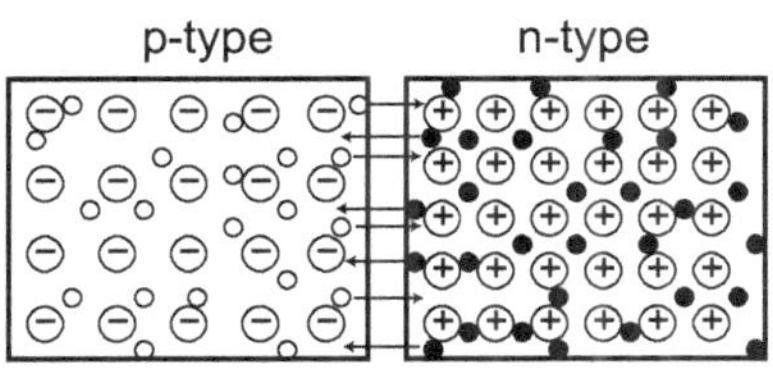


FIGURE 6.9 p-type and n-type semiconductors brought in close proximity.

n-type region is positively charged due to holes and donor ions. Then, a leftward (←) electric field is generated, and it operates to prevent further carrier diffusion. Eventually, carrier diffusion stops at a point where it is balanced by this electric field. Thus, an equilibrium p-n junction is formed. Note that the above is a thought experiment, and the actual procedure for making p-n junctions is quite different.

### 6.3.2 p-n Junctions in Equilibrium

Let us take a closer look at p-n junctions in equilibrium. In the following example, the pictures are drawn assuming that the acceptor ion density in the p-type region is lower than the donor ion density in the n-type region. This assumption affects the depletion layer thickness, described below (p. 186).

The leftward electric field mentioned at the end of §6.3.1 does not exist uniformly throughout the structure. For an electric field to exist, there must be positive charges, which are the sources of lines of electric force, and negative charges, which terminate the lines of electric force. The distribution of positive and negative charges is shown in Fig. 6.10(a). There are almost no carriers near the interface between the p-type and n-type regions, and the negative charges of the acceptor ions and the positive charges of the donor ions are exposed without being neutralized by carriers. This region, depleted of carriers, is called the *depletion layer* (because the region is a thin layer) or the *space-charge region*. "Space charge" here refers to fixed charges that are spatially distributed in a certain volume. The region beyond a certain distance from the junction interface is almost the same as the state before contact (Fig. 6.8) and is electrically neutral (except for the very thin transition regions next to the depletion layer), and is called the *neutral region* or the *quasi-neutral region*. The reason why "quasi-" is sometimes prepended is that there is no clear boundary between the depletion layer and the neutral region (§5.9), and the transition region is somewhat charged.

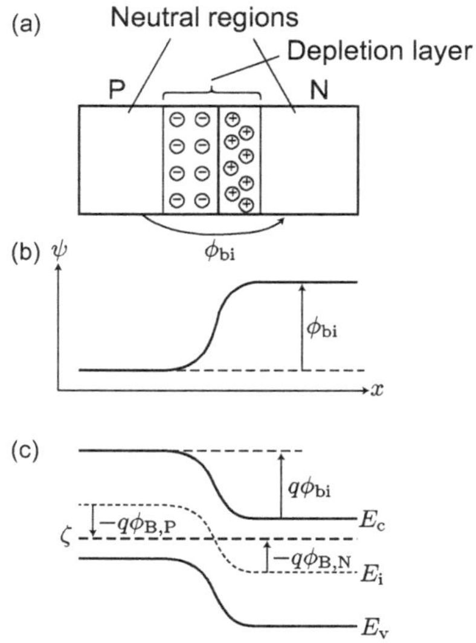


FIGURE 6.10 A p-n junction in equilibrium. (a) Structure. (b) Electrostatic potential. (c) Energy band diagram.

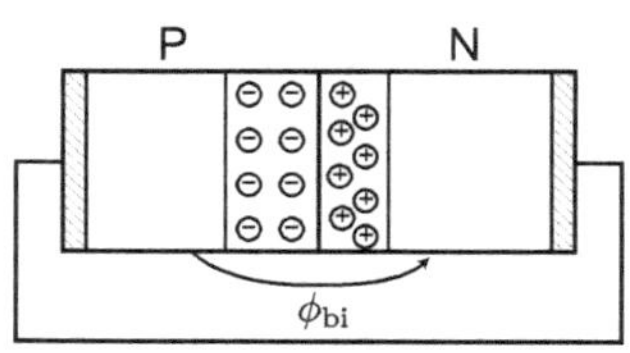


FIGURE 6.11 Short-circuited p-n junction.

As a result of the charge distribution described above, the position dependence of the electrostatic potential $\psi$ is as shown in Fig. 6.10(b). The potential gradient (i.e., the electrostatic field) exists only in the depletion layer. $\varphi_{bi}$ in Fig. 6.10(b) is the contact potential of the p-n junction and is called the *built-in potential*, *built-in voltage*, or *diffusion potential*. The corresponding energy band diagram is shown in Fig. 6.10(c). Naturally, the Fermi level $\zeta$ has a constant value throughout. Reading the carrier densities from this band diagram (§5.2.4), we see that a depletion layer is indeed formed as shown in Fig. 6.10(a). Since this state is an equilibrium state and no current is flowing, nothing should happen, even if electrodes are attached to both ends and short-circuited with a conducting lead as shown in Fig. 6.11.

The built-in potential can be expressed as follows using the bulk potential equations (6.8) and (6.6):

$$\varphi_{\mathrm{bi}} = \varphi_{\mathrm{B,P}} - \varphi_{\mathrm{B,N}} = \frac{kT}{q}\ln\left(\frac{p_{\mathrm{P0}}}{n_{\mathrm{i}}}\right) + \frac{kT}{q}\ln\left(\frac{n_{\mathrm{N0}}}{n_{\mathrm{i}}}\right)$$
$$\approx \frac{kT}{q}\ln\left(\frac{N_{\mathrm{A}}^{-}N_{\mathrm{D}}^{+}}{n_{\mathrm{i}}^{2}}\right). \qquad \text{(Built-in potential)} \qquad (6.10)$$

In a nondegenerate silicon p-n junction, $\varphi_{\mathrm{bi}}$ is often slightly less than 1 V (see Problem 6.3 on p. 213).

### 6.3.3 Biased p-n Junctions

As mentioned on p. 174, the potential difference in the circuit-theoretic sense measured by a voltmeter corresponds to the (quasi-)Fermi level difference. Therefore, when a bias voltage is applied to a p-n junction, the applied voltage appears in the energy band diagram as the difference between the quasi-Fermi levels of the majority carriers[2] in each region.

Fig. 6.12 shows a forward-biased p-n junction. The bias voltage $V$ here is assumed to be smaller than the built-in potential $\varphi_{\mathrm{bi}}$, that is, $0 < V < \varphi_{\mathrm{bi}}$. Then, the following equality holds.

$$\zeta_{\mathrm{nN}} - \zeta_{\mathrm{pP}} = qV(> 0) \qquad (6.11)$$

Again, the subscript "N" indicates a quantity in the n-type region, and the subscript "P" indicates a quantity in the p-type region. The difference in electrostatic potential between the p-type and the n-type regions is $\varphi_{\mathrm{bi}} - V$, which is smaller than that at zero bias. The depletion layer becomes thinner than at zero bias. Note that in the band diagram in Fig. 6.12, the quasi-Fermi levels for electrons and holes in the depletion layer are not drawn properly. We will see energy band diagrams with quasi-Fermi levels in the depletion layer properly drawn in §6.4 onward.

The reverse-biased case ($V < 0$) is shown in Fig. 6.13. Just as in the forward-biased case,

---

[2] We refer only to majority carriers here in view of the discussions on p. 125 and p. 150, which basically pointed out that the nonequilibrium majority carrier density is almost equal to the equilibrium value, whereas the nonequilibrium minority carrier density may be very different from the equilibrium value.

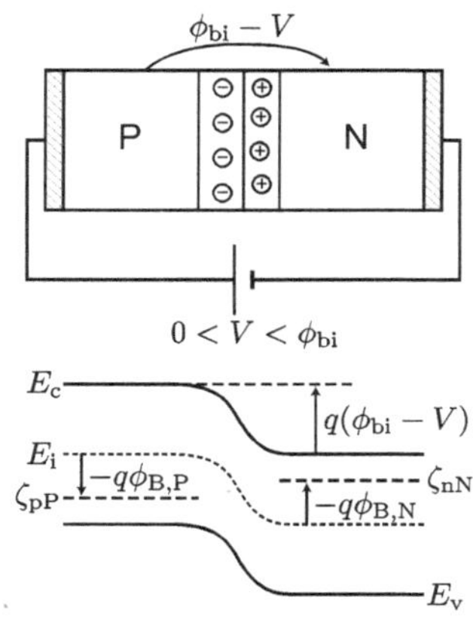


FIGURE 6.12 Forward-biased p-n junction.

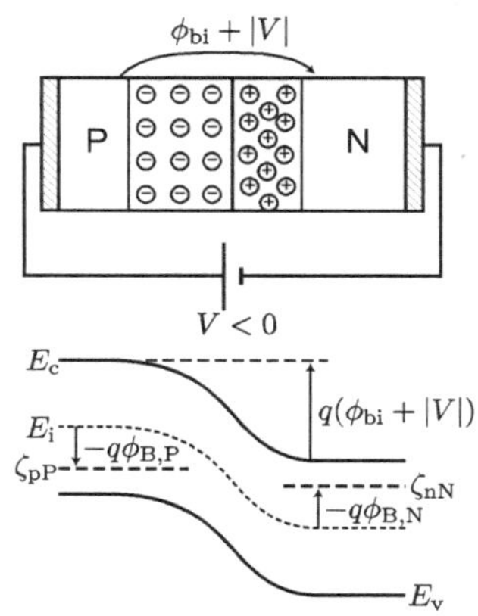


FIGURE 6.13 Reverse-biased p-n junction.

$$\zeta_{nN} - \zeta_{pP} = qV \quad (< 0) \tag{6.12}$$

holds. The difference in electrostatic potential between the p-type and n-type regions is $\varphi_{bi} + |V|$ and the depletion layer becomes thicker than at zero bias. The quasi-Fermi levels in the depletion layer are not drawn in Fig. 6.13, either.

## 6.4 QUALITATIVE DESCRIPTION OF RECTIFICATION

Let us consider qualitatively why the p-n junction shows the rectifying action, as shown in Fig. 6.1 on p. 168, using energy band diagrams.

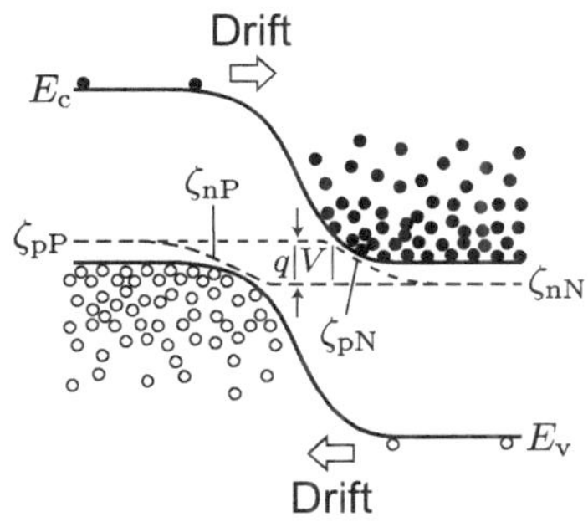


FIGURE 6.14 Energy band diagram of a forward-biased p-n junction with quasi-Fermi levels.

At zero bias ($V = 0$), carrier drift and diffusion are balanced (p. 175), and no net current flows. Under forward bias ($V > 0$), the potential barriers to carriers attempting to diffuse to the opposite region become lower for larger values of $V$, as we saw in Fig. 6.12. As a result, diffusion becomes dominant, and as shown in Fig. 6.14, the majority carriers in each region overcome the potential barrier and diffuse into the opposite region due to the density gradient. That is, the forward current through the depletion layer is the diffusion current. What happens then to the carriers that reach the opposite region? Since the electrons that have flowed into the p-type region by diffusion become minority carriers, they gradually recombine (§5.6) with majority carriers (i.e., holes) and disappear as they move deeper into the p-type region. Similarly, holes that have entered the n-type region are minority carriers, so they gradually recombine with electrons and disappear. Since recombination produces a gradient in minority carrier density, the diffusion current that flows is sometimes referred to as the *recombination current*.

Incidentally, since the band diagram cartoon in Fig. 6.14 shows electrons with • and holes with ∘, you might have easily accepted the above explanation that "diffusion is dominant." But what if these • and ∘ were not drawn? It is not surprising that you would find it difficult to accept the explanation that electrons and holes climb up the slopes of $E_c$ and $E_v$ against the potential gradient. However, if we remember that it is the gradient of the quasi-Fermi level, not the potential gradient, that determines the direction of carrier motion (p. 131), we should be able to see the directions of carrier motion from the slopes of the quasi-Fermi levels in Fig. 6.14, even if • and ◦ were not drawn. In the depletion layer, both $\zeta_n$ and $\zeta_p$ appear to be almost flat, but this is due to the logarithmic transform of carrier densities (§5.2.4), and in fact, there is a slight slope. Electrons go down the slope of $\zeta_n$ in Fig. 6.14, and the holes "go down" the slope of $\zeta_p$ when viewed upside-down.

What about electron and hole densities? Recalling the discussion on p. 123, if the quasi-Fermi levels are drawn on an energy band diagram, we can read the carrier densities (even if • and ◦ are not drawn). $E_c(x) - \zeta_n(x)$ in Fig. 6.14 shows that the electron density in the p-type side of the depletion layer and its vicinity is considerably higher than that at equilibrium. Similarly, $\zeta_p(x) - E_v(x)$ in Fig. 6.14 shows that the hole density is much higher than at equilibrium in the n-type side of the depletion layer and its vicinity. The forward current density is determined by these minority carrier densities, which are higher than at equilibrium (§6.8.2). However, the minority carrier densities are still significantly lower than the majority carrier densities. The injection of minority carriers in such a way is said to be *low injection*. We were able to read the above from Fig. 6.14 because the quasi-Fermi levels $\zeta_n(x)$ and $\zeta_p(x)$ were drawn in the energy band diagram. In contrast, it is difficult to read the desired information from an energy band diagram like that in Fig. 6.12. It is very important for the understanding of device operation that the quasi-Fermi levels are drawn in energy band diagrams.

The response of a p-n junction to a reverse bias ($V < 0$) is completely different from that to a forward bias. Drift becomes dominant in the depletion layer, and minority carriers drift to the opposite region, as shown in Fig. 6.15. So, the reverse current through the depletion layer is the drift current.[3] However, this reverse current is orders of magnitude smaller than the forward current. This is because, in the depletion layer, $\zeta_n < \zeta_p$ in the expression for the effective intrinsic carrier density, given by (5.14) on p. 124 (see Fig. 6.15), and the minority carrier density is smaller than at equilibrium. Since current density is primarily governed by the carrier density (p. 131), the reverse current density is very small despite the steep potential gradient in the depletion layer.

From the above, the p-n junction is a nonlinear resistor with rectifying action, as shown in Fig. 6.1.

[3] The current in the neutral regions is the diffusion current.

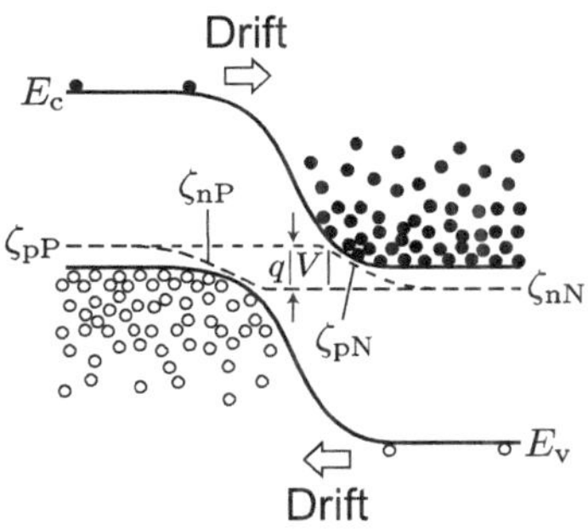


FIGURE 6.15 Energy band diagram of a reverse-biased p-n junction with quasi-Fermi levels.

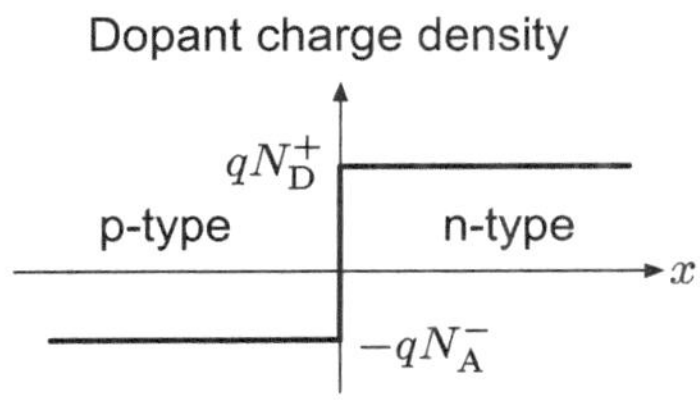


FIGURE 6.16 Ionized dopant charge density distribution in an abrupt junction.

## 6.5 ANALYSIS OF ABRUPT JUNCTIONS

In the previous sections, we have tried to explain the physics of p-n junctions as qualitatively as possible, using only the formulas derived up to Chapter 5. From this section onward, we will perform a detailed analysis of p-n junctions in which the ionized acceptor density in the p-type region and the ionized donor density in the n-type region are both constant, and the ionized dopant charge density changes abruptly at the junction interface, as shown in Fig. 6.16.

### 6.5.1 Zero-Bias Abrupt Junctions

A p-n junction with the ionized dopant charge density distribution, as shown in Fig. 6.16, is called a *step junction* or an *abrupt junction*. The actual charge density distribution of ionized dopants in a p-n junction is a little more complicated, but assuming an abrupt junction greatly simplifies the analysis, and various equations can be derived analytically.

First, let us write the Poisson equation (p. 153) for the zero-bias case.

$$\frac{d^2\psi(x)}{dx^2} = -\frac{\rho(x)}{\epsilon_{Si}}, \qquad \text{(One-dimensional Poisson equation)} \quad (6.13)$$

where the semiconductor is assumed to be silicon. The charges in silicon are holes, electrons, and donor and acceptor ions. The charge density distribution, $\rho(x)$, is therefore given by

$$\rho(x) = q\left[p(x) - n(x) + N_D^+(x) - N_A^-(x)\right]. \quad (6.14)$$

If the ionized dopant charge density, $N_D^+(x)$ or $N_A^-(x)$, changes spatially, it takes a distance longer than the Debye length (p. 158) for the carrier density to follow the change. In other words, the carrier density varies gently. However, the Debye length itself is usually shorter compared with device dimensions and depletion layer thicknesses (see Table 5.3 on p. 159 for numerical examples). Therefore, we apply the *depletion approximation* to (6.14), which assumes that the carrier density is negligible in the region where the band is bent near the junction interface. As a result, the charge density distribution becomes as shown in Fig. 6.17(a), with the (quasi-)neutral region being completely neutral and the depletion layer being charged with the ionized dopant charges. In Fig. 6.17, the p-n junction interface is assumed to be located at $x = 0$, and the edges of the depletion layer are at $-x_P$ and $x_N$. Note that $-x_P$ and $x_N$ are unknowns at this point. We will determine their values later (p. 185).

If we write down the charge density distribution (6.14) in accordance with Fig. 6.17(a), we find

$$\rho(x) = \begin{cases} 0 & (x \le -x_P) \\ -qN_A^- & (-x_P \le x \le 0) \\ qN_D^+ & (0 \le x \le x_N) \\ 0 & (x_N \le x) \end{cases}. \qquad \text{(Charge density distribution)} \quad (6.15)$$

Inserting (6.15) into the Poisson equation (6.13) yields

$$-\frac{d^2\psi(x)}{dx^2} = \begin{cases} 0 & (x \le -x_P) \\ -\frac{qN_A^-}{\epsilon_{Si}} & (-x_P \le x \le 0) \\ \frac{qN_D^+}{\epsilon_{Si}} & (0 \le x \le x_N) \\ 0 & (x_N \le x) \end{cases}. \qquad \text{(Poisson equation)} \quad (6.16)$$

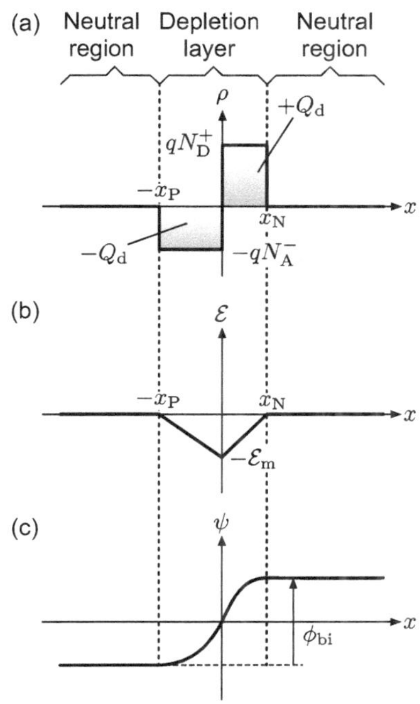


FIGURE 6.17 (a) Charge distribution of an abrupt junction. (b) Electric field. (c) Electrostatic potential.

Integrating (6.16) with $x$ yields the electric field:

$$\mathcal{E}(x) = -\frac{\mathrm{d}\psi(x)}{\mathrm{d}x}. \tag{6.17}$$

The constant of integration is determined so that the charge neutrality condition is satisfied. Specifically, according to Gauss' law, if $\mathcal{E}(x) = 0$ for $x \leq -x_\mathrm{P}$ and $x_\mathrm{N} \leq x$, then the p-n junction is electrically neutral. The result of the integration is

$$\mathcal{E}(x) = \int^{x} \frac{\rho(x')}{\epsilon_\mathrm{Si}}\mathrm{d}x' = \begin{cases} 0 & (x \leq -x_\mathrm{P}) \\ -\frac{qN_\mathrm{A}^-}{\epsilon_\mathrm{Si}}(x + x_\mathrm{P}) & (-x_\mathrm{P} \leq x \leq 0) \\ \frac{qN_\mathrm{D}^+}{\epsilon_\mathrm{Si}}(x - x_\mathrm{N}) & (0 \leq x \leq x_\mathrm{N}) \\ 0 & (x_\mathrm{N} \leq x) \end{cases}. \tag{6.18}$$

This is the electric field distribution. A plot of $\mathcal{E}(x)$ is shown in Fig. 6.17(b). The maximum magnitude of the electric field, $\mathcal{E}_{\mathrm{m}}$, is given by

$$\mathcal{E}_{\mathrm{m}} = |\mathcal{E}(0)| = \frac{qN_{\mathrm{A}}^{-}x_{\mathrm{P}}}{\epsilon_{\mathrm{Si}}} = \frac{qN_{\mathrm{D}}^{+}x_{\mathrm{N}}}{\epsilon_{\mathrm{Si}}} \tag{6.19}$$

$$= \frac{Q_{\mathrm{d}}}{\epsilon_{\mathrm{Si}}}, \text{(Maximum magnitude of electric field)} \tag{6.20}$$

where

$$Q_{\mathrm{d}} \equiv qN_{\mathrm{A}}^{-}x_{\mathrm{P}} = qN_{\mathrm{D}}^{+}x_{\mathrm{N}} \tag{6.21}$$

is the magnitude of the depletion charge per unit area on one side of the junction.

Since integrating the electric field and reversing the sign yields the electrostatic potential, we integrate sign-reversed (6.18) with $x$. There is arbitrariness in the value of the constant of integration (the reference point for the electrostatic potential). Here we choose it such that $\psi(0) = 0$.

$$\psi(x) = -\int^{x} \mathcal{E}(x')\,\mathrm{d}x' = \begin{cases} -\dfrac{qN_{\mathrm{A}}^{-}}{2\epsilon_{\mathrm{Si}}}x_{\mathrm{P}}^{2} & (x \le -x_{\mathrm{P}}) \\ \dfrac{qN_{\mathrm{A}}^{-}}{2\epsilon_{\mathrm{Si}}}\left(x^{2} + 2x_{\mathrm{P}}x\right) & (-x_{\mathrm{P}} \le x \le 0) \\ -\dfrac{qN_{\mathrm{D}}^{+}}{2\epsilon_{\mathrm{Si}}}\left(x^{2} - 2x_{\mathrm{N}}x\right) & (0 \le x \le x_{\mathrm{N}}) \\ \dfrac{qN_{\mathrm{D}}^{+}}{2\epsilon_{\mathrm{Si}}}x_{\mathrm{N}}^{2} & (x_{\mathrm{N}} \le x) \end{cases}. \tag{6.22}$$

Equation (6.22) gives the electrostatic potential distribution of the abrupt junction and is shown in Fig. 6.17(c).

From (6.22), the values of the electrostatic potential at the edges of the depletion layer are

$$\psi(-x_{\mathrm{P}}) = -\frac{qN_{\mathrm{A}}^{-}}{2\epsilon_{\mathrm{Si}}}x_{\mathrm{P}}^{2}, \tag{6.23}$$

$$\psi(x_{\mathrm{N}}) = \frac{qN_{\mathrm{D}}^{+}}{2\epsilon_{\mathrm{Si}}}x_{\mathrm{N}}^{2}. \tag{6.24}$$

Using these results, the built-in potential can be written as

$$\varphi_{\mathrm{bi}} = \psi(x_{\mathrm{N}}) - \psi(-x_{\mathrm{P}}) \qquad \text{(Built-in potential)} \tag{6.25}$$

$$= \frac{Q_d^2}{2q\epsilon_{Si}}\left(\frac{1}{N_D^+} + \frac{1}{N_A^-}\right), \tag{6.26}$$

where (6.21) was used to derive (6.26). $\varphi_{bi}$ can also be written as follows:

$$\varphi_{bi} = -\int_{-x_P}^{x_N} \mathcal{E}(x)\,dx = \frac{\mathcal{E}_m (x_P + x_N)}{2}. \tag{6.27}$$

The right-hand side of (6.27) represents the area of the triangle in Fig. 6.17(b) on p. 183. $\varphi_{bi}$ was already obtained in (6.10) on p. 177, but here we expressed it in relation to the depletion layer of the abrupt junction.

So far, we have written many equations using the coordinates of the depletion layer edges, $-x_P$ and $x_N$, but $x_P$ and $x_N$ can also be written in terms of the ionized dopant densities, $N_A^-$ and $N_D^+$, as we did in (6.10) on p. 177. To do so, first solve (6.26) for $Q_d$.

$$Q_d = \sqrt{2q\epsilon_{Si}\varphi_{bi}\frac{N_A^- N_D^+}{N_A^- + N_D^+}}. \tag{6.28}$$

Inserting (6.28) into (6.20) and solving for $x_P$ and $x_N$ using (6.19), we obtain

$$x_P = \sqrt{\frac{2\epsilon_{Si}\varphi_{bi}}{qN_A^-}}\sqrt{\frac{N_D^+}{N_A^- + N_D^+}}, \tag{6.29}$$

$$x_N = \sqrt{\frac{2\epsilon_{Si}\varphi_{bi}}{qN_D^+}}\sqrt{\frac{N_A^-}{N_A^- + N_D^+}}. \tag{6.30}$$

Equations (6.29) and (6.30) involve $\varphi_{bi}$, but since $\varphi_{bi}$ can be written as in (6.10) in terms of dopant densities, we have, in effect, written $x_P$ and $x_N$ in terms of dopant densities, too. From (6.29) and (6.30), we get

$$\frac{x_P}{x_N} = \frac{N_D^+}{N_A^-} = \frac{1/N_A^-}{1/N_D^+}. \tag{6.31}$$

Equation (6.31) shows that the depletion layer thickness on each side is inversely proportional to the ionized dopant density. From (6.29) and (6.30), the total depletion layer thickness is given by (see Problem 6.4 on p. 213)

$$d_{\text{dep}} = x_{\text{P}} + x_{\text{N}} = \sqrt{\frac{2\epsilon_{\text{Si}}\left(N_{\text{A}}^{-} + N_{\text{D}}^{+}\right)\varphi_{\text{bi}}}{qN_{\text{A}}^{-}N_{\text{D}}^{+}}}. \tag{6.32}$$

Of the built-in potential (6.25), the potential drop on each side is given by

$$\psi(x_{\text{N}}) = \frac{N_{\text{A}}^{-}}{N_{\text{A}}^{-} + N_{\text{D}}^{+}}\varphi_{\text{bi}}, \tag{6.33}$$

$$\psi(-x_{\text{P}}) = \frac{N_{\text{D}}^{-}}{N_{\text{A}}^{-} + N_{\text{D}}^{+}}\varphi_{\text{bi}}. \tag{6.34}$$

Similarly to (6.31), the potential drop across each side of the depletion layer is inversely proportional to the ionized dopant density.

### 6.5.2 Biased Abrupt Junctions

As explained in §6.4, a current flows when a bias voltage is applied to a p-n junction. When a current flows, there should be a voltage drop (or change in quasi-Fermi levels) in the neutral regions, too, because they have finite resistance. However, let us assume that the magnitude of the voltage drop there is negligibly small compared with the voltage drop across the depletion layer. This assumption is justified by the fact that there are almost no carriers in the depletion layer. Since the resistivity (p. 139) is inversely proportional to the carrier density, the resistivity of the depletion layer is large, whereas that of the neutral regions with a large number of carriers is much smaller. If the same current flows throughout, the voltage drop is proportional to the resistivity, so the voltage drop in the neutral region is negligible. This assumption was also reflected in Fig. 6.12 (p. 178) and Fig. 6.13 (p. 178). The voltage drop occurs almost entirely in the depletion layer, and the basic physics is similar to the zero-bias case. Strictly speaking, the presence of the conduction current in the depletion layer contradicts the depletion approximation, which assumes that there are *no* carriers in the depletion layer. The important point here is that in the low injection condition, the low electron and hole densities in the depletion layer contribute nonzero but orders of magnitude smaller charges to the right-hand side of the Poisson equation (5.106) on p. 158 or (6.16) on p. 182, compared with the ionized dopant densities. So the approximation is quite accurate.

With a bias voltage $V$ applied, the built-in potential $\varphi_{\mathrm{bi}}$ in each of the equations derived in §6.5.1 should be replaced with $(\varphi_{\mathrm{bi}} - V)$ (see Fig. 6.12 on p. 178). Note that the bias voltage $V$ must be lower than the built-in potential ($V < \varphi_{\mathrm{bi}}$) because we derived all equations under that assumption. Now, for example, the depletion charge formula (6.28) is modified as follows:

$$Q_{\mathrm{d}} = \sqrt{2q\epsilon_{\mathrm{Si}}\left(\varphi_{\mathrm{bi}} - V\right)\frac{N_{\mathrm{A}}^{-}N_{\mathrm{D}}^{+}}{N_{\mathrm{A}}^{-} + N_{\mathrm{D}}^{+}}}. \tag{6.35}$$

The depletion layer thickness formula (6.32) is modified as follows:

$$d_{\mathrm{dep}} = \sqrt{\frac{2\epsilon_{\mathrm{Si}}\left(N_{\mathrm{A}}^{-} + N_{\mathrm{D}}^{+}\right)\left(\varphi_{\mathrm{bi}} - V\right)}{qN_{\mathrm{A}}^{-}N_{\mathrm{D}}^{+}}}. \tag{6.36}$$

From (6.29) and (6.30), we obtain

$$x_{\mathrm{P}} = \sqrt{\frac{2\epsilon_{\mathrm{Si}}\left(\varphi_{\mathrm{bi}} - V\right)}{qN_{\mathrm{A}}^{-}}}\sqrt{\frac{N_{\mathrm{D}}^{+}}{N_{\mathrm{A}}^{-} + N_{\mathrm{D}}^{+}}}, \tag{6.37}$$

$$x_{\mathrm{N}} = \sqrt{\frac{2\epsilon_{\mathrm{Si}}\left(\varphi_{\mathrm{bi}} - V\right)}{qN_{\mathrm{D}}^{+}}}\sqrt{\frac{N_{\mathrm{A}}^{-}}{N_{\mathrm{A}}^{-} + N_{\mathrm{D}}^{+}}}. \tag{6.38}$$

From (6.36) to (6.38), we can see that the depletion layer becomes thinner for $V > 0$ (forward bias) and thicker for $V < 0$ (reverse bias). These equations also suggest that the depletion layer disappears at $V = \varphi_{\mathrm{bi}}$. The electrostatic potentials at the edges of the depletion layer are, from (6.33) and (6.34), given by

$$\psi\left(x_{\mathrm{N}}\right) = \frac{N_{\mathrm{A}}^{-}}{N_{\mathrm{A}}^{-} + N_{\mathrm{D}}^{+}}\left(\varphi_{\mathrm{bi}} - V\right), \tag{6.39}$$

$$\psi\left(-x_{\mathrm{P}}\right) = \frac{N_{\mathrm{D}}^{-}}{N_{\mathrm{A}}^{-} + N_{\mathrm{D}}^{+}}\left(\varphi_{\mathrm{bi}} - V\right). \tag{6.40}$$

## 6.6 CAPACITANCE OF P-N JUNCTIONS

### 6.6.1 Depletion Capacitance

Let us look again at the charge density distribution of the p-n junction in Fig. 6.17(a) (p. 183). Since there is a charge of $-Q_d$ per unit area on the p-type side of the depletion layer and $+Q_d$ on the n-type side, the p-n junction can be regarded as a kind of capacitor in which a charge $Q_d$ is stored per unit area. The incremental capacitance of this capacitor is called the *depletion capacitance*. The depletion capacitance per unit area of the abrupt junction can be found by differentiating the depletion charge (6.35) by the applied voltage and is given by

$$C_d \equiv \frac{dQ_d}{d(-V)} = \frac{dQ_d}{d(\varphi_{bi} - V)} \tag{6.41}$$

$$= \sqrt{\frac{q\epsilon_{Si} N_A^- N_D^+}{2(N_A^- + N_D^+)}} \frac{1}{\sqrt{\varphi_{bi} - V}}. \tag{6.42}$$

Here we assumed $\varphi_{bi} - V > 0$ (see §6.5.2). In (6.41), we differentiated $Q_d$ by $-V$ so that the capacitance is positive ($C_d > 0$). From (6.42), we can see that the depletion capacitance increases under a forward bias and decreases under a reverse bias.

Comparing the depletion layer thickness equation (6.36) with the depletion capacitance equation (6.42), we see that the following equality holds.

$$C_d = \frac{\epsilon_{Si}}{d_{dep}}. \qquad \text{(Depletion capacitance per unit area)} \tag{6.43}$$

Equation (6.43) has the same form as the expression for capacitance per unit area of a parallel-plate capacitor with dielectric constant $\epsilon_{Si}$ and thickness $d_{dep}$.

It is clear from the depletion capacitance equation (6.42) that $Q_d \neq C_d\,|V|$, so the depletion layer is a nonlinear capacitor (see §2.2.2). Nonlinear capacitors are also called *varactors* (= variable capacitors), and p-n junction diodes used as capacitors are called *varactor diodes*.

To summarize, the p-n junction is not a pure nonlinear resistor but has a nonlinear capacitor connected in parallel.

### 6.6.2 Diffusion Capacitance

Forward-biased p-n junctions have a large capacitance component in addition to the depletion capacitance. This capacitance is known as the *diffusion capacitance*, and it increases rapidly with the bias voltage $V$. Under a forward bias, majority carriers diffuse into the opposite regions, and a large number of minority carriers build up, as can be read from Fig. 6.14 (p. 179). These minority carriers are the charge stored in the diffusion capacitance.

The diffusion capacitance has a significant impact on the transient response of the p-n junction. In particular, when switching suddenly from forward bias to reverse bias, the current does not immediately become nearly zero as expected from Fig. 6.2 (p. 168), but a large reverse current flows for a short time. This is due to the discharging of the stored minority carriers. It is, therefore, practically important to consider the diffusion capacitance.

## 6.7 ONE-SIDED ABRUPT JUNCTIONS

A p-n junction in which one of the p-type and n-type regions is degenerately doped and the other is lowly doped is called a *one-sided junction*. The one-sided junction is a practically important structure found in various semiconductor devices such as MOSFETs, which will be discussed in Chapter 7. For a *one-sided abrupt junction*, approximate analytical treatment is possible by using the results of the analysis of the abrupt junction discussed in §6.5.

As an example, let us consider a one-sided abrupt junction whose n-type region is degenerately doped. In this case, $N_D^+ \gg N_A^-$ holds for the ionized dopant densities. From (6.29) and (6.30) on p. 185 for the depletion layer thickness on each side, we see that $x_P \gg x_N$. In other words, the depletion layer thickness is basically determined by the lowly doped p-type side, and the energy band on the degenerately doped side hardly bends. The depletion layer is formed practically only on the p-type side. From (6.32) on p. 186, the total depletion layer thickness is given approximately by

$$d_{\text{dep}} \approx x_P = \sqrt{\frac{2\epsilon_{\text{Si}}\,(\varphi_{\text{bi}} - V)}{qN_A^-}}, \tag{6.44}$$

where $\varphi_{\text{bi}}$ is given by (6.10) on p. 177. A numerical example of $d_{\text{dep}}$ for a zero-biased silicon one-sided abrupt junction is shown in Fig.

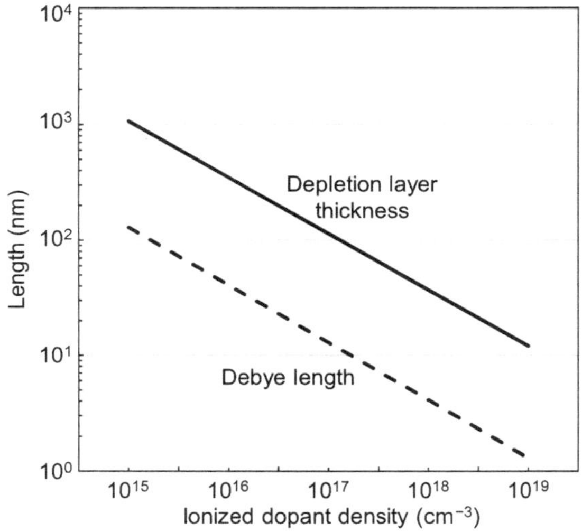


FIGURE 6.18 Depletion layer thickness and Debye length of a silicon one-sided abrupt junction.

6.18. The ionized donor density in the n-type region is assumed to be $N_D^+ = 10^{20}\,\text{cm}^{-3}$. In Fig. 6.18, the Debye length ((5.108) on p. 158) is also plotted for reference. The Debye length is about an order of magnitude shorter than the zero-bias depletion layer thickness (see Problem 6.5 on p. 213).

From (6.21) on p. 184 and (6.37) on p. 187, the depletion charge of a one-sided junction per unit area is given by

$$Q_d = qN_A^- x_P = \sqrt{2q\epsilon_{Si} N_A^- (\varphi_{bi} - V)}. \tag{6.45}$$

## 6.8 CURRENT-VOLTAGE CHARACTERISTICS OF P-N JUNCTIONS

### 6.8.1 Equation of Current-Voltage Characteristics

In §6.8, we derive the DC current-voltage characteristics of the abrupt p-n junction. Since the derivation is a lengthy process, we give the resulting equation first. The current density under a voltage bias $V$ is given by

$$J = J_s \left[ \exp\left(\frac{qV}{kT}\right) - 1 \right] \qquad (V < \varphi_{bi}). \tag{6.46}$$

Equation (6.46) is, in essence, the expression for an exponential function translated such that it goes through the origin of the *V-I* plane. At a forward bias lower than the built-in potential $\varphi_{bi}$, the current density increases exponentially with $V$. $J_s$ in (6.46) is the absolute value of the current density when $V$ is negative (reverse bias) and the exponential term is negligible (i.e., $qV/kT \lesssim -3$), and is called the *reverse saturation current density*.

$$J_s = q\left(\frac{D_n}{L_n} n_{P0} + \frac{D_p}{L_p} p_{N0}\right). \tag{6.47}$$

Symbols that appear in (6.47) are as follows:

$L_n$ : Electron diffusion length (p. 194)
$L_p$ : Hole diffusion length (p. 194)
$D_n$ : Electron diffusion coefficient (p. 140)
$D_p$ : Hole diffusion coefficient (p. 141)
$n_{P0}$ : Electron density in p-type neutral region (p. 149)
$p_{N0}$ : Hole density in n-type neutral region (p. 150)

### 6.8.2 Derivation of Current-Voltage Characteristics

The analytical derivation of the equation for current-voltage characteristics requires various assumptions. The assumptions of the analysis are listed below.

1. The p-n junction to be considered is an abrupt junction.
2. The applied bias voltage $V$ is assumed to be lower than the built-in potential $\varphi_{bi}$, that is, $V < \varphi_{bi}$.
3. Voltage drops in the neutral regions are assumed to be negligible.
4. The depletion approximation is to be applied. In other words, the carrier densities in the depletion layer are assumed to be negligibly small compared with the ionized dopant density.
5. Consistent with 1, assume the low injection condition (p. 180). That is to say, the minority carrier density flowing into the neutral region from the depletion layer is assumed to be significantly lower than the majority carrier density in the neutral region.

6. Both the p-type and n-type regions are assumed to be sufficiently thick (long enough in the $x$ direction) compared with the diffusion length (see p. 194).
7. The amount of change in the quasi-Fermi levels in the depletion layer is small compared with the thermal energy $kT$.
8. Carrier generation and recombination in the depletion layer are assumed to be negligible.

With Assumptions 1 to 4, the results of §6.5 can be used. Assumption 5 allows us to use the results of §5.6.4 regarding the behavior of minority carriers. We will try to find the DC current density at the edges of the neutral regions ($x = -x_\mathrm{P}$ and $x = x_\mathrm{N}$). In the depletion layer, there is a drift current due to the potential gradient, as well as a diffusion current, making it difficult to analyze. In contrast, in the neutral region, only a diffusion current exists due to Assumption 3, which simplifies the analysis. Assumption 7 says that the quasi-Fermi levels do not change in the depletion layer,[4] but if they really do not change at all, the current would be zero (see (5.33) and (5.34) on p. 131), so it says that there are small changes. Since we have Assumption 8, if we find the DC current density somewhere, that is the current density that flows through the whole system. To obtain the diffusion current density in the neutral regions, we find the $x$-dependence of the carrier densities in those regions. This is the outline of the analysis.

From here on, we will apply the continuity equations for charge and current, (5.102) and (5.103) on p. 157, to minority carriers. Rewriting the generation and recombination rates, $U_\mathrm{n}$ and $U_\mathrm{p}$, using (5.97) and (5.100) on p. 156, respectively, we obtain

$$\frac{\partial p_\mathrm{N}}{\partial t} = -\frac{p_\mathrm{N} - p_\mathrm{N0}}{\tau_\mathrm{p}} - \frac{1}{q}\frac{\partial J_\mathrm{p}}{\partial x}, \quad \text{(Continuity equation for electrons)} \tag{6.48}$$

$$\frac{\partial n_\mathrm{P}}{\partial t} = -\frac{n_\mathrm{P} - n_\mathrm{P0}}{\tau_\mathrm{n}} + \frac{1}{q}\frac{\partial J_\mathrm{n}}{\partial x}. \quad \text{(Continuity equation for holes)} \tag{6.49}$$

Thus, we have rewritten the equations using the electron and hole lifetimes, $\tau_\mathrm{n}$ and $\tau_\mathrm{p}$, respectively.

[4] This assumption is also known as the *quasi-equilibrium* assumption [12].

By Assumption 3, the bands are flat in the neutral regions ($x \leq -x_P$ and $x \geq x_N$), so drift current is zero and only diffusion currents exist. Therefore, from the current density equations (5.61) and (5.59) on p. 141, we obtain

$$J_p = -qD_p \frac{dp_N}{dx}, \qquad \text{(Hole diffusion current density in n-type)} \tag{6.50}$$

$$J_n = qD_n \frac{dn_P}{dx}. \qquad \text{(Electron diffusion current density in p-type)} \tag{6.51}$$

Putting these equations in the continuity equations (6.48) and (6.49) yields

$$\frac{\partial p_N}{\partial t} = -\frac{p_N - p_{N0}}{\tau_p} + D_p \frac{\partial^2 p_N}{\partial x^2}, \tag{6.52}$$

$$\frac{\partial n_P}{\partial t} = -\frac{n_P - n_{P0}}{\tau_n} + D_n \frac{\partial^2 n_P}{\partial x^2}. \tag{6.53}$$

Since $\partial p_N/\partial t = 0$ and $\partial n_P/\partial t = 0$ in a steady state, the continuity equations become

$$\frac{d^2 p_N}{dx^2} = \frac{p_N - p_{N0}}{D_p \tau_p}, \tag{6.54}$$

$$\frac{d^2 n_P}{dx^2} = \frac{n_P - n_{P0}}{D_n \tau_n}. \tag{6.55}$$

These differential equations are of the same form as the wave equation in (5.105) on p. 157. The solutions of these differential equations have the following form:

$$\Delta p = p_N - p_{N0} = c_1 \exp\left(-\frac{x}{L_p}\right) + c_2 \exp\left(\frac{x}{L_p}\right), \tag{6.56}$$

$$\Delta n = n_P - n_{P0} = c_3 \exp\left(-\frac{x}{L_n}\right) + c_4 \exp\left(\frac{x}{L_n}\right), \tag{6.57}$$

where $c_1$ through $c_4$ are unknown constants that are to be determined using boundary conditions, discussed shortly. $L_p$ and $L_n$ are called the *diffusion lengths* and can be written as follows:

$$L_p \equiv \sqrt{D_p \tau_p}, \qquad \text{(Hole diffusion length)} \tag{6.58}$$

$$L_n \equiv \sqrt{D_n \tau_n}, \qquad \text{(Electron diffusion length)} \tag{6.59}$$

where $D_p$ and $D_n$ are the diffusion coefficients (p. 140), and $\tau_p$ and $\tau_n$ are the minority carrier lifetimes (p. 149). Equations (6.56) and (6.57) indicate that the densities of excess minority carriers diffusing into the neutral regions vary (decrease) exponentially as functions of $x$. As discussed in §5.6.4, minority carriers recombine and disappear after a while.

The first set of boundary conditions for determining the unknown constants are that the excess minority carrier densities must go to zero at a sufficient distance from the junction interface:

$$\Delta p\,(+\infty) \to 0, \qquad \text{(Boundary condition at far right)} \tag{6.60}$$

$$\Delta n(-\infty) \to 0. \qquad \text{(Boundary condition at far left)} \tag{6.61}$$

These come from Assumption 6 on p. 192. Of course, the actual values of coordinate $x$ at which the excess minority carrier density is practically zero do not have to be $\pm\infty$. That is to say, "long enough in the $x$ direction" in Assumption 6 means long enough in the sense that $\Delta p \to 0$ and $\Delta n \to 0$ hold at certain points. From the above, we immediately obtain $c_2 = 0$ and $c_3 = 0$. Thus, the equations for the excess minority carrier densities, (6.56) and (6.57), are simplified as follows.

$$\Delta p = p_N - p_{N0} = c_1 \exp\left(-\frac{x}{L_p}\right) \quad (x \geq x_N), \tag{6.62}$$

$$\Delta n = n_P - n_{P0} = c_4 \exp\left(\frac{x}{L_n}\right) \quad (x \leq -x_P). \tag{6.63}$$

The second set of boundary conditions required to determine $c_1$ and $c_4$ are less obvious. These are needed to write down the minority

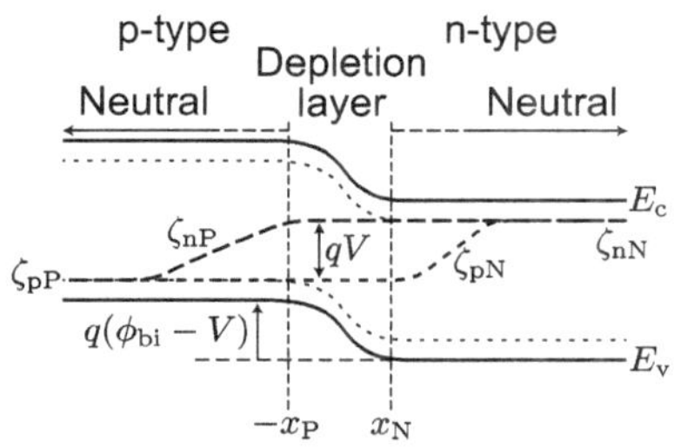


FIGURE 6.19 Quasi-Fermi levels in a forward-biased p-n junction.

carrier densities at the depletion layer edges ($x = -x_P$ and $x = x_N$)[5]. Specifically, we postulate the following:

- The hole quasi-Fermi level $\zeta_p$ in the entire depletion layer takes on nearly the same value as in the p-type neutral region, where holes are the majority carriers.
- The electron quasi-Fermi level $\zeta_n$ in the entire depletion layer takes on nearly the same value as in the n-type neutral region, where electrons are the majority carriers.

These are Assumption 7 on p. 192. Fig. 6.19 visualizes these in an energy band diagram.

Under Assumption 7, the quasi-Fermi levels in the depletion layer satisfy

$$\zeta_n - \zeta_p \approx qV. \qquad \text{(Vertical quasi Fermi level opening)} \qquad (6.64)$$

See also (5.13) on p. 124 and Fig. 6.15 (p. 181). The latter shows the reverse-biased case. Equation (6.64) says that the vertical opening between $\zeta_n$ and $\zeta_p$ is determined by the applied bias voltage, $V$. Incidentally, the horizontal opening is determined by the minority carrier lifetimes or diffusion lengths.

The boundary condition at $x = x_N$ can be written as follows:

$$p_N(x_N) \approx p_{P0} \exp\left[-\frac{q(\varphi_{bi} - V)}{kT}\right] = p_{N0} \exp\left(\frac{qV}{kT}\right). \qquad (6.65)$$

Equation (6.65) is the condition for the minority carrier (i.e., hole) density at the depletion layer edge in the n-type region. To write

[5] These are also the edges of the neutral regions.

down (6.65), we used the fact that the hole density can be variously expressed in the form of (5.26) on p. 129. To be more specific, to derive the middle equation in (6.65), we set the "reference density" in (5.26) to be the hole density $p_{\mathrm{P0}}$ in the p-type neutral region. The "reference energy" is the "relative energy" of $\zeta_{\mathrm{pP}}$ in the p-type neutral region (see p. 128). Then for $x \leq -x_{\mathrm{P}}$, the exponent of (5.26) becomes 0, and the hole density is given by $p_{\mathrm{P}}(x) = p_{\mathrm{P0}}$. Considering now the hole density at $x = x_{\mathrm{N}}$, since the band is bent down by $q(\varphi_{\mathrm{bi}} - V)$, the "reference energy" position is also bent down parallel to $E_{\mathrm{v}}$ by the same amount. $\zeta_{\mathrm{pN}}$ at $x = -x_{\mathrm{P}}$, therefore, is quite far from the "reference energy," and the hole density becomes correspondingly low. This explains why the hole density at $x = x_{\mathrm{N}}$ is given by the middle equation in (6.65). The right-hand side of (6.65) rewrites the same quantity with the "reference density" being the hole density $p_{\mathrm{N0}}$ in the n-type neutral region. In this case, the "reference energy" is the "relative energy" of $\zeta_{\mathrm{pN}}$ in the n-type neutral region. At $x = x_{\mathrm{N}}$, $\zeta_{\mathrm{pN}}$ deviates from the "reference energy" by $qV$, so the exponential factor on the right-hand side of (6.65) is needed.

Similarly, the boundary condition at $x = -x_{\mathrm{P}}$ can be written as

$$n_{\mathrm{P}}(-x_{\mathrm{P}}) \approx n_{\mathrm{N0}} \exp\left[-\frac{q(\varphi_{\mathrm{bi}} - V)}{kT}\right] = n_{\mathrm{P0}} \exp\left(\frac{qV}{kT}\right). \qquad (6.66)$$

This is the condition for the minority carrier (i.e., electron) density at the depletion layer edge in the p-type region. The derivation of (6.66) is similar to the derivation of (6.65) (see Problem 6.6 on p. 213).

Inserting $x = x_{\mathrm{N}}$ and (6.65) into (6.62) yields

$$\Delta p(x_{\mathrm{N}}) = p_{\mathrm{N0}} \exp\left(\frac{qV}{kT}\right) - p_{\mathrm{N0}} = c_1 \exp\left(-\frac{x_{\mathrm{N}}}{L_{\mathrm{p}}}\right). \qquad (6.67)$$

Equation (6.67) can be solved for $c_1$ as follows:

$$c_1 = p_{\mathrm{N0}}\left[\exp\left(\frac{qV}{kT}\right) - 1\right] \exp\left(\frac{x_{\mathrm{N}}}{L_{\mathrm{p}}}\right). \qquad (6.68)$$

Putting $c_1$ in (6.62) on p. 194, the excess hole density distribution in the n-type region is finally found to be

$$\Delta p(x) = p_{\mathrm{N0}}\left[\exp\left(\frac{qV}{kT}\right) - 1\right] \exp\left(-\frac{x - x_{\mathrm{N}}}{L_{\mathrm{p}}}\right) \quad (x \geq x_{\mathrm{N}}). \qquad (6.69)$$

Likewise, to find the excess electron density distribution in the p-type region, insert $x = -x_{\rm P}$ and (6.66) into (6.63): on p. 194.

$$n_{\rm P0}\exp\left(\frac{qV}{kT}\right) - n_{\rm P0} = c_4\exp\left(-\frac{x_{\rm P}}{L_{\rm n}}\right). \tag{6.70}$$

Solving this equation for $c_4$ gives

$$c_4 = n_{\rm P0}\left[\exp\left(\frac{qV}{kT}\right) - 1\right]\exp\left(\frac{x_{\rm P}}{L_{\rm n}}\right). \tag{6.71}$$

Putting $c_4$ into (6.63) finally yields

$$\Delta n(x) = n_{\rm P0}\left[\exp\left(\frac{qV}{kT}\right) - 1\right]\exp\left[\frac{x-(-x_{\rm P})}{L_{\rm n}}\right] \quad (x \le -x_{\rm P}). \tag{6.72}$$

This is the excess electron density distribution in the p-type region.

We derived (6.69) and (6.72) to find the DC current density. Noting that $\partial p_{\rm N0}/\partial x = 0$ (§5.1), insert (6.69) into the current density expression (6.50) on p. 193. The hole current density in the n-type neutral region is thus found to be

$$\begin{aligned} J_{\rm p}(x) &= -qD_{\rm p}\frac{{\rm d}p_{\rm N}(x)}{{\rm d}x} = -qD_{\rm p}\frac{{\rm d}\Delta p(x)}{{\rm d}x} \\ &= q\frac{D_{\rm p}}{L_{\rm p}}p_{\rm N0}\left[\exp\left(\frac{qV}{kT}\right) - 1\right]\exp\left(-\frac{x-x_{\rm N}}{L_{\rm p}}\right) \quad (x \ge x_{\rm N}). \end{aligned} \tag{6.73}$$

Equation (6.73) shows that the hole current density decreases exponentially toward the right.

Similarly, insert (6.72) in (6.51) on p. 193, noting that $\partial n_{\rm P0}/\partial x = 0$. The electron current density in the p-type neutral region is found to be

$$\begin{aligned} J_{\rm n}(x) &= qD_{\rm n}\frac{{\rm d}n_{\rm P}(x)}{{\rm d}x} = qD_{\rm n}\frac{{\rm d}\Delta n(x)}{{\rm d}x} \\ &= q\frac{D_{\rm n}}{L_{\rm n}}n_{\rm P0}\left[\exp\left(\frac{qV}{kT}\right) - 1\right]\exp\left[\frac{x-(-x_{\rm P})}{L_{\rm n}}\right] \quad (x \le -x_{\rm P}). \end{aligned} \tag{6.74}$$

Equation (6.74) shows that the electron current density decreases exponentially toward the left.

Both electron and hole currents contribute to the total current that flows in a p-n junction. So, it would be desirable for the current density at $x$ to be written as

$$J(x) = J_{\rm p}(x) + J_{\rm n}(x), \tag{6.75}$$

but (6.73) and (6.74) only cover different ranges of $x$. Now we can invoke Assumption 8 on p. 192, which says that neither the hole current density nor the electron current density changes in the depletion layer. Thanks to this assumption, for the hole current density, the value at $x = x_N$ can be adopted, and for the electron current density, the value at $x = -x_P$ can be adopted. Since we are considering DC current here, the sum in (6.75) is constant regardless of the value of $x$. The total current density, therefore, is given by

$$J = J_p(x_N) + J_n(-x_P). \qquad \text{(Current density of abrupt junction)} \tag{6.76}$$

Substituting (6.73) and (6.74) for the first and second terms of (6.76), respectively, we finally obtain

$$\begin{aligned} J &= q\left(\frac{D_n}{L_n} n_{P0} + \frac{D_p}{L_p} p_{N0}\right)\left[\exp\left(\frac{qV}{kT}\right) - 1\right] \\ &= J_s\left[\exp\left(\frac{qV}{kT}\right) - 1\right] \qquad (V < \varphi_{bi}). \end{aligned} \tag{6.77}$$

This completes the derivation of the DC current-voltage characteristics (6.46) on p. 191 and the reverse saturation current density (6.47).

## SHOCKLEY'S P-N JUNCTION THEORY

The theory for the current-voltage characteristics of p-n junctions described in this section was developed by William Shockley, one of the founders of the field of semiconductor devices and the inventor of the bipolar transistor [27]. In the above derivation, the most puzzling and seemingly unwarranted assumption would be Assumption 7 on p. 192. and the boundary conditions (6.65) and (6.66), which embody this assumption. In fact, it is not so easy to justify equations (6.65) and (6.66) theoretically. A lengthy discussion is required to do so [31]. Moreover, when the magnitude of the reverse bias voltage, $|V|$, is about 0.1 V or higher, it is known that these equations do not hold [31]! Nevertheless, thanks to the bold assumption, we were able to derive the equation (6.77) for the current density. Geniuses do tricks like this that would not occur to those who can think only logically (see Problem 6.7 on p. 213). Such a bold move has undoubtedly helped the development of semiconductor electronics in the early days.

### 6.8.3 Additional Notes on p-n Junctions

#### *6.8.3.1 Length Scales*

First, let us summarize length scales relevant to semiconductor devices. The diffusion lengths are given by (6.58) and (6.59) on p. 194. To see how long diffusion lengths actually are, we estimated the electron diffusion length $L_n$ and hole diffusion length $L_p$ by referring to the mobilities in Table 1.3 (p. 5) and the minority carrier lifetimes given in Table 5.2 (p. 122). Einstein's relations (5.62) and (5.63) on p. 141 were used to express the diffusion coefficient in terms of mobility. The results are shown in Table 6.1 for several different lifetimes. Note that the number of significant digits in the table should be considered to be about one.

The depletion layer thickness in silicon is usually less than 1 μm, as we saw in the depletion layer thickness graph (Fig. 6.18) on p. 190. The diffusion lengths in Table 6.1 are generally longer than the depletion layer thickness. That is, the following inequality holds:

$$\text{(diffusion length)} \gtrsim \text{(depletion layer thickness)} > \text{(Debye length)}. \quad (6.78)$$

If Assumption 6 on p. 192 is satisfied and therefore the excess minority carrier densities go to zero at the left and right ends of the p-n junction diode (as in (6.60) and (6.61) on p. 194), the lengths of the

TABLE 6.1 Diffusion Lengths in Silicon

| Minority carrier lifetime (s) | $L_n$ (μm) | $L_p$ (μm) |
|---|---|---|
| $10^{-10}$ | 0.62 | 0.36 |
| $10^{-9}$ | 2.0 | 1.1 |
| $10^{-8}$ | 6.2 | 3.6 |
| $10^{-7}$ | 20 | 11 |
| $10^{-6}$ | 62 | 36 |
| $10^{-5}$ | 196 | 113 |
| $10^{-4}$ | 621 | 358 |

p-type and n-type regions must be at least be about an order of magnitude longer than the diffusion length. This is because the diffusion length is the length at which the minority carrier density is $1/e \simeq 0.37$ times the excess minority carrier density, whereas the minority carrier density must decrease by several orders of magnitude for the excess minority carrier density to become negligible. Therefore, the depletion layer is usually negligibly thin compared with the total length of the p-n junction diode. In other words, p-n junction pictures in most books, including this one, either exaggerate the depletion layer thickness or draw the lengths of the p-type and n-type regions too short.

Note that it is not at all true that actual p-n junctions must be long enough so that Assumption 6 on p. 192 is satisfied. It is common in real devices for the p-type and n-type regions to be much shorter than the diffusion length. For example, the base region of a bipolar transistor (p. 46) is made as short as possible compared with the diffusion length. In such a case, the boundary conditions change, and we have $c_2 \neq 0$ and $c_3 \neq 0$ in (6.56) and (6.57). Then, a different form of the current-voltage characteristics equation is derived than (6.77) on p. 198.

### *6.8.3.2 Carrier Generation-Recombination in the Depletion Layer*

Regarding Assumption 8 on p. 192, in actual silicon p-n junctions, it is known that generation and recombination in the depletion layer cannot be neglected, and Assumption 8 does not hold in practice. As a result, the current-voltage characteristic is somewhat different from that expressed by (6.77) (see Fig. 6.20 on p. 201).

### *6.8.3.3 High Forward Bias*

If Assumption 2 on p. 191 regarding the bias voltage $V$ does not hold and $V \geq \varphi_{\mathrm{bi}}$, the depletion layer disappears and the entire band of the p-n junction diode slopes. In this case, the p-type and n-type regions behave like conductive solids with a certain resistivity, and the voltage drop occurs across the entire structure. This requires a completely different treatment from that in §6.8.2. We will see numerical examples covering $V \geq \varphi_{\mathrm{bi}}$ in §6.8.3.4 and §6.9.1.

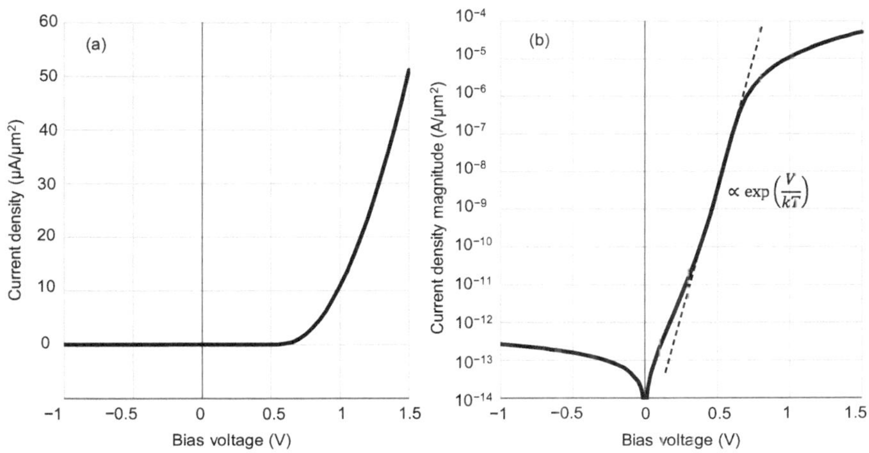


FIGURE 6.20 An example of numerical analysis of current-voltage characteristics of a p-n junction.

### *6.8.3.4 Example of Numerical Analysis*

Fig. 6.20 shows an example of current-voltage characteristics obtained by numerical analysis (i.e., device simulation or "TCAD"), including generation-recombination and the bias conditions for $V \geq \varphi_{\mathrm{bi}}$, which were not covered in our analysis. The vertical axis of Fig. 6.20(a) uses a linear scale, and that of Fig. 6.20(b) uses a logarithmic scale. The conditions for the device simulation are shown in Table 6.2. The lengths of the p-type and n-type regions are 5 μm each. The doping densities are kept quite low to make the depletion layer relatively thick and easy to see in the energy band diagram. In order to limit the diffusion lengths, the minority carrier lifetimes are made quite short. But still, Assumption 6 on p. 191 and $\Delta p$ (right end) $\to 0$ and $\Delta n$ (left end) $\to 0$, corresponding to (6.60) and (6.61) on p. 194, do not hold. Indirect generation and recombination are allowed to occur throughout the structure, including in the depletion layer, according to the specified lifetimes. Therefore, Assumption 8 does not hold either. Thus, we do not intend here to reproduce or verify what we did in §6.8.2. The built-in potential is $\varphi_{\mathrm{bi}} \simeq 0.7\,\mathrm{V}$.

Fig. 6.20(a), where the vertical axis uses a linear scale, shows that there appears to be almost no current flow at $V \lesssim \varphi_{\mathrm{bi}}$. To see the exponential characteristic described by the current-voltage characteristic

TABLE 6.2 Conditions for Device Simulation of a p-n Junction

| Material | Silicon |
|---|---|
| Length of p-type region | 5 μm |
| Length of n-type region | 5 μm |
| Acceptor ion density $N_A^-$ in p-type region | $1 \times 10^{16}$ cm$^{-3}$ |
| Donor ion density $N_D^+$ in n-type region | $2 \times 10^{16}$ cm$^{-3}$ |
| Electron lifetime $\tau_n$ | $10^{-9}$ s |
| Hole lifetime $\tau_p$ | $10^{-9}$ s |
| Electron mobility $\mu_n$ | 1450 cm$^2$/(V·s) |
| Hole mobility $\mu_n$ | 500 cm$^2$/(V·s) |

equation (6.46) on p. 191, the vertical axis must be set to a logarithmic scale, as shown in Fig. 6.20(b).

According to (6.46), if the vertical axis is a logarithmic scale, a straight line should appear at low forward bias ($0 < V < \varphi_{bi} \simeq 0.7\,\text{V}$). However, in Fig. 6.20(b), two slopes are observed below 0.7 V.[6] The gentler slope seen at $0.1 \lesssim V \lesssim 0.4\,\text{V}$ is related to the exponential factor $\exp(qV/2kT)$ of (5.14) for the effective intrinsic carrier density[7] $n_i'$ on p. 124 and recombination in the depletion layer (see p. 200). Fig. 6.20 also has other features different from (6.46), but a detailed description is beyond the scope of this book. However, as long as $V < \varphi_{bi}$, it should be fair to say that (6.46) captures the most important features of the p-n junction—the rectifying action and the exponential variation of the current density.

### *6.8.3.5 Breakdown*

As the reverse bias voltage $|V|$ is made larger beyond a certain point, the current increases rapidly, as shown in Fig. 6.21. This phenomenon is known as *breakdown*. The *breakdown voltage* depends on the forbidden bandwidth $E_g$ of the material as well as the doping densities. The larger $E_g$ is, the larger the absolute value of the breakdown voltage. Because of this, materials with large $E_g$, such as silicon carbide (SiC) and gallium nitride (GaN), are often used in high-voltage devices (see Problem 4.3 on p. 113).

[6] The slope seen at $0 < V \lesssim 0.1\,\text{V}$ is due to the fact that (6.46) passes through the origin and should be ignored.

[7] Note that (6.64) on p. 195 was used.

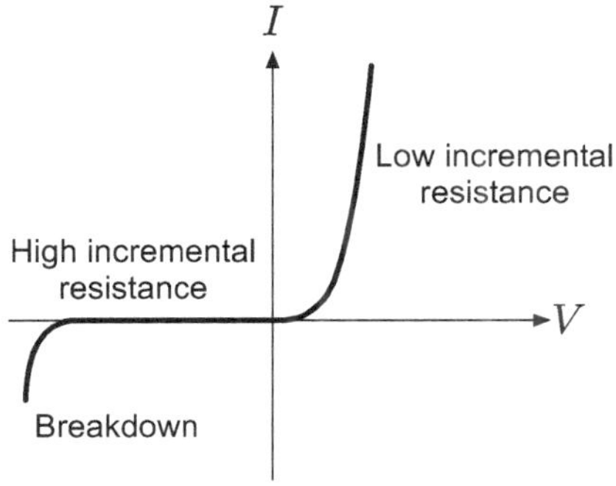


FIGURE 6.21 *I-V* curve of a p-n junction showing breakdown.

## 6.9 READING ENERGY BAND DIAGRAMS OF P-N JUNCTIONS

We already read carrier densities and rectifying action from p-n junction energy band diagrams in §6.4. In this section, as a more practical exercise, we will read energy band diagrams of the p-n junction diode obtained by numerical analysis using a device simulator (§5.7).

### 6.9.1 Bias Voltage Dependence

The conditions for device simulation are the same as in Fig. 6.20 (p. 201) and are shown in Table 6.2 (p. 202). The interface between the p-type and n-type regions is at $x = 5\,\mu\text{m}$.

Fig. 6.22 shows the band diagram at zero bias ($V = 0\,\text{V}$). The p-type depletion layer is slightly thicker than the n-type depletion layer because the p-type side is doped at a lower concentration (see (6.29) and (6.30) on p. 185 and Table 6.2 on p. 202). Since both quasi-Fermi levels $\zeta_\text{n}$ and $\zeta_\text{p}$ are flat, no current flows.

Next, let us look at band diagrams at reverse bias. First, Fig. 6.23 (p. 204) shows a band diagram at a low reverse bias voltage, $V = -0.1\,\text{V}$. For the electron quasi-Fermi level $\zeta_\text{n}$ at the right end ($x = 10\,\mu\text{m}$) and the hole quasi-Fermi level $\zeta_\text{p}$ at the left end ($x = 0\,\mu\text{m}$),

$$\zeta_\text{n}\,(10\mu\text{m}) - \zeta_\text{p}\,(0\mu\text{m}) = qV(< 0) \tag{6.79}$$

holds (see also (6.12) on p. 178). This equation is imposed by the device simulator as a boundary condition. In and around the depletion layer, $\zeta_\text{n}$ and $\zeta_\text{p}$ are split in and around the depletion layer, indicating

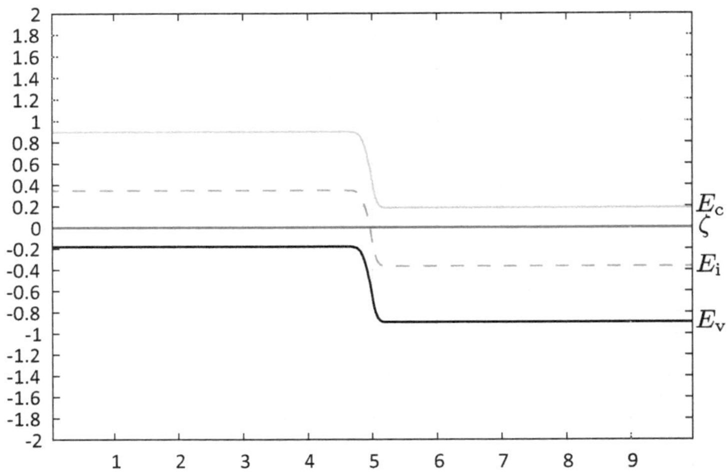


FIGURE 6.22 A TCAD-drawn energy band diagram of a p-n junction diode. Lifetimes: $\tau_n = \tau_p = 10^{-9}$ s, Voltage bias: $V = 0$ V.

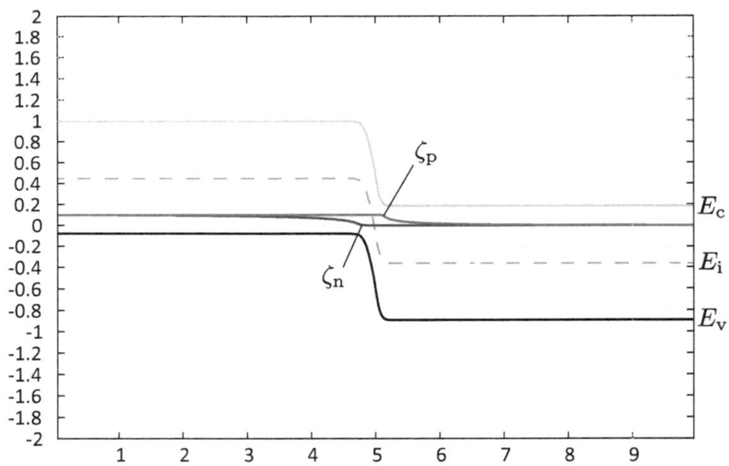


FIGURE 6.23 A TCAD-drawn energy band diagram of a p-n junction diode. Lifetimes: $\tau_n = \tau_p = 10^{-9}$ s, Voltage bias: $V = -0.1$ V.

a larger deviation from equilibrium (see p. 124) than in the neutral regions, where $\zeta_n \simeq \zeta_p$. Since $\zeta_n < \zeta_p$ in the depletion layer, the effective intrinsic carrier density is greater than the intrinsic carrier density

($n_i' < n_i$), and therefore carrier generation is dominant over recombination (see p. 151). In Fig. 6.23, it appears that the vertical opening of quasi-Fermi levels is given, indeed, by (6.64) on p. 195. However, if we check $\zeta_n$ and $\zeta_p$ at the edges of the depletion layer carefully, it seems that Assumption 7 on p. 192 does not hold completely.

The direction of carrier motion is determined by the gradient of the corresponding quasi-Fermi level (see Fig. 5.3 (p. 122) and (5.31) and (5.32) on p. 131). Electrons move rightward as they descend the slope of $\zeta_n$ in Fig. 6.23 (p. 204). Holes move leftward as they descend the upside-down slope of $\zeta_p$. The carriers generated in the depletion layer also contribute to the current. This current is sometimes called the *generation current*. To determine the magnitude of the current density, one can look at the corresponding carrier density (electron density for a slope of $\zeta_n$ or hole density for a slope of $\zeta_p$) at a position where the gradient of the quasi-Fermi level is steep (see p. 131). We can see that at such positions in Fig. 6.23, both electron density and hole density are extremely small (which is natural, since $n_i' < n_i$). Therefore, we can conclude that the current density is very small.

When the reverse bias voltage is set to $V = -0.5$ V, the energy band diagram is as shown in Fig. 6.24 (p. 206). Now it is no longer clear that the expression (6.64) for the vertical opening of the quasi-Fermi levels on p. 195 holds in the depletion layer. Obviously, Assumption 7 on p. 192 does not hold (see the Box on p. 198). Therefore, the boundary conditions (6.65) and (6.66) on p. 195 do not hold either. For (6.65) and (6.66) to hold, $\zeta_p$ must go into the conduction band of the n-type region and $\zeta_n$ must go into the valence band of the p-type region (see Problem 6.7 on p. 213). Since the slopes of $E_c$ and $E_v$ and the slopes of $\zeta_n$ and $\zeta_p$ have the same sign in the depletion layer, it is clear that the drift current is dominant over the diffusion current (see p. 119). In §6.8.2, we calculated the current through the p-n junction only from the diffusion current at the edges of neutral regions. In the reverse-biased case, however, a drift current flows first due to the steep potential gradient in the depletion layer, resulting in a decrease in the minority carrier densities at the edges of the neutral regions, which, in turn, is compensated for by the diffusion currents in the neutral regions. Basically, the same thing is happening in Fig. 6.25 (p. 206), where the reverse bias is increased in magnitude to $V = -1$ V.

The band diagram for a forward bias voltage of $V = 0.5$ V is shown in Fig. 6.26 (p. 207). At forward bias, (6.11) on p. 177 holds for the hole quasi-Fermi level $\zeta_p$ at the left end ($x = 0\,\mu\text{m}$) and the electron

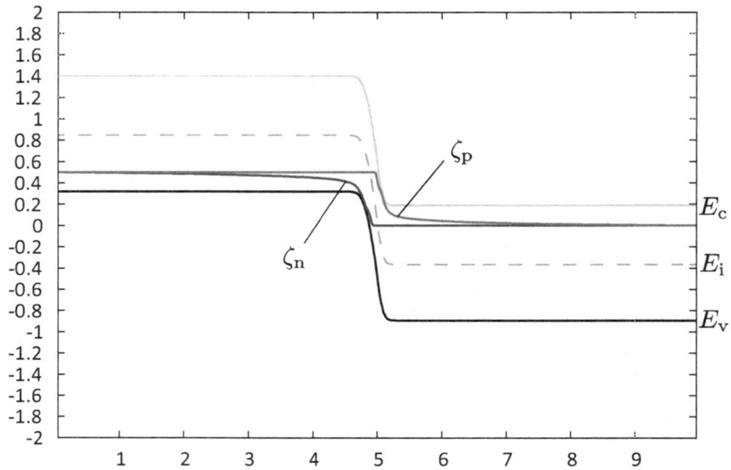


FIGURE 6.24 TCAD-drawn energy band diagram of a p-n junction diode. Lifetimes: $\tau_n = \tau_p = 10^{-9}$ s, Voltage bias: $V = -0.5$ V.

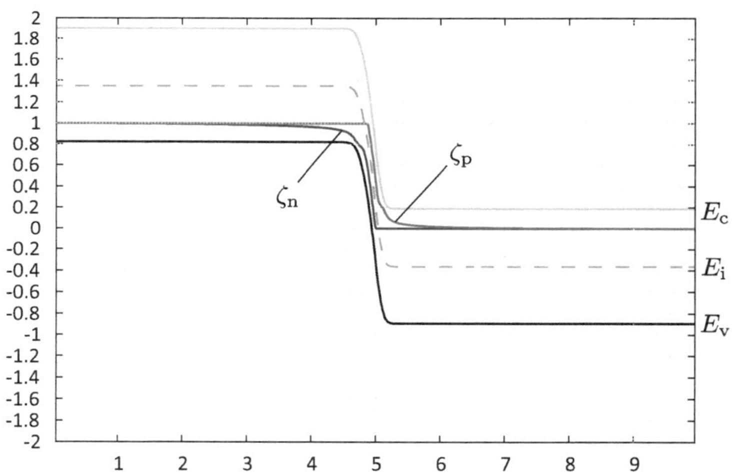


FIGURE 6.25 TCAD-drawn energy band diagram of a p-n junction diode. Lifetimes: $\tau_n = \tau_p = 10^{-9}$ s, Voltage bias: $V = -1$ V.

quasi-Fermi level $\zeta_n$ at the right end ($x = 10\,\mu$m). $\zeta_n$ and $\zeta_p$ are split over the entire structure due to the relatively long diffusion length. This is a different situation from Assumption 6 on p. 192, but it is quite common in real devices. Since $\zeta_n > \zeta_p$, $n_i' > n_i$ holds (see p.

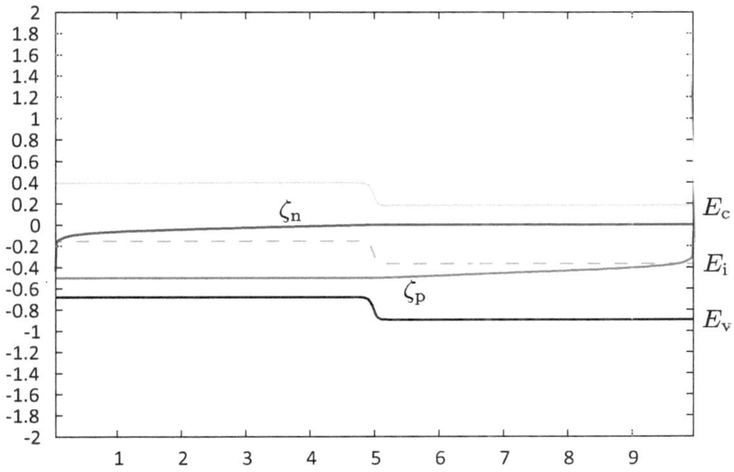


FIGURE 6.26 TCAD-drawn energy band diagram of a p-n junction diode. Lifetimes: $\tau_n = \tau_p = 10^{-9}$ s, Voltage bias: $V = 0.5$ V.

151), carrier recombination is dominant (see p. 179). Since $\zeta_n > E_i$ in the p-type region of Fig. 6.26, the electron (minority carrier) density is higher than the intrinsic carrier density $n_i$ as if it were an n-type semiconductor.[8] Similarly, since $\zeta_p < E_i$ in most parts of the n-type region, the hole (minority carrier) density is higher than $n_i$ as if it were a p-type semiconductor.[9]

The current density is by far greater than at reverse bias (see Fig. 6.20 on p. 201) because much higher densities of minority carriers contribute to the current density. Since the electron quasi-Fermi level $\zeta_n$ slopes down toward the left, electrons diffuse leftward against the potential gradient in the depletion layer. Since the hole quasi-Fermi level $\zeta_p$ slopes up toward the right, holes diffuse rightward also against the potential gradient in the depletion layer. At both ends ($x = 0\,\mu m$ and $x = 10\,\mu m$), in addition to (6.79) on p. 203, the boundary condition that the quasi-Fermi levels for electrons and holes coincide is imposed:

$$\zeta_n\,(0\mu m) = \zeta_p\,(0\mu m)\,, \tag{6.80}$$

8 The hole (majority carrier) density is still higher than the electron density.
9 The electron (majority carrier) density is still higher than the hole density.

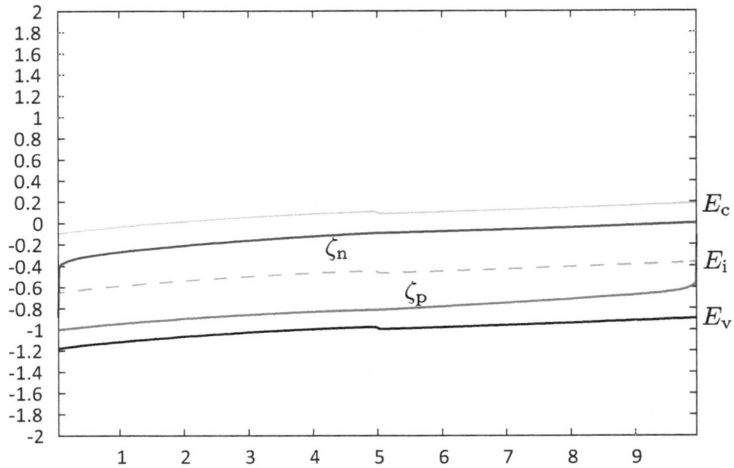


FIGURE 6.27 TCAD-drawn energy band diagram of a p-n junction diode. Lifetimes: $\tau_n = \tau_p = 10^{-9}$ s, Voltage bias: $V = 1$ V.

$$\zeta_n\,(10\mu\text{m}) = \zeta_p\,(10\mu\text{m})\,. \tag{6.81}$$

This is the reason for the (barely visible) sudden changes in minority carrier quasi-Fermi levels at the edges.

Fig. 6.27 (p. 208) shows the results for forward bias $V = 1$ V. Since $V > \varphi_{bi} \simeq 0.7$ V, it goes beyond the scope of the discussion in §6.8. Gradients of electrostatic potential, electron quasi-Fermi level, and hole quasi-Fermi level can be seen in the entire region. In addition, both electron and hole densities are high throughout. This is because the long diffusion lengths (or equivalently, the long minority carrier lifetimes) make the carriers drift away before recombination takes place.

## 6.9.2 Lifetime Dependence

Now, according to the discussion on p. 121, the reason why quasi-Fermi levels must be defined separately for electrons and holes is because of the following relationship:

$$\text{(minority carrier lifetime)} \gg \text{(dielectric relaxation time)}. \tag{6.82}$$

Generation and recombination work to bring electrons and holes into equilibrium ($n_i' \to n_i$), but if minority carriers have a long lifetime, the effectiveness of generation-recombination will be poor. Conversely, if we somehow shorten the minority carrier lifetime (by, for example, introducing traps), the system should get closer to equilibrium (see Problem 5.1 on p. 164). The separation between electron and hole quasi-Fermi levels, $|\zeta_n - \zeta_p|$, represents the degree of deviation from equilibrium (see (5.13) on p. 124), and the generation or recombination becomes dominant depending on $\zeta_n \lesseqgtr \zeta_p$ (p. 151).

Let us take a look at the results of analyzing the same structure with $\tau_n = \tau_p = 10^{-7}$ s or $\tau_n = \tau_p = 10^{-11}$ s, longer or shorter than the minority carrier lifetime given in Table 6.2 ($\tau_n = \tau_p = 10^{-9}$ s). Fig. 6.28 and Fig. 6.29 are the results for reverse bias $V = -0.5$ V, and Fig. 6.30 and Fig. 6.31 are the results for forward bias $V = 0.5$ V. In all cases, the opening of quasi-Fermi levels is larger when the lifetime is longer and smaller when the lifetime is shorter. When $\tau_n = \tau_p = 10^{-11}$ s, the diffusion lengths are sufficiently short, so that in Fig. 6.31 exponential decreases in minority carrier densities (linear changes in quasi-Fermi levels) are observed just as in (6.69) and (6.72) on p. 196. As minority carriers move along the gradients of $\zeta_n$ and $\zeta_p$, the minority carrier densities are decreasing rapidly due to recombination. When net generation or recombination is occurring at a high rate as in Fig. 6.29 and Fig. 6.31, the electron current density and the hole current density are not separately conserved,[10] and the current density is somewhat more difficult to read from an energy band diagram.

Incidentally, such large changes in minority carrier lifetime will, of course, affect the current-voltage characteristics. Fig. 6.32 shows the corresponding current-voltage characteristics of the p-n junction (see Problem 6.9 on p. 214).

## P-N JUNCTIONS THAT DO NOT RECTIFY

In this chapter, we looked at the rectifying action of the p-n junction diode. However, it is not always the case that p-n junctions in semiconductor devices exhibit rectifying action.

For example, the bipolar transistor in Fig. 2.19 (p. 46) has an n-p-n or p-n-p structure. One of the two p-n junctions can have

[10] The sum of the two current densities, (6.75) on p. 197, is conserved when considering a DC current.

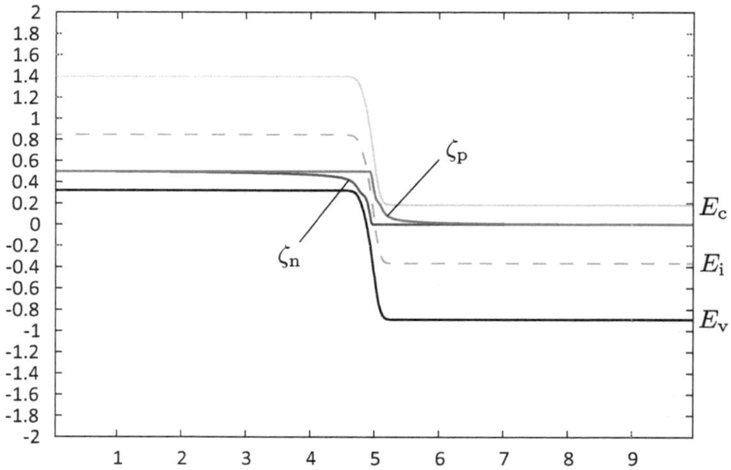


FIGURE 6.28 TCAD-drawn energy band diagram of a p-n junction diode. Lifetimes: $\tau_n = \tau_p = 10^{-7}$ s, Voltage bias: $V = -0.5$ V.

a large reverse current flow. The reason is related to the fact that the base region is made thinner than the diffusion length, and therefore the boundary conditions are different from those in §6.8.2. This can be understood by looking at the energy band diagram of a bipolar transistor.

Another example is the MOSFET. The nMOS transistor also has an n-p-n structure as shown in Fig. 7.3 (p. 217). By a mechanism different from that in bipolar transistors, a large reverse current can flow from the degenerately doped n-type drain to the p-type region just below the gate oxide film.

## 6.10 SUMMARY

In this chapter, we considered the physics and characteristics of the p-n junction.

- The difference in electrostatic potential between solid substances of different properties, when they are brought into contact with each other, is called the contact potential.

- The contact potential between the p-type and n-type regions of a p-n junction is called the built-in potential.

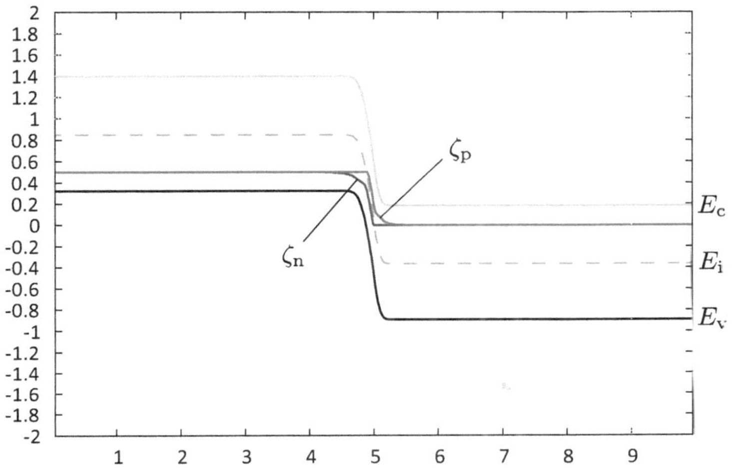


FIGURE 6.29 TCAD-drawn energy band diagram of a p-n junction diode. Lifetimes: $\tau_n = \tau_p = 10^{-11}$ s, Voltage bias: $V = -0.5$ V.

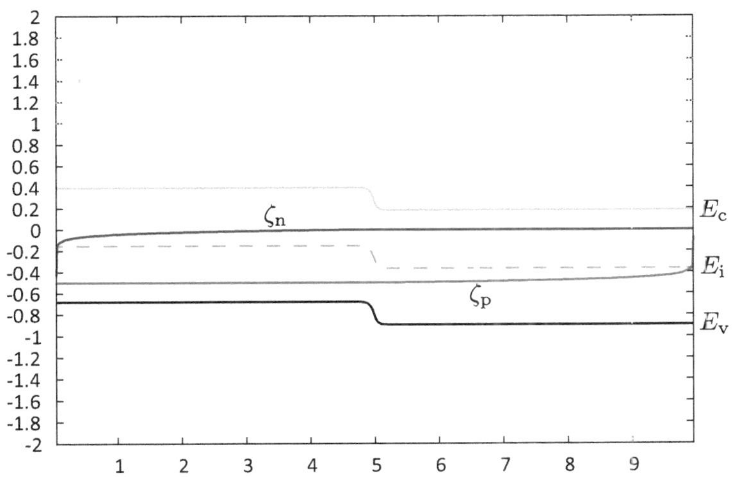


FIGURE 6.30 TCAD-drawn energy band diagram of a p-n junction diode. Lifetimes: $\tau_n = \tau_p = 10^{-7}$ s, Voltage bias: $V = 0.5$ V.

- A depletion layer with very few carriers is formed around the p-n junction interface, and it has associated depletion capacitance.
- The p-n junction diode has a rectifying effect.
- Properties of the one-sided junction are determined by the lowly doped side.
- The DC current-voltage characteristics and the reverse saturation current of the abrupt junction can be derived analytically under several assumptions.
- (Diffusion length) $\gtrsim$ (Depletion layer thickness) $>$ (Debye length).
- The physics of p-n junctions can be understood to a large extent from energy band diagrams with quasi-Fermi levels.

## 6.11 PROBLEMS

6.1 Why is (6.3) on p. 170 not $q\varphi_{\mathrm{AB}} = q\left(\varphi_{\mathrm{W,A}} - \varphi_{\mathrm{W,B}}\right)$?

6.2 The energy band diagram in Fig. 6.10(c) for a zero-biased p-n junction on p. 176 does not depict electrons, holes, or dopant

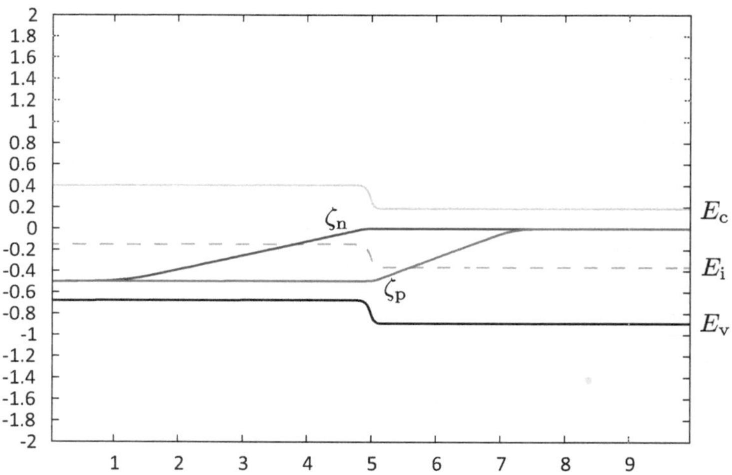


FIGURE 6.31 TCAD-drawn energy band diagram of a p-n junction diode. Lifetimes: $\tau_{\mathrm{n}} = \tau_{\mathrm{p}} = 10^{-11}$ s, Voltage bias: $V = 0.5$ V.

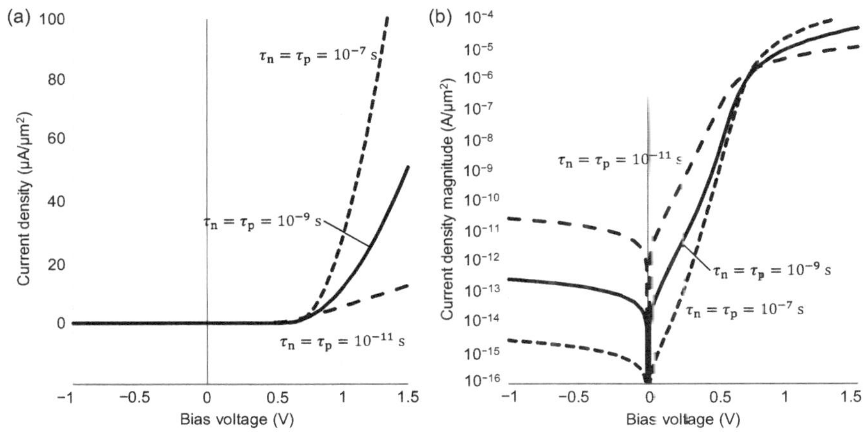


FIGURE 6.32 Lifetime dependence of p-n junction current-voltage characteristics.

atoms. Draw an energy band diagram including electrons, holes, and dopant atoms (both ionized and non-ionized ones).

6.3 The built-in potential of a p-n junction made of silicon is often somewhat below 1 V (see p. 177). Explain why, referring to Table 1.3.

6.4 Derive the depletion layer thickness equation (6.32) on p. 186 for a zero-biased p-n junction.

6.5 Comparing equation (6.44) on p. 189 for the depletion layer thickness $d_{\text{dep}}$ of a one-sided abrupt junction with the Debye length equation (5.108) on p. 158, the Debye length is equal to $d_{\text{dep}}$ when $\varphi_{\text{bi}} - V = kT/q$. On this basis, how can we understand the Debye length?

6.6 Read the energy band diagram shown in Fig. 6.19 on p. 195 and explain (or derive) the boundary condition (6.66) on p. 196.

6.7 An energy band diagram of a p-n junction with a relatively high reverse bias (more than a few hundred millivolts), satisfying Assumption 7 on p. 192 and the boundary conditions (6.65) on p. 195 and (6.66) at the edges of the depletion layer, is shown in Fig. 6.33 (see also Fig. 6.15 (p. 181), Box on p. 198, and Fig. 6.24 (p. 206)). Band diagrams looking similar to Fig. 6.33 can be found in several books [22, 25, 30, 31]. However, there actually is something qualitatively wrong with this band diagram. Read this

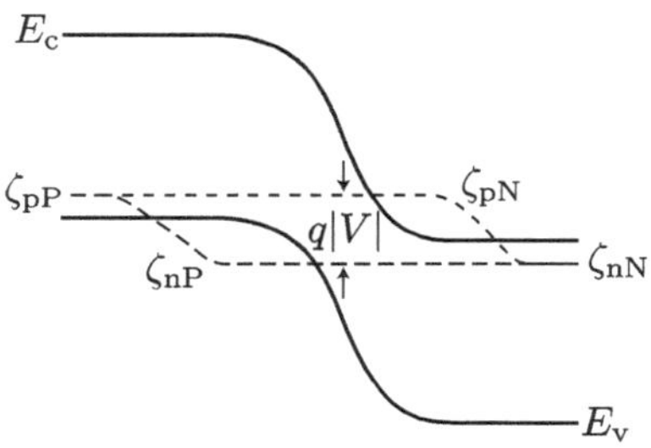


FIGURE 6.33 Energy band diagram of a reverse-biased p-n junction satisfying (6.65) and (6.66).

energy band diagram and consider what is wrong. Hints: p. 15, p. 119, and pp. 178–181.

6.8 Looking at Fig. 6.19 (p. 195) and Fig. 6.31 (p. 212), which are energy band diagrams at forward bias, the lengths of the slopes of $\zeta_n$ and $\zeta_p$ are different (i.e., the slope of $\zeta_p$ is steeper and shorter). Explain the reason.

6.9 Looking at the lifetime dependence of the current-voltage characteristics of the p-n junction diode (Fig. 6.32 on p. 213), the shorter the lifetime, the larger the current density at reverse bias. Read the corresponding energy band diagrams (Fig. 6.28 (p. 210) and Fig. 6.29 (p. 211)) and explain the reason.

CHAPTER 7

# MOS Transistors

MOS transistors contain a metal-oxide-semiconductor structure (MOS structure), which is a basic structure with a p-n junction. In this chapter, the physics of MOS transistors is studied and DC current-voltage characteristics are derived.

## 7.1 MOSFET STRUCTURE AND BASIC CHARACTERISTICS

### 7.1.1 Structure of MOSFETs

When we mentioned MOS transistors or MOSFETs (metal-oxide-semiconductor field-effect transistors) on p. 44, we described them as three-terminal devices. However, traditional MOSFETs, commonly known as *planar bulk MOSFETs*, are four-terminal devices as shown in Fig. 7.1. The *back gate* terminal may sometimes be omitted from the schematic symbol, as shown in Fig. 2.17 on p. 44. If it is omitted, the back gate of the nMOS transistor is connected to ground, and the back gate of the pMOS transistor is connected to a supply voltage.

Fig. 7.2 shows examples of MOSFET schematic symbols. Various other symbols are also used. Between the *drain* and the *source* is a nonlinear variable resistor, and the *gate* is the control terminal. In the middle symbol in Fig. 7.2, the source has an arrow indicating the direction of the current flow. Somewhat strangely, the left and the right symbols in Fig. 7.2 do not distinguish between drain and source per se, but drain and source are marked as such. Structurally, the drain and the source of a MOSFET for digital circuits are often fabricated exactly the same, and this fact is reflected in those symbols. Usually, connections to a power supply and a ground determine which is the drain and which is the source. In an nMOS transistor (nMOSFET), the DC current flows into the drain and exits from the source. In a pMOS transistor (pMOSFET), the DC current flows into the source and exits

DOI: 10.1201/ 9781003439417-7

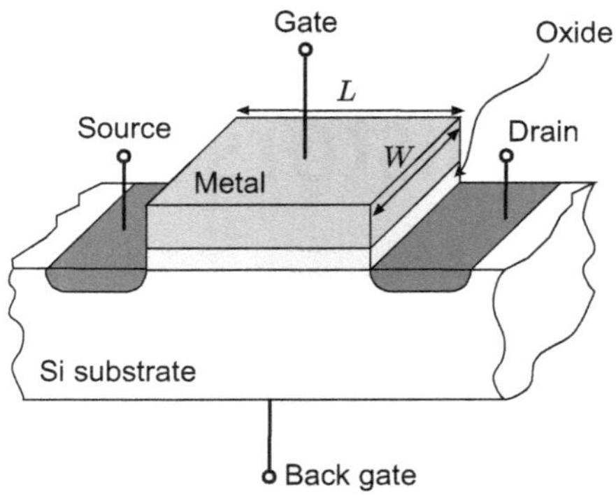


FIGURE 7.1 Structure of a planar bulk MOSFET.

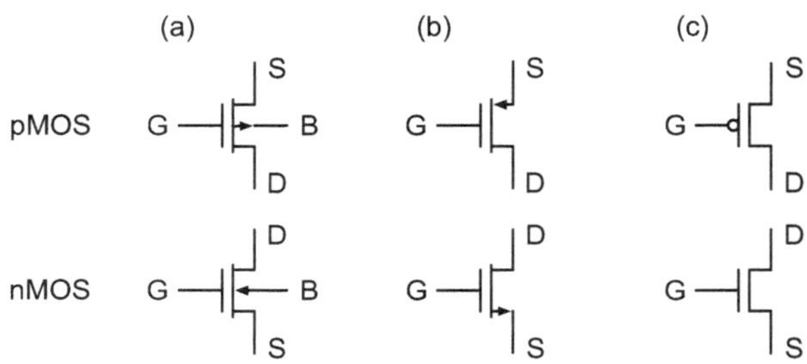


FIGURE 7.2 Schematic symbols of pMOSFET and nMOSFET. (a) Often used by analog circuit designers. (b) The back gate terminal is not drawn. (c) Often used by digital circuit designers.

from the drain. The schematic symbols in Fig. 7.2 are drawn on the assumption that the DC current passes from top to bottom. Actual circuit schematics are often drawn in such a way.

The term "MOS" comes from the *metal-oxide-semiconductor* structure contained in a MOSFET. Of these three materials, the material of the semiconductor *substrate* is usually silicon (Si). Regarding the oxide, silicon dioxide (SiO2), with a relative permittivity of about 3.9, had long been the only material used. But today, insulators (not necessarily oxides) with higher permittivities are also used. Therefore, it is technically more accurate to call it a *metal-insulator-semiconductor (MIS)* structure. The gate is made of metal material, but instead of metal, degenerately doped polycrystalline silicon (polysilicon), which exhibits properties similar to metal, is also used. It is common to use the term "MOS structure" even when the gate

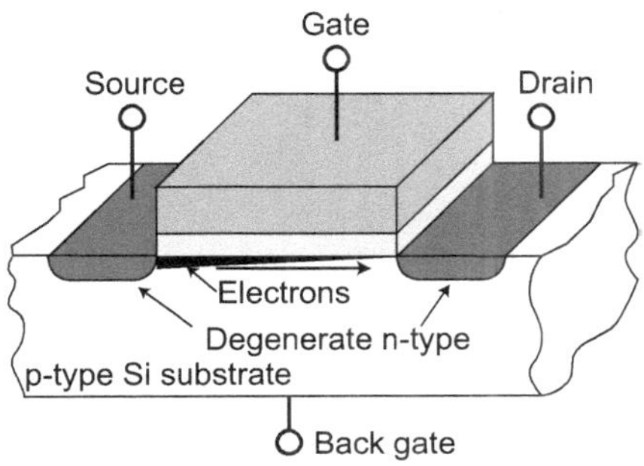


FIGURE 7.3 Structure of an nMOSFET.

insulator is not SiO2 and/or when the gate material is not metal. Fig. 7.1 shows the simplest type of *planar MOSFET*, but there are planar MOSFETs with more complex structures and also nonplanar MOSFETs.

The difference between an nMOSFET and a pMOSFET is the polarity of doping. In the nMOSFET, as shown in Fig. 7.3, two degenerate n-type regions are formed on a p-type silicon substrate and used as the source and drain. In the nMOSFET in an on state (p. 218), there is a high density of electrons just below the gate insulator, even though it is a p-type region, and these electrons are responsible for electrical conduction. This current path consisting of high-density carriers is called a *channel*. The current in an nMOSFET flows from drain to source. The black elongated triangle labeled as "Electrons" in Fig. 7.3 indicates that the density of electrons is high near the source and decreases toward the drain. On the other hand, the channel thickness is thin near the source and thickens as it approaches the drain [37].

A pMOSFET is the opposite, with two degenerate p-type regions formed in a wide n-type region as a source and drain, as shown in Fig. 7.4. The reason why we referred to the "wide n-type region" and not an "n-type substrate" is that this n-type region (or *n-well*) is usually formed in a p-type substrate. This allows nMOSFETs and pMOSFETs to coexist in a CMOS (complementary MOS) configuration (p. 44). A detailed discussion of device structures is beyond the scope of this book, so we will not go into it further. The channel is formed in the n-type region just below the gate insulator, and the current is carried by holes (in spite of being in the n-type region). The current in a pMOSFET flows from source to drain.

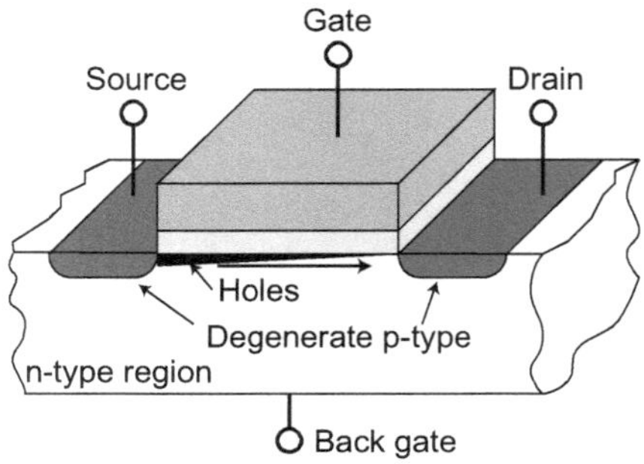


FIGURE 7.4 Structure of a pMOSFET.

The terms "source" and "drain" come from the fact that the former provides carriers to the channel and the latter serves as the drain of carriers. The arrows in the left schematic symbols in Fig. 7.2 (p. 216) indicate the direction of the forward current in the p-n junction formed between the substrate (or n-well) and the source (and drain) regions [22]. In fact, the same is true for the arrow in the middle schematic symbols in Fig. 7.2.

### 7.1.2 Basic Characteristics of MOSFETs

#### *7.1.2.1 nMOSFETs*

Fig. 7.5 shows how MOSFETs are usually biased. Fig. 2.18 (p. 45) showed the rough sketch of current-voltage characteristics of the nMOSFET. Its $I_{DS}$-$V_{DS}$ characteristics are shown in Fig. 7.6. The region where the drain current $I_{DS}$ increases nearly linearly with the drain-source voltage $V_{DS}$ is called the *linear region*, *triode region*, or *nonsaturation region*. The region where $I_{DS}$ does not depend on $V_{DS}$ is called the *pentode region* or *saturation region* (see Problem 7.1 on p. 264).

Fig. 7.7 shows the $I_{DS}$-$V_{GS}$ characteristics of nMOSFET when $V_{DS}$ is chosen to be in the saturation region. The value of $V_{GS}$, which is the boundary between the *on state* where current flows and the *off state* where little current flows, is called the *threshold voltage* of the MOSFET. Here, the threshold voltage is denoted by $V_T$. Note that the concept of threshold voltage is somewhat sloppy because some leakage current, also known as subthreshold current, flows even when the $V_{GS}$ is applied below the threshold value. The corresponding region

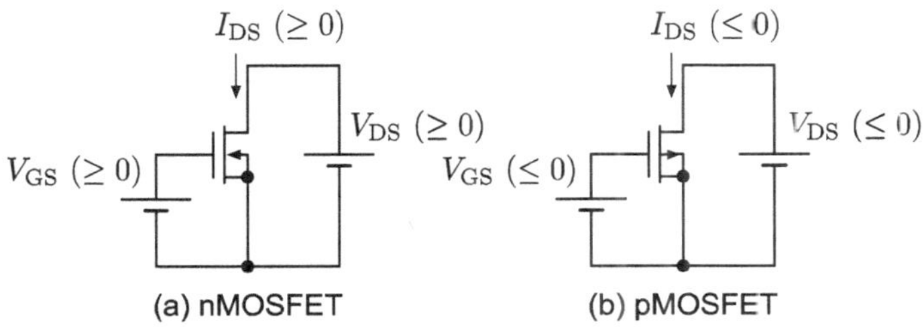


FIGURE 7.5 Biasing nMOSFET and pMOSFET.

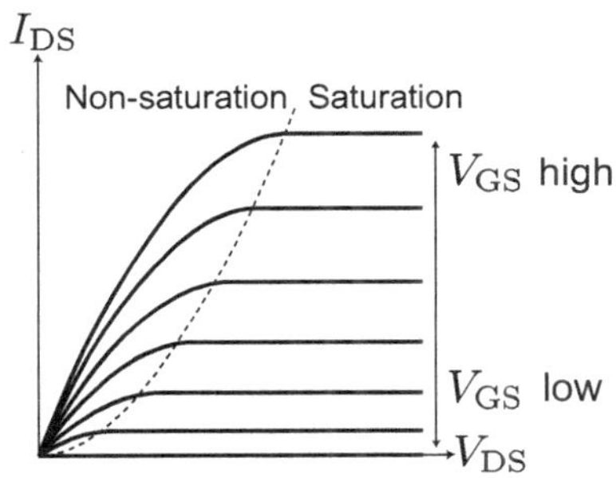


FIGURE 7.6 $I_{DS}$-$V_{DS}$ characteristics of nMOSFETs.

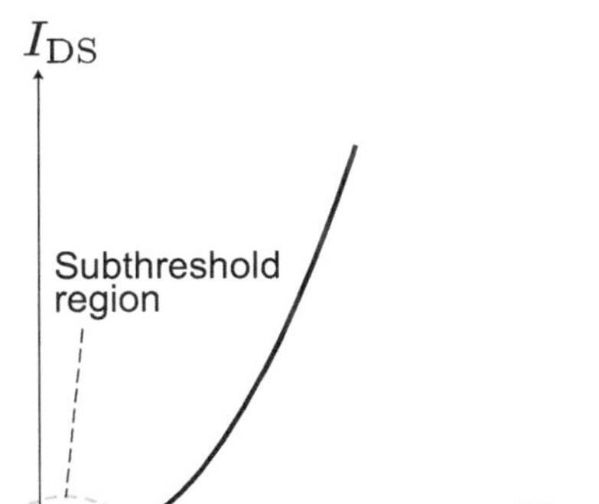


FIGURE 7.7 $I_{DS}$-$V_{GS}$ characteristics of nMOSFETs.

of operation is called the *subthreshold region*. To see the subthreshold currents, the vertical axis must be set to a log scale (see Problem 7.11 on p. 267).

Normal nMOSFETs are made so that the threshold voltage is positive ($V_T > 0$). A field-effect transistor (FET, p. 44) like this, which can be turned off by setting $V_{GS} = 0$, is called a *normally-off* FET. Conversely, a FET with a negative threshold voltage ($V_T < 0$) that remains on when $V_{GS} = 0$ is called a *normally-on* FET. Normally off MOSFETs are essential for implementing low-power digital circuits. Whether a FET is normally-off or normally-on (i.e., the sign of $V_T$) is related to the work function difference between the material used for the gate and the semiconductor substrate (§7.4).

The DC current-voltage characteristics in the nonsaturation region in Fig. 7.6 can be written as

$$I_{DS} = \frac{\mu_n W C_{ox}}{L}\left[(V_{GS} - V_T)V_{DS} - \frac{1}{2}V_{DS}^2\right] \quad (0 \le V_{DS} \le V_{GS} - V_T), \tag{7.1}$$

where $\mu_n$ is the electron mobility (§5.3.2). Actually, the value of $\mu_n$ in (7.1) is considerably smaller than that given in Table 1.3 (p. 5)—the bulk mobility. The reason is that the current flows at the surface of the substrate (interface with the gate insulator), where there is more carrier scattering due to surface roughness. This results in lower mobility than in the crystal far from the interface (i.e., bulk). $W$ in (7.1), shown in Fig. 7.1 (p. 216), is called the *gate width* or *channel width*. $L$, also shown in Fig. 7.1, is called the *gate length* or *channel length*. The symbols $W$ and $L$ are universally used in this field as symbols for the MOSFET gate width and gate length. $C_{ox}$ in (7.1) is the capacitance per unit area of the gate oxide. Since (7.1) is a quadratic function with respect to $V_{DS}$, the characteristics of the nonsaturation region in Fig. 7.6 (p. 219) are parabolic. A goal of this chapter is to understand the process of deriving (7.1) and the associated device physics.

The equation for the saturation region characteristic where the drain voltage is $V_{DS} \ge V_{GS} - V_T$ is given by (Problem 7.2 on p. 265)

$$I_{DSsat} = \frac{\mu_n W C_{ox}}{2L}(V_{GS} - V_T)^2 \quad (V_{DS} \ge V_{GS} - V_T). \tag{7.2}$$

The current in the saturation region (saturated drain current) $I_{DSsat}$ is independent of $V_{DS}$ (see Fig. 7.6).

#### *7.1.2.2 pMOSFETs*

The current and voltage of a pMOSFET are customarily defined to be negative as shown in Fig. 7.5 (p. 219). Actual current flows upward from source to drain. By defining the voltage and current of pMOSFETs in this way, their current-voltage characteristics are like Fig. 7.6 (p. 219) and Fig. 7.7 (p. 219) rotated 180° around the origin. However, even if the values of $W$ and $L$ are the same as those of an nMOSFET, the current in a pMOSFET is smaller than that in an nMOSFET because hole mobility is smaller than electron mobility (p. 136). This often necessitates that pMOS gate width $W$ be made larger than nMOS gate width (see (7.2)).

From now on, we will mainly consider nMOSFETs.

### 7.1.3 Outline of Analyzing MOSFETs

Analysis of the MOSFET operation is more difficult than commonly perceived. Modeling upon the lucid approach developed by [33, 34], we will first look at a simpler structure—the *MOS capacitor*—and increase the complexity in the following order.

1. Two-terminal MOS structure (MOS capacitor). Terminals: Gate and back gate.
2. Three-terminal MOS structure (Gated diode). Terminals: Gate, back gate, and drain.
3. Four-terminal MOSFET. Terminals: Gate, back gate, drain, and source.

## 7.2 MOS CAPACITOR

### 7.2.1 Structure of MOS Capacitors

A structure consisting of only the gate and the back gate of a MOSFET, as shown in Fig. 7.8, is called a *MOS capacitor* or *MOS diode*. "Diode" here means a two-terminal element. As noted in §7.1.1, the gate material may be metal or degenerately doped polycrystalline silicon (poly-Si), but hereafter it will be simply referred to as "metal" for simplicity. We will also refer to the gate insulator as the "gate oxide" or simply "oxide" regardless of the actual material type. The back side of the silicon substrate (opposite the gate) is assumed to be covered with the same metal as the gate, serving as an electrode. Since we want to analyze an nMOSFET, the silicon substrate is assumed to be p-type.

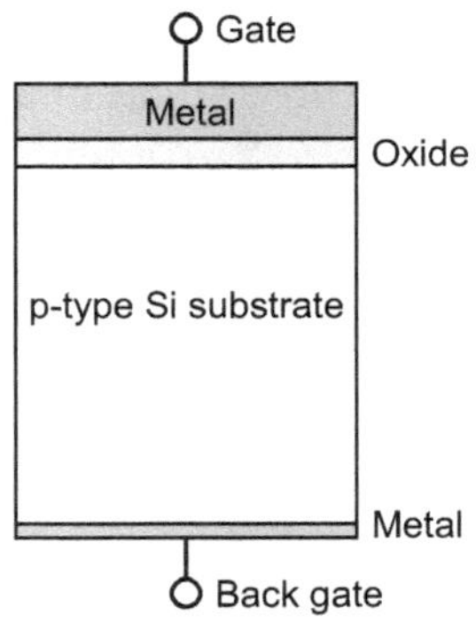


FIGURE 7.8 Structure of a MOS capacitor.

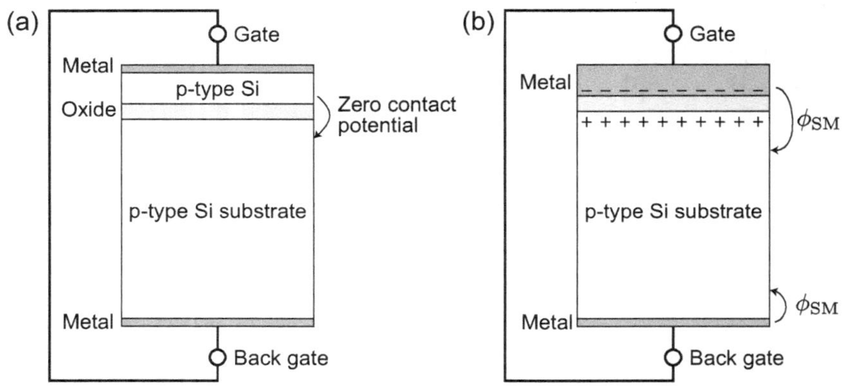


FIGURE 7.9 (a) A MOS capacitor built of a p-type silicon gate and a substrate. (b) A MOS capacitor built of a metal gate and a p-type silicon substrate.

## 7.2.2 Analysis of MOS Capacitors

### *7.2.2.1 A MOS Capacitor under Zero Bias*

First, as a simpler structure than that in Fig. 7.8, consider the case where the gate is also made of p-type silicon with the same dopant density as the substrate, as shown in Fig. 7.9(a). In this cartoon, a loop structure (p. 173) is formed using a conducting wire, and the entire structure is in thermal and diffusive equilibrium. Since the gate and substrate materials are the same, the contact potential between them is, of course, 0.

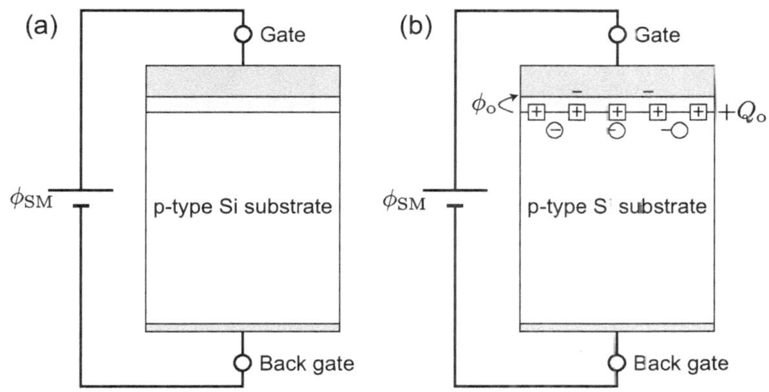


FIGURE 7.10 (a) When the gate voltage is increased by $\varphi_{SM}$, the charge stored in the MOS capacitor becomes zero. (b) In practice, the capacitor charge may not become zero due to the fixed charge $Q_o$ in the oxide or at the interface between the substrate and the oxide.

Let us now go back to the original structure (Fig. 7.8). Since the gate and substrate materials are different, a nonzero contact potential $\varphi_{SM}$ ($\neq 0$)is generated as shown in Fig. 7.9(b) (the subscript "SM" is from semiconductor-metal). Positive and negative charges then appear on one and the other side of the oxide layer, respectively. In other words, some charge is stored in the MOS capacitor. The gate and back gate electrodes are short-circuited by a conducting wire, but since two different materials form the capacitor, charge is stored due to the difference in the work functions of the two materials (see also Fig. 4.10 (p. 110) and Fig. 4.11 (p. 111)). The sign of $\varphi_{SM}$ and the polarity of charge stored on each side depend on the combination of materials. Fig. 7.9(b) is a cartoon for $\varphi_{SM} > 0$.

### 7.2.2.2 Flat-Band Condition

Next, let us consider canceling the charges that appeared on both sides of the gate oxide in Fig. 7.9(b). This can be done by connecting a voltage source and raising the gate voltage by $\varphi_{SM}$ (p. 111). Fig. 7.10(a) shows this situation.

In practice, the voltage required to cancel the charge stored in the MOS capacitor may deviate slightly from $\varphi_{SM}$. The reasons may be as follows:

- Fixed charge in the gate oxide.

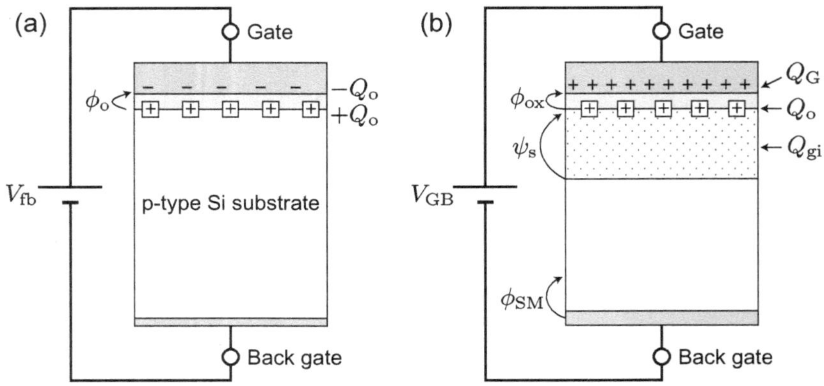


FIGURE 7.11 (a) The gate voltage at which the charge stored in the MOS capacitor just cancels out is the flat-band voltage $V_{fb}$. (b) A biased MOS capacitor.

- Charge trapped and fixed at the interface between the oxide and the silicon substrate (p. 144).

Assuming that all fixed charges are at the interface between the oxide and the substrate, let $Q_o$ be the amount of charge per unit area (the subscript "o" is from the offset charge). Fig. 7.10(b) shows a drawing with the assumption that $Q_o > 0$. Due to the charge neutrality condition of the system, a negative charge, $-Q_o$, is induced around the substrate surface, including the bottom surface of the gate. Let $\varphi_o$ be the voltage (electrostatic potential difference) across the oxide in this state.

$$\varphi_o = -\frac{Q_o}{C_{ox}}. \qquad \text{(Voltage across gate oxide due to} Q_o\text{)} \qquad (7.3)$$

Equation (7.3) has a minus sign on the right-hand side because the voltage here is substrate-referenced. To cancel $Q_o$, the bias voltage must be changed from $\varphi_{SM}$. The gate voltage required to cancel, as shown in Fig. 7.11(a), both the contact potential $\varphi_{SM}$ and the potential difference $\varphi_o$ due to the fixed charge is called the *flat-band voltage* $V_{fb}$.

$$V_{fb} \equiv \varphi_{SM} + \varphi_o = \varphi_{SM} - \frac{Q_o}{C_{ox}}. \qquad \text{(Flat-band voltage)} \qquad (7.4)$$

#### *7.2.2.3 A MOS capacitor under General Gate Bias*

Suppose now that a certain gate voltage $V_{GB}$ is applied without limiting it to a specific value. This situation is depicted in Fig. 7.11(b). In this cartoon, $V_{GB} > V_{fb}$ is assumed, and the charge stored in the gate (per unit area), $Q_G$, is drawn as being positive. $Q_{gi}$ in the cartoon is the charge per unit area (depletion charge, etc.) induced in the silicon substrate by the gate voltage (the subscript "gi" is from gate-induced).

From the charge neutrality condition of the system, we have

$$Q_G + Q_o + Q_{gi} = 0. \qquad \text{(Charge neutrality condition)} \qquad (7.5)$$

Also, since the sum of the electrostatic potential differences (i.e., contact potentials) around a loop is zero (p. 173),

$$V_{GB} = \varphi_{ox} + \psi_s + \varphi_{SM} \qquad \text{(Potential balance equation)} \qquad (7.6)$$

holds, where $\psi_s$ is the *surface potential*, that is, the electrostatic potential at the surface of the silicon substrate. $\varphi_{ox}$ is the voltage (i.e., electrostatic potential) across the oxide. The *datum point* for the electrostatic potential is the region deep in the silicon substrate that is neutral and has no potential gradient. In other words, it is the region below the starting point of the arrow showing $\psi_s$ in Fig. 7.11(b).

If $Q_o$ takes a fixed value independent of the gate voltage $V_{GB}$, then the following equation holds for the changes in $Q_G$ and $Q_{gi}$, that is, $\Delta Q_G$ and $\Delta Q_{gi}$:

$$\Delta Q_G + \Delta Q_{gi} = 0. \qquad (7.7)$$

Therefore, when $V_{GB}$ is varied, $\Delta Q_G$ and $\Delta Q_{gi}$ change by the same absolute value with opposite signs. In other words, the gate voltage can be used to control the amount of charge induced near the surface of the silicon substrate.

### 7.2.3 Classification of Surface Conditions of MOS Capacitors

The surface condition of the silicon substrate of the MOS capacitor depends on the applied gate voltage $V_{GB}$. In the following, we will classify the conditions of the substrate surface using energy band diagrams. However, for the purpose of simplifying the energy band diagrams, we assume that there is no work function difference (or contact potential difference) between the gate metal and the silicon

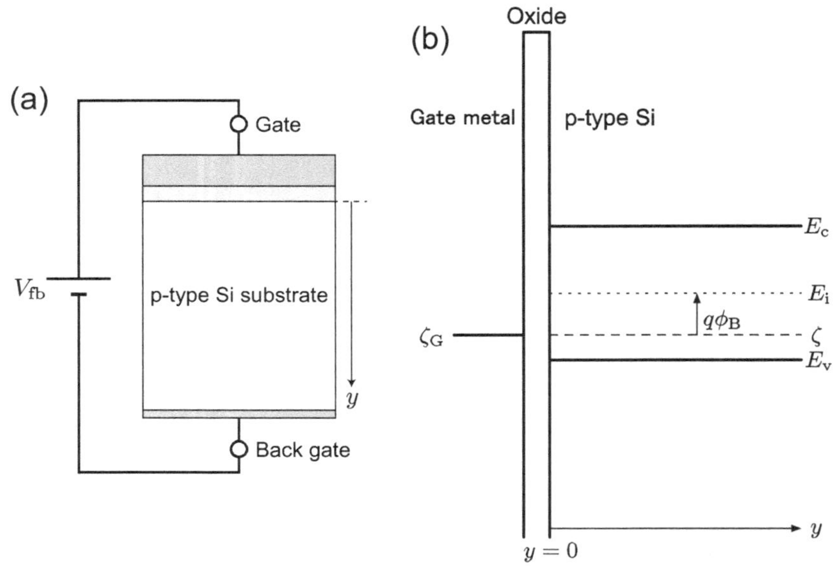


FIGURE 7.12 A MOS capacitor in a flat-band condition.

substrate ($\varphi_{SM} = 0$) and that no fixed charge exists on the oxide surface ($Q_o = 0$). Then, the Fermi level $\zeta_G$ of the gate metal coincides with the Fermi level $\zeta$ of the silicon substrate. From (7.4) we obtain $V_{fb} = 0$. However, since the purpose is to make the band diagram simple and easy to read, we will not assume $\varphi_{SM} = 0$ and $Q_o = 0$ in equations, and $V_{fb}$ will be left as it is (see Problem 7.4 on p. 265).

### 7.2.3.1 Flat Band

The state realized when the gate voltage $V_{GB}$ equals the flat-band voltage $V_{fb}$, given by (7.4) on p. 224, is called the *flat-band condition* or simply *flat band*. In this condition, there is no induced charge in the semiconductor ($Q_{gi} = 0$) as shown in Fig. 7.12, and therefore, by Gauss' law, the energy band is completely flat (no band bending).

$\varphi_B$ in the energy band diagram of Fig. 7.12 is the bulk potential of (6.4) on p. 226. Since the oxide is an insulator, only the top of the forbidden band is drawn in the band diagram for it (see Fig. 4.3 on p. 91). Note, however, that for the convenience of drawing, this energy is written at a much lower position than it should be. Let us set the constant term of (4.23) on p. 98 to 0, and map $E_i$ to $\psi$. In Fig. 7.12, $\psi = 0$ at the right end of the band diagram. Let $p_s$ and $n_s$ denote the hole density and the electron density at the surface of the silicon substrate, respectively.

$$Q_{gi} = 0 \qquad \text{(Induced charge in flat band)} \qquad (7.8)$$

$$\psi_s = 0 \qquad \text{(Surface potential in flat band)} \qquad (7.9)$$

$$p_s = p_{P0} \gg n_i \qquad \text{(Surface hole density in flat band)} \qquad (7.10)$$

$$n_s = n_{P0} \ll n_i \qquad \text{(Surface electron density in flat band)} \qquad (7.11)$$

Here $p_{P0}$ is the hole density in equilibrium p-type semiconductors without band bending, and $n_{P0}$ is the electron density in equilibrium p-type semiconductors without band bending.

#### *7.2.3.2 Accumulation*

When the gate voltage is lower than the flat-band voltage ($V_{GB} < V_{fb}$), the gate is negatively charged, and more majority carriers (holes in this case) accumulate on the p-type silicon substrate surface than in the flat-band condition, as shown in Fig. 7.13. Such a condition, where majority carriers have accumulated more than in the flat-band condition, is called *accumulation*. As can be seen from the band diagram in Fig. 7.13, the surface hole density increases as $E_v$ approaches $\zeta$ at the substrate surface (see (4.17) on p. 95). The surface hole density in this case can be written as (7.14). The lines of electric force emitted from the holes on the substrate surface are terminated by negative charges in the gate. In the cartoon on the left of Fig. 7.13, the negative charges on the gate and the positive charges in the silicon substrate are equal in number, representing the charge neutrality condition (likewise in Figs. 7.14–7.16).

$$Q_{gi} > 0 \qquad \text{(Induced charge in accumulation)} \qquad (7.12)$$

$$\psi_s < 0 \qquad \text{(Surface potential in accumulation)} \qquad (7.13)$$

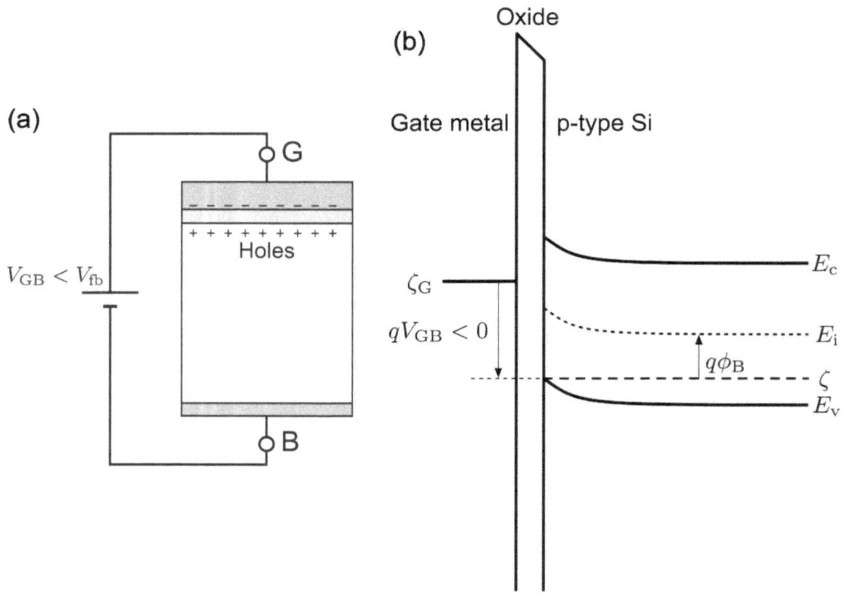


FIGURE 7.13 A MOS capacitor in an accumulation condition.

$$p_s > p_{P0} \quad \text{(Surface hole density in accumulation)} \tag{7.14}$$

$$n_s = \frac{n_i^2}{p_s} < n_{P0} \quad \text{(Surface electron density in accumulation)} \tag{7.15}$$

The reason why $Q_{gi} > 0$ in (7.12) is that there are more holes at the substrate surface than in the flat-band condition. Equations (7.14) and (7.15) can be confirmed by reading the surface carrier densities from the band diagram in Fig. 7.13 (see p. 228).

### *7.2.3.3 Depletion*

When the gate voltage is set higher than the flat-band voltage ($V_{GB} > V_{fb}$), the gate is positively charged, and as shown in Fig. 7.14, majority carriers (holes) on the substrate surface are driven away, creating a depletion layer. This condition is called *depletion*. The surface hole density is reduced because $E_v$ has gone far away from the Fermi level

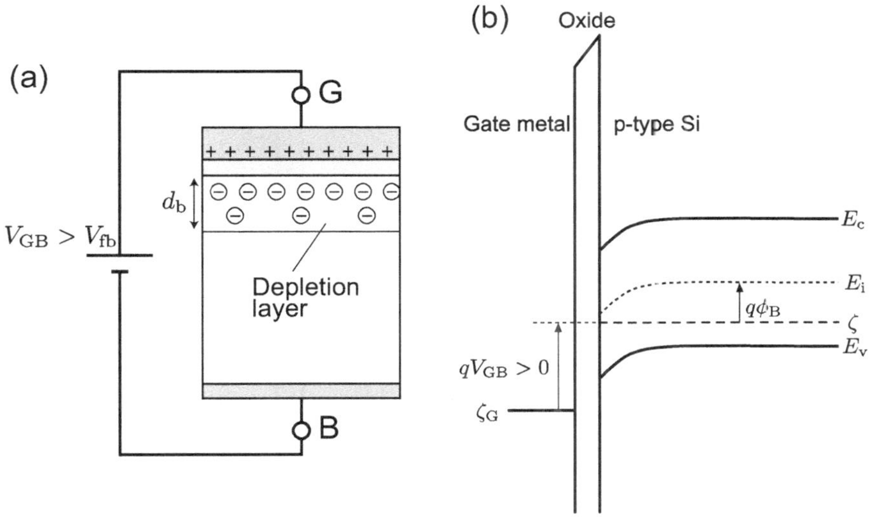


FIGURE 7.14 A MOS capacitor in a depletion condition.

$\zeta$ at the substrate surface. The lines of electric force emitted from the positive charges on the gate surface are terminated by depletion charges (acceptor ions) in the substrate. The formula for surface hole density, from which the name "depletion" is derived, is (7.18). In the cartoon of Fig. 7.14, the depletion layer thickness is denoted by $d_b$ (the subscript "b" is from body, which means substrate).

$$Q_{gi} < 0 \qquad \text{(Induced charge in depletion)} \qquad (7.16)$$

$$0 < \psi_s \leq \varphi_B \qquad \text{(Surface potential in depletion)} \qquad (7.17)$$

$$n_i < p_s < p_{P0} \qquad \text{(Surface hole density in depletion)} \qquad (7.18)$$

$$n_s = \frac{n_i^2}{p_s} > n_{P0} \qquad \text{(Surface electron density in depletion)} \qquad (7.19)$$

The negative charge in (7.16) is due to acceptor ions, which are now exposed because of the absence of holes near the substrate surface. Read the band diagram in Fig. 7.14 and confirm that the surface carrier densities are as given by (7.18) and (7.19).

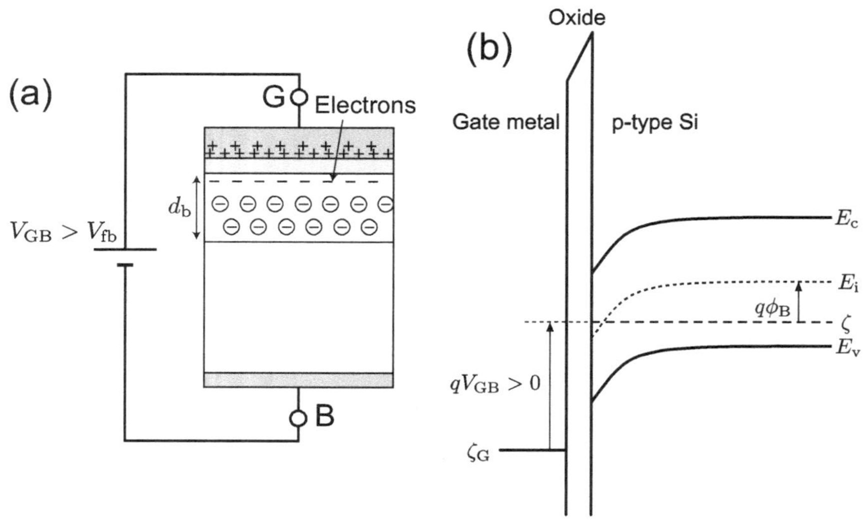


FIGURE 7.15 MOS capacitor in the weak inversion condition.

### *7.2.3.4 Weak Inversion*

$V_{GB} > V_{fb}$ as in depletion, but at higher gate voltages, the gate is more strongly positively charged. This not only strongly bends the band near the substrate surface downwards, driving away holes and expanding the depletion layer, but also induces electrons, so that the surface electron density $n_s$ exceeds the intrinsic carrier density $n_i$. As shown in the energy band diagram in Fig. 7.15, the substrate surface is effectively n-type because $\zeta > E_i$ at the substrate surface (see (4.99) on p. 99). This is the condition called *weak inversion*. The lines of electric force from the positive charges on the gate surface are terminated in the depletion charges and electrons in the substrate. The surface electron density, from which the name "weak inversion" is derived, is given by (7.23).

$$Q_{gi} < 0 \qquad \text{(Induced charge in weak inversion)} \qquad (7.20)$$

$$\varphi_B \leq \psi_s \leq 2\varphi_B \qquad \text{(Surface potential in weak inversion)} \qquad (7.21)$$

$$p_s < n_i \qquad \text{(Surface hole density in weak inversion)} \qquad (7.22)$$

$$n_s = \frac{n_i^2}{p_s} > n_i \quad \text{(Surface electron density in weak inversion)} \quad (7.23)$$

The negative charge in (7.20) is due to the acceptor ions and electrons. Read and check the surface carrier densities (7.22) and (7.23) from the band diagram in Fig. 7.15.

In the traditional treatment, weak inversion was included in depletion, but later it was recognized that weak inversion was important for understanding the subthreshold characteristics of MOSFETs (p. 219), and it is now treated separately from depletion.

#### *7.2.3.5 Strong Inversion*

If the gate voltage is increased beyond that in the weak inversion condition, *strong inversion* is reached, as shown in Fig. 7.16. The surface electron density $n_s$ at strong inversion even exceeds the majority carrier density (i.e., hole density) $p_{P0}$ deep in the substrate where the energy band is flat. The surface electron density can be written as (7.27). A very thin minority-carrier layer formed at the substrate surface by strong inversion is called an *inversion layer*. The lines of electric force emitted from the positive charges on the gate surface are terminated by the depletion charges in the substrate and the electrons constituting the inversion layer. Once strong inversion is reached, a further increase in $V_{GB}$ does not affect the band bending very much. This is because a further increase in $V_{GB}$ only induces further inversion charges (i.e., electrons) at the substrate surface, and not many more depletion charges deeper in the substrate. Thus, the depletion layer thickness $d_b$ no longer depends very much on the gate voltage $V_{GB}$ (see Problem 7.3 on p. 265).

$$Q_{gi} < 0 \quad \text{(Induced charge in strong inversion)} \quad (7.24)$$

$$2\varphi_B \le \psi_s \quad \text{(Surface potential in strong inversion)} \quad (7.25)$$

$$p_s \le n_{P0} \ll n_i \quad \text{(Surface hole density in strong inversion)} \quad (7.26)$$

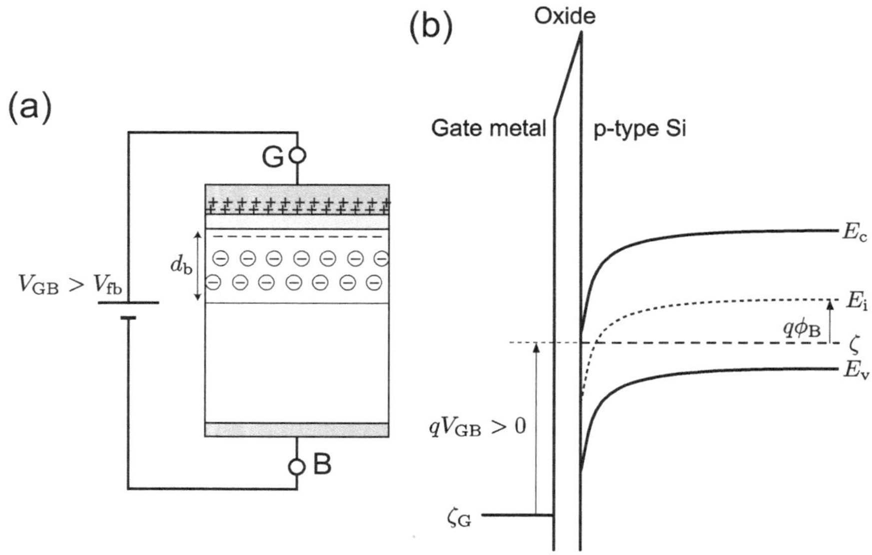


FIGURE 7.16 A MOS capacitor in a strong inversion condition.

$$n_s = \frac{n_i^2}{p_s} \geq p_{P0} \gg n_i \quad \text{(Surface electron density in strong inversion)} \tag{7.27}$$

In the traditional treatment, strong inversion was simply called *inversion*. Fig. 7.17 shows the relationship between surface electron density $n_s$ and surface potential $\psi_s$. In the parentheses are the traditional terms. $n_s = p_s = n_i$ when $\psi_s = \varphi_B$.

### 7.2.4 Surface Electron Density and Surface Potential

Here we discuss the surface conditions of MOS capacitors considered in §7.2.3 in relation to the general form of carrier density expressions discussed in §5.2.5. The channel of a MOSFET is an inversion layer formed at the substrate surface of the MOS capacitor. So, here we write the surface electron density $n_s$ of a p-type MOS capacitor in the form of (5.23) on p. 128.

Using the bulk potential $\varphi_B$ defined in (6.4) on p. 170 instead of the Fermi level $\zeta$, the surface electron density $n_s$ can be expressed in

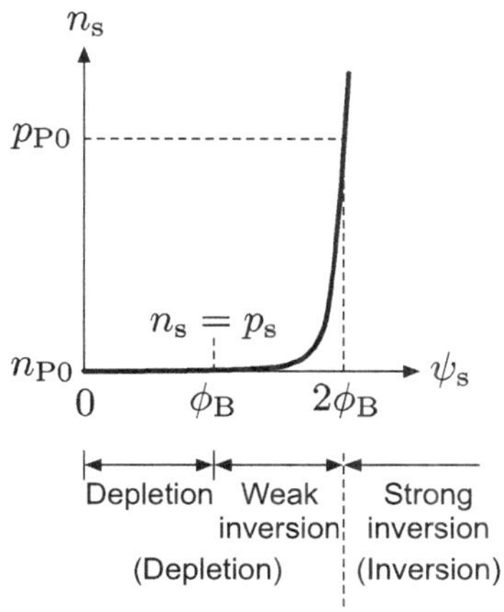


FIGURE 7.17 The surface electron density of a MOS capacitor with a p-type substrate.

different ways as follows:

$$n_s = n_{P0} \exp\left(\frac{q\psi_s}{kT}\right) \qquad \text{(Flat-band referenced)} \qquad (7.28)$$

$$= n_i \exp\left[\frac{q(\psi_s - \varphi_B)}{kT}\right] \qquad \text{(Weak-inversion-onset referenced)} \qquad (7.29)$$

$$= p_{P0} \exp\left[\frac{q(\psi_s - 2\varphi_B)}{kT}\right] \qquad \text{(Strong-inversion-onset referenced)} \qquad (7.30)$$

In the flat-band condition, $\psi_s = 0$ from (7.9) on p. 227, so (7.28) follows immediately from (7.11). Equation (7.29), which is based on the condition of the onset of weak inversion (the left equality in (7.21) on p. 230 holds), is the same as (4.27) on p. 99. Putting the surface potential at the onset of strong inversion (the equality in (7.25) holds) into (7.30), we obtain the surface electron density at the onset of strong inversion (the equality in (7.27) holds). The above should be understood in conjunction with Table 7.1, which summarizes the results of §7.2.3. In Table 7.1, the "boxed" entries indicate the "boundaries" where the values of the exponents in (7.28) through (7.30) become 0 when the equality in the table holds.

TABLE 7.1 Surface Potential and Surface Carrier Densities of a MOS Capacitor with a p-Type Substrate

| | Accumu. | Flat-band | Depletion | Weak inv. | Strong inv. |
|---|---|---|---|---|---|
| Surface potential $\psi_s$ | $< 0$<br>$< \varphi_B$<br>$< 2\varphi_B$ | $= 0$<br>$< \varphi_B$<br>$< 2\varphi_B$ | $> 0$<br>$\leq \varphi_B$<br>$< 2\varphi_B$ | $> 0$<br>$\geq \varphi_B$ $(> 0)$<br>$\leq 2\varphi_B$ | $> 0$<br>$> \varphi_B$<br>$\geq 2\varphi_B$ |
| Surface hole density $p_s$ | $> p_{P0}$<br>$> n_i$<br>$> n_{P0}$ | $= p_{P0}$<br>$> n_i$<br>$> n_{P0}$ | $< p_{P0}$<br>$\geq n_i$<br>$> n_{P0}$ | $< p_{P0}$<br>$\leq n_i$<br>$\geq n_{P0}$ | $< p_{P0}$<br>$< n_i$<br>$\leq n_{P0}$ |
| Surface electron density. $n_s$ | $< n_{P0}$<br>$< n_i$<br>$< p_{P0}$ | $= n_{P0}$<br>$< n_i$<br>$< p_{P0}$ | $> n_{P0}$<br>$\leq n_i$<br>$< p_{P0}$ | $> n_{P0}$<br>$\geq n_i$<br>$\leq p_{P0}$ | $> n_{P0}$<br>$> n_i$<br>$\geq p_{P0}$ |

### 7.2.5 Relation between Gate Voltage and Inversion Charge

Since inversion charge (electrons in nMOS and holes in pMOS) is responsible for electrical conduction in MOSFETs, let us look at the relationship between the gate voltage $V_{GB}$, which is the control voltage, and inversion charge density. If the surface conditions to be considered are limited to weak inversion and strong inversion, the charge $Q_{gi}$ induced on the silicon substrate surface consists of the inversion charge $Q_{inv}$ and the depletion charge $Q_b$ (the subscript "b" is from body, which means substrate).

$$Q_{gi} = Q_{inv} + Q_b. \qquad \text{(Per-unit-area induced charge at the surface)} \tag{7.31}$$

Therefore, the goal here is to find the relationship between $Q_{inv}$ and $V_{GB}$.

First, the per-unit-area gate charge $Q_G$ can be written using $\varphi_{ox}$ in Fig. 7.11(b) (p. 224) as follows:

$$Q_G = C_{ox}\varphi_{ox}. \qquad \text{(Per-unit-area gate charge)} \tag{7.32}$$

Putting (7.32) into (7.5) on p. 225 yields

$$C_{ox}\varphi_{ox} + Q_o + Q_{gi} = 0. \tag{7.33}$$

Eliminating $\varphi_{ox}$ using equation (7.6) yields

$$C_{ox}(V_{GB} - \psi_s - \varphi_{SM}) + Q_o + Q_{gi} = 0. \tag{7.34}$$

Rewriting (7.34) using the flat-band voltage $V_{fb}$ defined in (7.4) on p. 224, we obtain

$$C_{ox}(V_{GB} - \psi_s - V_{fb}) + Q_{gi} = 0. \tag{7.35}$$

Putting (7.31) into (7.35) and solving for $Q_{inv}$ yields

$$Q_{inv} = -C_{ox}\left(V_{GB} - V_{fb} - \psi_s + \frac{Q_b}{C_{ox}}\right). \tag{7.36}$$

This is the relationship between inversion charge $Q_{inv}$ and the gate voltage $V_{GB}$.

We will now apply the *depletion approximation* (p. 182) to write down the equation for the depletion charge $Q_b$. Then the depletion layer thickness at the p-type substrate surface is given by

$$d_b = \sqrt{\frac{2\epsilon_{Si}\psi_s}{qN_A^-}}. \tag{7.37}$$

Equation (7.37) can be derived in the same way as we considered the depletion layer of the abrupt junction in Fig. 6.17 (p. 183). Equation (7.37) is also similar to the depletion layer thickness equation, (6.44) on p. 189, for a one-sided abrupt junction. Using (7.37), the depletion charge per unit area can be written as

$$Q_b = -qN_A^- d_b = -\sqrt{2q\epsilon_{Si}N_A^-\psi_s}. \tag{7.38}$$

Eliminating $Q_b$ in (7.36) using (7.38), we obtain

$$Q_{inv} = -C_{ox}\left(V_{GB} - V_{fb} - \psi_s - \frac{\sqrt{2q\epsilon_{Si}N_A^-\psi_s}}{C_{ox}}\right) \tag{7.39}$$

$$= -C_{ox}\left(V_{GB} - V_{fb} - \psi_s - \gamma\sqrt{\psi_s}\right), \tag{7.40}$$

where we introduced the *body-effect coefficient* $\gamma$.

$$\gamma \equiv \frac{\sqrt{2q\epsilon_{Si}N_A^-}}{C_{ox}}. \qquad \text{(Body-effect coefficient)} \tag{7.41}$$

Here, "body" refers to the silicon substrate, and specifically, "body effect" refers to the effect originating from the dopant ions in the substrate (acceptor ions, in this p-type substrate case). Thus, if we know the surface potential $\psi_s$ when the gate voltage $V_{GB}$ is given, then we can find the inversion charge $Q_{inv}$ from (7.40). So, we next need to find $\psi_s$ from $V_{GB}$.

## 7.2.6 Relation between Gate Voltage and Surface Potential

### *7.2.6.1 Approximation for strong Inversion*

As mentioned on p. 231, the depletion layer thickness depends only weakly on the gate voltage and takes a nearly fixed value in the strong inversion condition, so the surface potential takes a nearly fixed value as well. Therefore, we can approximate the surface potential as follows in strong inversion:

$$\psi_s \simeq \psi_{sT}, \quad \text{(Fixed-value approximation of surface potential)} \tag{7.42}$$

where

$$\psi_{sT} \equiv 2\varphi_B + 3\varphi_{th}, \qquad \text{(Approximate surface potential)} \tag{7.43}$$

and

$$\varphi_{th} \equiv \frac{kT}{q} \qquad \text{(Thermal voltage)} \tag{7.44}$$

$$\simeq 26\text{mV} \qquad (\text{Thermal voltage at } T = 300\,\text{K}) \tag{7.45}$$

is the *thermal voltage* (see Problem 1.3 on p. 26). Equation (7.43) has a slightly larger value than the condition for the onset of strong inversion in (7.25) on p. 231. Coefficient 3 in the second term of (7.43) is an approximate value, so an appropriate value should be chosen depending on the situation.

Under this approximation, the inversion charge equation (7.40) becomes

$$Q_{inv} = \begin{cases} -C_{ox}(V_{GB} - V_{T0}) & (V_{GB} \geq V_{T0}) \\ 0 & (V_{GB} < V_{T0}) \end{cases}, \tag{7.46}$$

where $V_{T0}$ is the *threshold voltage* of the MOS capacitor.

$$V_{T0} \equiv V_{fb} + \psi_{sT} + \gamma\sqrt{\psi_{sT}}. \qquad \text{(MOS capacitor threshold voltage)} \tag{7.47}$$

$V_{T0}$ is somewhat different from the threshold voltage $V_T$ that appeared in Fig. 7.7 (p. 219). Note that (7.46) assumes that $Q_{inv} = 0$ at the threshold (around the boundary between strong inversion and weak inversion), and electrons in the weak inversion condition are neglected due to the second line of (7.46).

### 7.2.6.2 General case

The surface potential $\psi_s$ cannot be obtained analytically for the general case, not limited to strong inversion. However, it is possible to derive an equation (not a differential or integral equation) that can be used to obtain $\psi_s$ numerically using a computer. The starting point for the derivation is the Poisson equation (p. 153). Suppose that the $y$-axis is oriented toward the depth of the substrate (Fig. 7.12 on p. 226) and that there are no donors. Then the Poisson equation is given by

$$\frac{d^2\psi(y)}{dy^2} = -\frac{q}{\epsilon_{Si}}[p(y) - n(y) - N_A^-]. \qquad \text{(Poisson equation)} \tag{7.48}$$

Note that the acceptor ions are assumed to be uniformly distributed in the substrate. The carrier densities at depth $y$ are given by

$$p(y) = p_{P0} \exp\left[-\frac{q\psi(y)}{kT}\right], \qquad \text{(Hole density)} \tag{7.49}$$

$$n(y) = n_{P0} \exp\left[\frac{q\psi(y)}{kT}\right] = \frac{n_i^2}{p_{P0}} \exp\left[\frac{q\psi(y)}{kT}\right]. \qquad \text{(Electron density)} \tag{7.50}$$

Equation (7.50) is the same as (7.28) on p. 233. Equation (7.49) can also be understood in a similar manner. Also, using the charge neutrality condition deep in the substrate (or in the flat-band condition), the acceptor ion density can be written as

$$N_A^- = p_{P0} - n_{P0} = p_{P0} - \frac{n_i^2}{p_{P0}}. \qquad \text{(Acceptor ion density)} \tag{7.51}$$

The Poisson equation (7.48) can be rewritten using (7.49) through (7.51) as follows.

$$\begin{aligned}\frac{d^2\psi}{dy^2} &= -\frac{q}{\epsilon_{Si}}\left(p_{P0}e^{-q\psi/kT} - \frac{n_i^2}{p_{P0}}e^{q\psi/kT} - p_{P0} + \frac{n_i^2}{p_{P0}}\right)\\ &= -\frac{q}{\epsilon_{Si}}\left[p_{P0}\left(e^{-q\psi/kT} - 1\right) - \frac{n_i^2}{p_{P0}}\left(e^{q\psi/kT} - 1\right)\right]\\ &\approx -\frac{q}{\epsilon_{Si}}\left[N_A^-\left(e^{-q\psi/kT} - 1\right) - \frac{n_i^2}{N_A^-}\left(e^{q\psi/kT} - 1\right)\right]. \qquad (7.52)\end{aligned}$$

This differential equation can be analytically integrated once by transforming the variables using the electrostatic field $\mathcal{E} = -d\psi/dy$ as follows:

$$\frac{d^2\psi}{dy^2} = \frac{d}{dy}\frac{d\psi}{dy} = \left(\frac{d}{d\psi}\frac{d\psi}{dy}\right)\frac{d\psi}{dy} = \frac{d(-\mathcal{E})}{d\psi}(-\mathcal{E}) = \mathcal{E}\frac{d\mathcal{E}}{d\psi} \qquad (7.53)$$

Putting (7.53) into (7.52), the variable of integration on the left-hand side changes to $\mathcal{E}$ and that on the right-hand side to $\psi$. Then, we can integrate the resulting equation from a point deep in the substrate ($y = \infty$, $\psi = 0$) to the surface of the substrate ($y = 0$, $\psi = \psi_s$) as follows:

$$\int_0^{\mathcal{E}_s}\mathcal{E}d\mathcal{E} = -\frac{q}{\epsilon_{Si}}\int_0^{\psi_s}\left[N_A^-\left(e^{-q\psi/kT} - 1\right) - \frac{n_i^2}{N_A^-}\left(e^{q\psi/kT} - 1\right)\right]d\psi, \quad (7.54)$$

where $\mathcal{E}_s$ is the electric field at the substrate surface. The result of the integration is as follows.

$$\begin{aligned}\frac{\mathcal{E}_s^2}{2} &= -\frac{qN_A^-}{\epsilon_{Si}}\left[-\frac{kT}{q}(e^{-q\psi_s/kT} - 1) - \psi_s\right.\\ &\quad\left.-\left(\frac{n_i}{N_A^-}\right)^2\frac{kT}{q}(e^{q\psi_s/kT} - 1 - \frac{q\psi_s}{kT})\right]\\ &= \frac{kTN_A^-}{\epsilon_{Si}}\left[(e^{-q\psi_s/kT} - 1 + \frac{q\psi_s}{kT})\right.\\ &\quad\left.+\left(\frac{n_i}{N_A^-}\right)^2(e^{q\psi_s/kT} - 1 - \frac{q\psi_s}{kT})\right]. \qquad (7.55)\end{aligned}$$

We now apply Gauss' law to the induced charge $Q_{gi}$ at the substrate surface (Fig. 7.11(b) on p. 224). Since there is no transverse electric field (perpendicular to the $y$ axis), we only need to integrate over the upper and lower planes. The electric field at the surface ($y = 0$) is $\mathcal{E}_s$, and that deep in the substrate ($y = \infty$) is 0. Thus, the induced charge is given by

$$Q_{gi} = -\epsilon_{Si}\mathcal{E}_s \qquad \text{(Per-unit-area induced charge)}$$

$$= \pm\sqrt{2\epsilon_{Si}kTN_A^-}\left[\left(e^{-q\psi_s/kT} - 1 + \frac{q\psi_s}{kT}\right) + \left(\frac{n_i}{N_A^-}\right)^2\left(e^{q\psi_s/kT} - 1 - \frac{q\psi_s}{kT}\right)\right]^{1/2}. \qquad (7.56)$$

The minus sign of the double sign in (7.56) corresponds to depletion and inversion, and the plus sign to accumulation. Putting this $Q_{gi}$ in (7.35) on p. 235 gives the equation for numerically finding $\psi_s$ for a given $V_{GB}$ (see Problem 7.5 on p. 266).[1] Putting the surface potential $\psi_s$ thus obtained into (7.40) on p. 235, we can finally obtain the inversion charge $Q_{inv}$.

We have explained the flow of the calculation of the surface potential $\psi_s$ and the inversion charge $Q_{inv}$, but it might have been difficult to understand what was being done because of the mathematical technicalities involved. Let us take a look back at what we did.

The Poisson equation (7.48) is a second-order differential equation, and finding its solution means that the electrostatic potential $\psi(y)$ is obtained as a function of $y$. However, here we actually did not need $\psi(y)$. What we needed was only $\psi(0) = \psi_s$. So we changed the variables of integration using (7.53), evaluated integrals, and obtained the equation for finding $\psi_s$. Thus, we did not solve the Poisson equation in the literal sense. We obtained the information we needed without solving the given differential equation.

## METAL-SEMICONDUCTOR CONTACTS

One important topic that should have been covered but is omitted from this book is metal-semiconductor contacts. Metal-semiconductor junctions usually exhibit nonlinear characteristics—especially rectifying action. When the nonlinearity is weak and the resistance is small, the contact is called an *ohmic contact*. Fig. 1.8 (p. 14), Fig. 6.1 (p. 168), Fig. 7.1

[1] Typically, the Newton–Raphson method is used to numerically find $\psi_s$.

(p. 216), and so on included metal-semiconductor junctions, but we either ignored the details or implicitly assumed that they were very low-resistance ohmic contacts. A contact that is not ohmic is called a *Schottky contact*. Schottky contacts that exhibit marked rectifying action are sometimes used as diodes and are called Schottky-barrier diodes. Since only majority carriers are involved in the operation of Schottky barrier-diodes, their responses are much faster than those of p-n junction diodes. Therefore, Schottky-barrier diodes are often used in radio-frequency (RF) circuits.

The metal-semiconductor contact can be regarded as the extreme limit of a MOS capacitor with its oxide thinned to an infinitesimal thickness. If the oxide is very thin, a large tunnel current can flow. If any charge is stored in the MOS capacitor for some reason (the fixed charge $Q_o$ in Fig. 7.10 (p. 223) can be such charge), the potential drop across this negligibly thin oxide is not zero no matter how thin it is, so the electrostatic potential should change abruptly at the metal-(oxide-)semiconductor interface. On the other hand, since current flows through the infinitesimally thin oxide, the quasi-Fermi levels do not change across the oxide. This is the difference from the real MOS capacitor.

It is, in theory, possible to predict whether a metal-semiconductor contact is ohmic or Schottky based on the work function of the metal and the doping of the semiconductor. However, in practice, this simplistic theoretical prediction often fails. This does make it difficult to cover metal-semiconductor contacts in introductory books, such as this one. The main reason for the failure of the theory is considered to be the presence of traps on the semiconductor surface (corresponding to $Q_o$ in Fig. 7.10). To achieve an ohmic contact, the semiconductor should be highly doped, but this is not always sufficient. Ohmic contacts are often quite difficult to achieve. Realizing good ohmic contacts is an important research subject in the development of new semiconductor materials.

## 7.3 THREE-TERMINAL MOS STRUCTURES

A three-terminal MOS structure might sound like a MOSFET without a back gate. But what we consider here is a structure in which a region

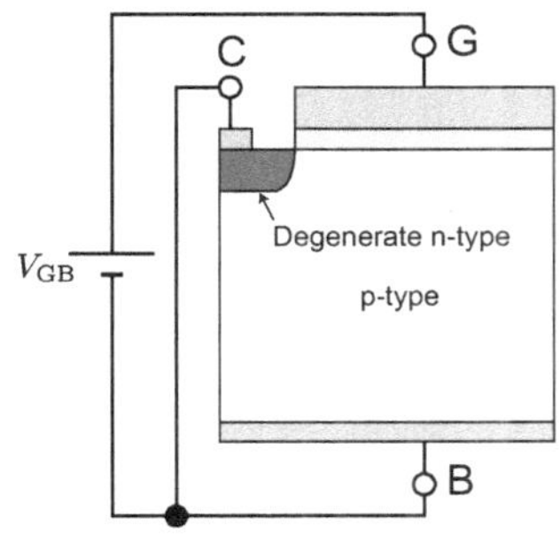


FIGURE 7.18 Three-terminal MOS structure.

corresponding to the drain (or source) is added to the substrate of a MOS capacitor.

## 7.3.1 Back-Gate-Referenced Analysis

A degenerate n-type region is added to the left end of the substrate surface, as shown in Fig. 7.18. Since an inversion layer is formed under an appropriate gate bias, and it becomes the *channel* of the MOSFET, let the terminal name be C. As a result, a one-sided p-n junction (p. 189) is formed between the p-type substrate and the degenerate n-type region. Therefore, this structure is also called a *gated diode*. If terminal C is directly connected to the back gate (terminal B) as shown in Fig. 7.18, the system remains in equilibrium, so the behavior of that part of the MOS capacitor, which is at a sufficient distance from the p-n junction, should be the same as in the two-terminal case.

### *7.3.1.1 Effect of Biasing the Channel*

Up to this point, we have only discussed the structure, and we did not say much about the value of the gate voltage $V_{GB}$. Next, let us consider the case where $V_{GB} = V_{fb}$, as shown in Fig. 7.19. Since the MOS capacitor is in the flat-band condition, no charge is stored. Since the p-n junction is zero-biased, the electrostatic potential difference between the p-type and n-type regions equals the built-in potential $\varphi_{bi}$ (p. 177). Since it is a one-sided junction, a depletion layer is formed only on the p-type side (p. 189).

Next, let us set $V_{GB} > V_{fb}$ and bring the MOS capacitor into strong inversion. Let us assume for convenience that the value of the surface

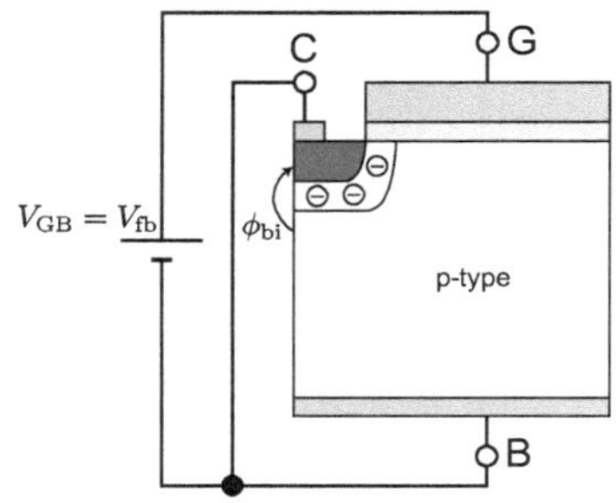


FIGURE 7.19 Three-terminal MOS structure in a flat-band condition.

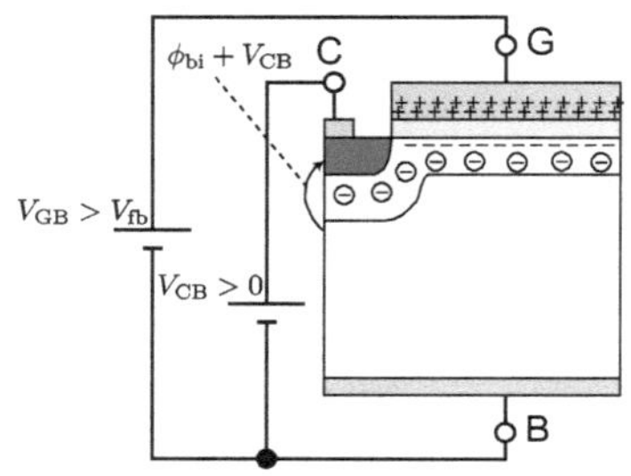


FIGURE 7.20 Three-terminal MOS structure in a strong inversion condition.

potential equals $\psi_{sT}$ in (7.43) on p. 236. Then, let us increase the voltage $V_{CB}$ at terminal C as shown in Fig. 7.20. $V_{CB}$ is called the *channel potential*.

Since the p-n junction is now reverse biased, the system is in a nonequilibrium state. Specifically, as shown in Fig. 6.15 (p. 181), $\zeta_{nP} < \zeta_{pP}$ at the substrate surface near the p-n junction. Because of $V_{CB} > 0$, electrons are attracted to the degenerate n-type region, so the degree of inversion of the MOS capacitor substrate surface is reduced. Since $\zeta_{nP} < \zeta_{pP}$, the surface electron density is lower than at equilibrium ($V_{CB} = 0$), as discussed on p. 124. Since the channel is formed from a large number of electrons and has lower resistance than the rest of the silicon substrate, the entire channel can be considered to remain equipotential.

In order to keep the degree of inversion at the substrate surface the same as it was at $V_{CB} = 0$, the surface potential must be increased from $\psi_{sT}$ to $\psi_{sT} + V_{CB}$ by increasing $V_{GB}$. So the expression for the surface electron density for $V_{CB} \neq 0$ can be written by replacing $\psi_s$ in equations (7.28) through (7.30) on p. 233 with $\psi_s - V_{CB}$ as

$$n_s = n_{P0} \exp\left[\frac{q(\psi_s - V_{CB})}{kT}\right] \qquad \text{(Flat-band referenced)} \qquad (7.57)$$

$$= n_i \exp\left[\frac{q(\psi_s - V_{CB} - \varphi_B)}{kT}\right] \qquad \text{(Weak-inversion-onset referenced)} \qquad (7.58)$$

$$= p_{P0} \exp\left[\frac{q(\psi_s - V_{CB} - 2\varphi_B)}{kT}\right]. \qquad \text{(Strong-inversion-onset ref.)} \qquad (7.59)$$

The value of $\psi_s$ in (7.59), for example, must be larger than that in the two-terminal MOS capacitor case by $V_{CB}$ so that $n_s = p_{P0}$ holds. That is why $V_{CB}$ in (7.59) has a minus sign.

The above derivation is similar to that in the analysis of p-n junctions in Chapter 6, where the zero-bias case was first examined (§6.5.1), and then the equation for the biased case was derived by variable substitution (§6.5.2). This is no coincidence because Fig. 7.20 includes a reverse-biased p-n junction.

#### *7.3.1.2 Approximation for Strong Inversion*

In the case of strong inversion, the surface potential $\psi_s$ on the left-hand side of (7.42) on p. 236 can be replaced by $\psi_s - V_{CB}$ to obtain the following approximation.

$$\psi_s \simeq \psi_{sT} + V_{CB}. \qquad \text{(Fixed-value approximation)} \qquad (7.60)$$

Substituting equation (7.60) into the depletion layer thickness equation (7.37) on p. 235 yields

$$d_b = \sqrt{\frac{2\epsilon_{Si}(\psi_{sT} + V_{CB})}{qN_A^-}}. \qquad \text{(Depletion layer thickness)} \qquad (7.61)$$

Equation (7.61) indicates that if $V_{CB} > 0$, the depletion layer at the substrate surface is thicker than in the two-terminal case. This is natural because a higher gate voltage is applied.

Likewise, putting (7.60) in (7.38) on p. 235 gives the approximate depletion charge in strong inversion.

$$Q_b = -qN_A^- d_b = -\sqrt{2q\epsilon_{Si}N_A^- (\psi_{sT} + V_{CB})}. \qquad \text{(Depletion charge)} \tag{7.62}$$

Replacing $\psi_{sT}$ in (7.47) on p. 237 for the two-terminal MOS capacitor threshold voltage with $\psi_{sT} + V_{CB}$, we obtain

$$V_{TB}(V_{CB}) \equiv V_{fb} + \psi_{sT} + V_{CB} + \gamma\sqrt{\psi_{sT} + V_{CB}}. \tag{7.63}$$

This is the threshold voltage of the three-terminal MOS structure. The left-hand side of (7.63) explicitly shows that $V_{TB}$ is a function of the channel potential $V_{CB}$. If $V_{CB} > 0$, the threshold becomes higher and it becomes more difficult to strongly invert the channel. The inversion charge can be expressed using the threshold voltage (7.63) as follows:

$$Q_{inv} = \begin{cases} -C_{ox}(V_{GB} - V_{TB}) & (V_{GB} \geq V_{TB}) \\ 0 & (V_{GB} < V_{TB}) \end{cases} \tag{7.64}$$

$$= \begin{cases} -C_{ox}\left(V_{GB} - V_{fb} - \psi_{sT} - V_{CB} - \gamma\sqrt{\psi_{sT} + V_{CB}}\right) & (V_{GB} \geq V_{TB}) \\ 0 & (V_{GB} < V_{TB}) \end{cases}. \tag{7.65}$$

The concept of "threshold voltage" is based on the idea that the inversion charge $Q_{inv}$ is controlled by the gate voltage $V_{GB}$. This is shown in Fig. 7.21. When $V_{GB}$ exceeds the threshold voltage $V_{TB}$, $|Q_{inv}|$ increases linearly. Actually, $V_{TB}$ depends on the channel potential $V_{CB}$ as well, according to (7.63), so it is more appropriate to express the inversion charge as a function of $V_{GB}$ and $V_{CB}$ as $Q_{inv}(V_{GB}, V_{CB})$.

### *7.3.1.3 Pinch-Off Voltage*

When a certain gate voltage $V_{GB}$ is given, the value of $V_{CB}$ at which the inversion charge becomes $Q_{inv} \simeq 0$ (the boundary between strong inversion and weak inversion) is called the *pinch-off voltage* and is denoted by $V_P$. Inserting $Q_{inv} = 0$ and $V_{CB} = V_P$ into (7.65) according to this definition, we obtain

$$0 = -C_{ox}\left(V_{GB} - V_{fb} - \psi_{sT} - V_P - \gamma\sqrt{\psi_{sT} + V_P}\right). \tag{7.66}$$

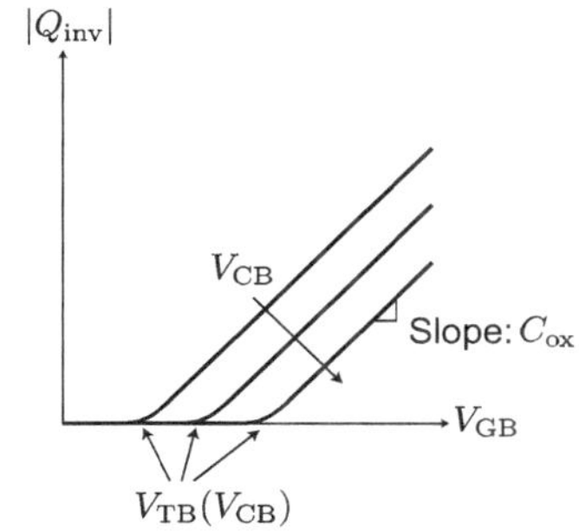


FIGURE 7.21 Controlling the inversion charge $Q_{\mathrm{inv}}$ by the gate voltage $V_{\mathrm{GB}}$.

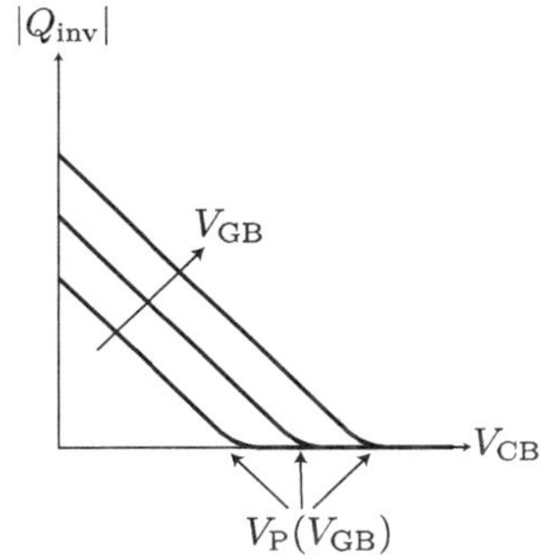


FIGURE 7.22 Controlling the inversion charge $Q_{\mathrm{inv}}$ by the channel potential $V_{\mathrm{CB}}$.

Solving (7.66) for $V_{\mathrm{P}}$ results in (Problem 7.6 on p. 266)

$$V_{\mathrm{P}}\left(V_{\mathrm{GB}}\right) = \left(-\frac{\gamma}{2} + \sqrt{\frac{\gamma^2}{4} + V_{\mathrm{GB}} - V_{\mathrm{fb}}}\right)^2 - \psi_{\mathrm{sT}}. \quad \text{(Pinch-off voltage)} \tag{7.67}$$

The left-hand side of (7.67) states that $V_{\mathrm{P}}$ is a function of $V_{\mathrm{GB}}$.

A graph similar to Fig. 7.21, with $V_{\mathrm{CB}}$ on the horizontal axis, is shown in Fig. 7.22. Both the threshold voltage and the pinch-off voltage are defined as the point where (strong) inversion begins/ends on the respective axes. The pinch-off voltage can be interpreted as a kind of "threshold voltage" when the inversion charge $Q_{\mathrm{inv}}$ is considered to be controlled by the channel potential $V_{\mathrm{CB}}$.

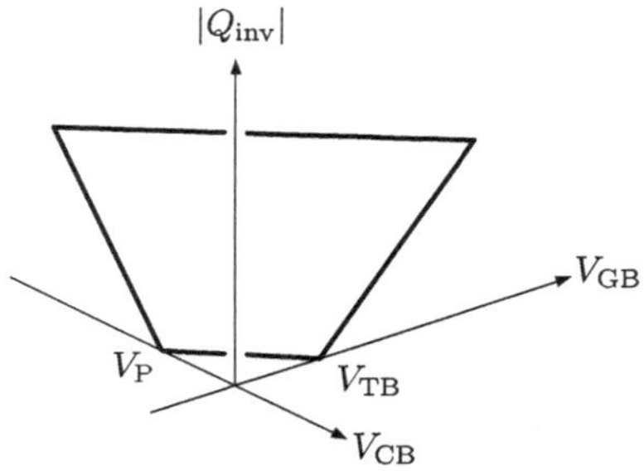


FIGURE 7.23 Dependence of the absolute value of inversion charge, $|Q_{inv}|$, on $V_{GB}$ and $V_{CB}$.

In short, the inversion charge $Q_{inv}(V_{GB}, V_{CB})$ can be controlled by either the gate voltage $V_{GB}$ or the channel potential $V_{CB}$, both of which are involved in the MOSFET operation. The two graphs in Figs. 7.21 and 7.22 can be combined into a single graph as in Fig. 7.23 (the point of intersection of the horizontal axes is not the "origin").

#### 7.3.1.4 Cases Other than Strong Inversion

The above discussion was for the case of strong inversion, but to include other cases than strong inversion, the surface potential $\psi_s$ can be found numerically by replacing $\psi_s$ in (7.56) on p. 239 with $\psi_s - V_{CB}$. The inversion charge $Q_{inv}$ can then be found from (7.40) on p. 235.

### 7.3.2 Channel-Terminal-Referenced Analysis

Here we will rewrite the equations we derived using terminal C (corresponding to the source of the MOSFET) as the *datum node* or *reference node*. The biasing is as shown in Fig. 7.24, and the following equality holds:

$$V_{GB} = V_{GC} + V_{CB}. \tag{7.68}$$

The approximate expression for the inversion charge in the strong inversion condition can be obtained by putting (7.68) into (7.64) on p. 244.

$$Q_{inv} = \begin{cases} -C_{ox}(V_{GC} - V_{TC}) & (V_{GC} \geq V_{TC}) \\ 0 & (V_{GC} < V_{TC}) \end{cases}, \tag{7.69}$$

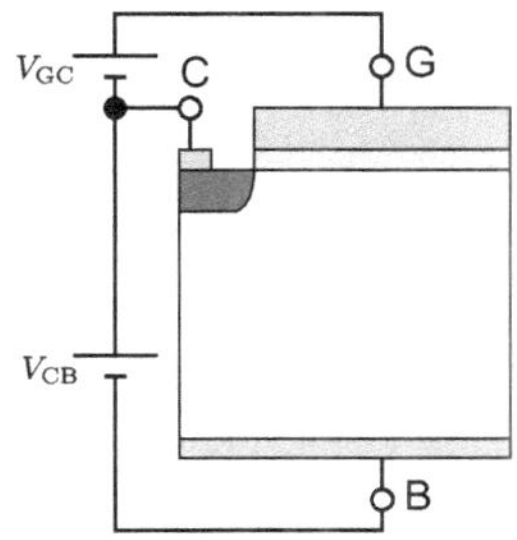


FIGURE 7.24 Channel-terminal-referenced three-terminal MOS structure.

where we introduced yet another threshold voltage, $V_{TC}$.

$$\begin{aligned} V_{TC} &\equiv V_{TB} - V_{CB} \\ &= V_{fb} + \psi_{sT} + \gamma\sqrt{\psi_{sT} + V_{CB}}. \end{aligned} \tag{7.70}$$

This is the channel-terminal-referenced threshold voltage. We will use this result when we consider the MOSFET threshold voltage (p. 257).

### *C-V* CHARACTERISTICS OF MOS CAPACITORS

Even though we are dealing with a structure called a "MOS capacitor," we have not discussed its capacitance, except that we referred to the per-unit-area gate oxide capacitance, $C_{ox}$. This is certainly not because the capacitance is unimportant.

MOS capacitors are clearly not linear capacitors. Differentiating the gate charge $Q_G$ by the gate voltage $V_{GB}$, the incremental capacitance of the MOS capacitor is obtained. A graph plotting the incremental capacitance on the vertical axis and the gate voltage on the horizontal axis is called the "*C-V* curve" of the MOS capacitor. "*C-V* measurement" is performed by applying a small sinusoidal signal on top of a DC bias to the gate [23]. Thus, this is a measurement of a sinusoidal steady state (p. 116). This incremental capacitance depends on the frequency of the superimposed sine wave, reflecting the lifetime of minority carriers and other factors at the substrate surface. The capacitance component due to the inversion charge and that due to the depletion charge can be measured separately by using a MOS structure

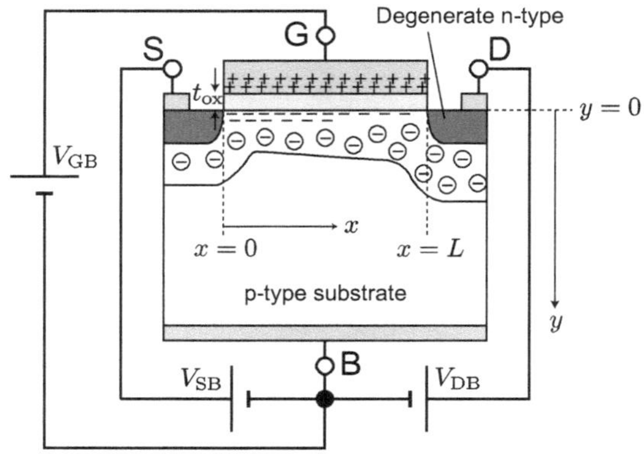


FIGURE 7.25 Back-gate-referenced four-terminal nMOSFET.

with three or more terminals (split C-V measurement). C-V measurement is an important tool for investigating the condition and quality of substrate surfaces.

## 7.4 FOUR-TERMINAL MOSFET

Let us finally add the other degenerate n-type region and analyze the four-terminal MOSFET. Traditionally, the terminal voltages of a MOSFET are measured with respect to its source, but using the back gate as the datum node makes the equations more symmetrical and much easier to understand [34].

### 7.4.1 Back-Gate-Referenced Analysis

#### *7.4.1.1 Four-Terminal MOSFET Structure*

Let us add a degenerate n-type region, similar to the one added in Fig. 7.18 (p. 241), on the right side as shown in Fig. 7.25. As before, we will proceed with the discussion using the back gate terminal as the voltage reference. This kind of treatment is also said to be body-referenced or bulk-referenced.

#### *7.4.1.2 Assumptions of the Analysis*

The assumptions of the analysis are listed below.

1. Since we consider an nMOSFET here, the drain voltage $V_{DB}$ is assumed to be higher than the source voltage $V_{SB}$ (see Fig. 7.5 on p. 219). More specifically, we assume the following inequality to hold:

$$0 \le V_{SB} < V_{DB}. \tag{7.71}$$

The source voltage is usually $V_{SB} = 0$ (Fig. 7.5). Under this condition, current flows from the drain to the source. Also, channel potential $V_{CB}$ and the degree of inversion depend on the position $x$ in the channel.

$$V_{SB} \le V_{CB}(x) \le V_{DB}. \tag{7.72}$$

2. The *gradual-channel approximation* is to be applied. The gate voltage $V_{GB}$ creates a $y$-direction electric field $\mathcal{E}_y$ in the channel. There also exists an $x$-direction electric field $\mathcal{E}_x$ due to the biasing stipulated in Assumption 1. Since $|V_{GB}|$ and $|V_{DB} - V_{SB}|$ are usually comparable, if the oxide thickness $t_{ox}$ (Fig. 7.25) is sufficiently thinner than the channel length $L$, then $|\mathcal{E}_x| \ll |\mathcal{E}_y|$ holds. Based on this observation, the approximation that ignores $\mathcal{E}_x$ is the gradual-channel approximation. MOSFETs for which this approximation is valid are called *long-channel MOSFETs*.

3. Gate and back gate currents are assumed to be zero. First, because of the insulating gate oxide, no DC current flows into the gate. The p-n junction diode between the substrate and the drain is reverse biased according to (7.71). Since the reverse current is very small, we can ignore it. In equations,

$$I_G = 0 \qquad \text{(Current flowing into gate)} \tag{7.73}$$

$$I_B = 0 \qquad \text{(Current flowing into back gate)} \tag{7.74}$$

$$I_D = -I_S = I_{DS} \qquad \text{(Drain current)} \tag{7.75}$$

4. Carrier generation and recombination are assumed to be negligible.

5. The electron mobility $\mu_n$ is assumed to be constant, independent of the position $x$ in the channel. In reality, the mobility may depend on $x$, but we ignore this for simplicity.

6. Approximation for strong inversion is to be used. We consider only the cases where the surface potential can be considered to take the following form from (7.60) on p. 243 with $x$-dependence.

$$\psi_s(x) = \psi_{sT} + V_{CB}(x), \qquad \text{(Approximate surface potential)} \quad (7.76)$$

where $\psi_{sT}$ is given by (7.43) on p. 236. The range of $V_{CB}(x)$ is as given in (7.72). Under this assumption, we can directly consider only the nonsaturation region of the on state. The current in the saturation region can be derived from it, but the off-state (subthreshold characteristics) cannot be treated.

7. The current is assumed to be carried only by electrons in the inversion layer. The equations for inversion charge and threshold voltage are also modified from (7.64) and (7.63) on p. 244 to account for $x$-dependence as follows.

$$Q_{inv}(x) = \begin{cases} -C_{ox}\left[V_{GB} - V_{TB}(x)\right] & (V_{GB} \geq V_{TB}) \\ 0 & (V_{GB} < V_{TB}) \end{cases}, \quad (7.77)$$

$$V_{TB}(x) = V_{fb} + \psi_{sT} + V_{CB}(x) + \gamma\sqrt{\psi_{sT} + V_{CB}(x)}. \quad (7.78)$$

*7.4.1.3 Channel Potential and Quasi-Fermi Levels*

The substrate in Fig. 7.25 (p. 248) is p-type, and the drain and the source are degenerate n-type, so one-sided p-n junctions are formed between the substrate and the drain/source. According to the discussion in §6.8.2 (especially Assumption 7 on p. 192), the hole quasi-Fermi level in the depletion layer of a p-n junction equals that in the p-type neutral region, and the electron quasi-Fermi level in the depletion layer equals that in the n-type neutral region (see Fig. 6.19 on p. 195). The voltage in the circuit-theoretic sense corresponds to the quasi-Fermi potential of majority carriers (§6.3.3). If so, then the hole quasi-Fermi potential $\psi_p$ just below the gate oxide in Fig. 7.25 (p. 248) is determined by the deep part of the p-type substrate (i.e., back gate). The electron quasi-Fermi potential $\psi_n(x)$ is determined by $V_{SB}$ at the source end ($x = 0$) of the channel and by $V_{DB}$ at the drain end ($x = L$):

$$\psi_n(0) = V_{SB}, \qquad \text{(Electron quasi Fermi potential at source end)} \quad (7.79)$$

$$\psi_{\text{n}}(L) = V_{\text{DB}}. \qquad \text{(Electron quasi Fermi potential at drain end)} \tag{7.80}$$

From the above, the channel potential at position $x$ in the channel is given approximately by

$$V_{\text{CB}}(x) \approx \psi_{\text{n}}(x) - \psi_{\text{p}} = \frac{\zeta_{\text{n}}(x) - \zeta_{\text{p}}}{-q}. \qquad \text{(Channel potential)} \tag{7.81}$$

This equation corresponds to (6.64) on p. 195. In words, (7.81) is saying that the channel potential $V_{\text{CB}}(x)$ is the quasi-Fermi potential $\psi_{\text{n}}(x)$ for the inversion charge, measured with respect to the hole quasi-Fermi potential $\psi_{\text{p}}$ of the substrate.

Actually, (7.81) is quite a questionable expression. This is because Assumption 7 on p. 192 is not valid for reverse-biased p-n junctions (unless the reverse bias is very small), as already discussed in the Box on p. 198, §6.9, and Problem 6.7 on p. 213. However, in the derivation of the current-voltage characteristics of the p-n junction diode, we used Assumption 7 because we could not move forward without it. The situation is basically the same. After all, our MOSFET is also a "gated diode" (p. 241). So although we know it is questionable, we will use (7.81) in the following.

### *7.4.1.4 Channel Potential and Current*

Consider the current $I_{\text{DS}}$ flowing through an infinitesimal channel section $\Delta x$ at position $x$. Note that $I_{\text{DS}}$ flows in the left $(-x)$ direction. Since the steady-state drain current $I_{\text{DS}}$ is independent of $x$ according to Assumption 4 on p. 249, the current is obtained by integrating the current density at a certain $x$ in the $y$-direction and multiplying it by the channel width $W$. Thus,

$$I_{\text{DS}} = -W \int_0^{y_{\text{C}}} \mu_{\text{n}} n(x, y) \frac{\Delta \zeta_{\text{n}}(x)}{\Delta x} \text{d}y \tag{7.82}$$

$$= -W\mu_{\text{n}} \frac{Q_{\text{inv}}(x)}{-q} \frac{\Delta \zeta_{\text{n}}(x)}{\Delta x} = -W\mu_{\text{n}} Q_{\text{inv}}(x) \frac{\Delta V_{\text{CB}}(x)}{\Delta x}. \tag{7.83}$$

In (7.82), we used (5.33) on p. 131 for the electron current density. The upper end of the integral in (7.82), $y_{\text{C}}$, is the depth of the channel or the thickness of the inversion layer, which itself depends on $x$ (p. 217). In (7.83), use was made of (7.81). From the above, we obtain

$$I_{DS}\Delta x = -W\mu_n Q_{inv}(x)\Delta V_{CB}(x). \tag{7.84}$$

Integrate both sides of (7.84) from the source end ($x = 0$, $V_{CB} = V_{SB}$) to the drain end ($x = L$, $V_{CB} = V_{DB}$).

$$(\text{Left-hand side}) = \int_0^L I_{DS}\mathrm{d}x = I_{DS}\int_0^L \mathrm{d}x = I_{DS}L, \tag{7.85}$$

$$(\text{Right-hand side}) = -W\mu_n \int_{V_{SB}}^{V_{DB}} Q_{inv}(V_{CB})\,\mathrm{d}V_{CB}, \tag{7.86}$$

where Assumption 5 on p. 249 was used. From the above, the drain current is given by

$$I_{DS} = -\frac{W}{L}\mu_n \int_{V_{SB}}^{V_{DB}} Q_{inv}(V_{CB})\,\mathrm{d}V_{CB}. \quad \text{(General form of drain current)} \tag{7.87}$$

We have not yet used the strong inversion approximation in our derivation so far.

*7.4.1.5 Drain Current in Strong Inversion (Nonsaturation Region)*

We now use Assumptions 6 and 7 on p. 250 to give the inversion charge by (7.77) on p. 250. Putting (7.77) in (7.87) and integrating, we obtain the following expression for the drain current (Problem 7.7 on p. 266).

$$I_{DS} = \frac{\mu_n W C_{ox}}{L}\left\{(V_{GB} - V_{fb} - \psi_{sT})(V_{DB} - V_{SB}) - \frac{1}{2}(V_{DB}^2 - V_{SB}^2) - \frac{2}{3}\gamma\left[(\psi_{sT} + V_{DB})^{3/2} - (\psi_{sT} + V_{SB})^{3/2}\right]\right\}. \tag{7.88}$$

Note that the drain voltage $V_{DB}$ needs to be limited so that the entire channel is strongly inverted. Recalling the pinch-off argument from p. 255, $V_{DB}$ must satisfy

$$0 \leq V_{SB} < V_{DB} \leq V_P, \quad \text{(Strong inversion and nonsaturation)} \tag{7.89}$$

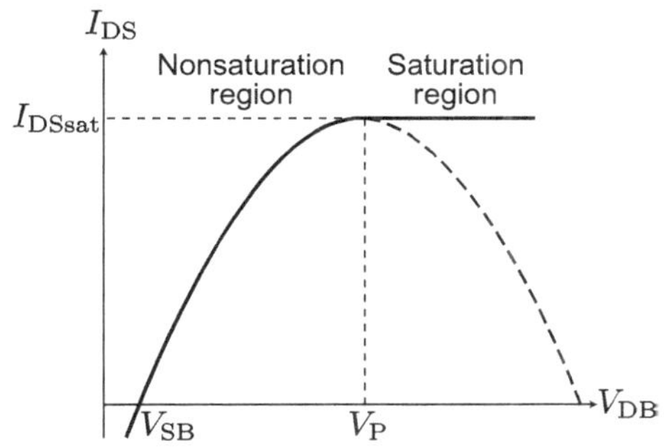


FIGURE 7.26 The $I_{DS}$-$V_{DB}$ characteristic of an nMOSFET.

which is more restrictive than (7.71) on p. 249 (see also Fig. 7.22 on p. 245). Equation (7.89) can be regarded as the definition of the nonsaturation region. A plot of $I_{DS}$ with $V_{DB}$ on the horizontal axis is shown in Fig. 7.26.

#### *7.4.1.6 Drain Current in the Saturation Region*

The region where the drain voltage $V_{DB}$ satisfies

$$V_{DB} \geq V_P \qquad \text{(Saturation region)} \tag{7.90}$$

is the saturation region. When $V_{DB} = V_P$, pinch-off occurs exactly at the drain end ($x = L$). When $V_{DB} > V_P$, the pinch-off onset point moves to a position $x = L' (< L)$, where $V_{CB}(L') = V_P$ holds. The larger $V_{DB}$ is, the smaller $L'$ is. This phenomenon, in which the *effective channel length* $L'$ changes depending on the applied drain voltage, is called *channel-length modulation*. The range $L' \leq x \leq L$ is weakly inverted. In this case, it is known that the current remains approximately the same as when $V_{DB} = V_P$ in (7.88) and is almost independent of $V_{DB} - V_{SB}$, *provided the channel length $L$ is sufficiently long*. The current in this saturation region is given by

$$I_{DSsat} = \frac{\mu_n W C_{ox}}{L} \left\{ (V_{GB} - V_{fb} - \psi_{sT})(V_P - V_{SB}) - \frac{1}{2}(V_P^2 - V_{SB}^2) - \frac{2}{3}\gamma \left[ (\psi_{sT} + V_P)^{3/2} - (\psi_{sT} + V_{SB})^{3/2} \right] \right\}. \tag{7.91}$$

The drain current that is independent of $V_{DB} - V_{SB}$ is convenient and important for realizing amplifier circuits.

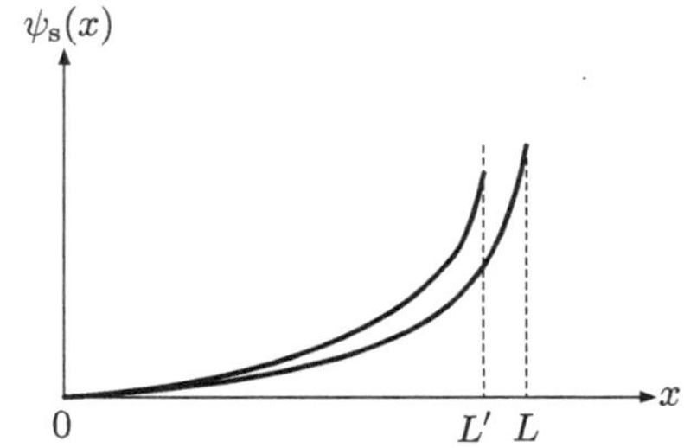


FIGURE 7.27 The slope of surface potential $\psi_s$ at the source end ($x = 0$) does not change significantly due to channel-length modulation.

The qualitative reason why the saturation drain current (7.91) does not depend on $V_{DB} - V_{SB}$ is as follows. In the pinched-off region ($L' \leq x \leq L$), the inversion charge is drastically reduced, so its resistance increases according to the conductivity equation (5.52) on p. 138. As a result, the voltage $V_{DB} - V_P$ applied beyond $V_P$ mostly drops across the pinched-off region. Then, the pinched-off region will also have a larger electric field in the $x$-direction, and the gradual-channel approximation will no longer hold in that part of the channel. However, it can be shown that the electric field intensity at the source end does not change very much by increasing $V_{DB}$ beyond $V_P$ (Fig. 7.27). Recall that the DC drain current $I_{DS}$ is independent of the position $x$ (p. 251). The insensitivity of the electric field at the source end to $V_{DB}$ dictates the almost constant drain current in the saturation region [2] (Fig. 7.26). The current is approximately determined by the region $0 \leq x \leq L'$ where the gradual-channel approximation holds. The saturation drain current is given by (7.91) with $L$ replaced by the effective channel length $L'$ ($< L$). But as long as $L$ is large and $L'/L$ is close to unity, (7.91) can be used as is.

#### *7.4.1.7 Approximation That Neglects Depletion Charge*

Although the formula for the back-gate-referenced drain current has been derived, the depletion charge is sometimes neglected to simplify the formula. This corresponds to setting the body-effect coefficient (p. 235) to $\gamma \to 0$ in previous equations. The theoretical basis for doing so is tenuous (unless the substrate is undoped), but the equations do become very simple. Simplifying the equations has the important advantage of making the concept of pinch-off easier to understand by looking at the equations.

Let us look at some equations in detail. When $\gamma = 0$, from (7.67) on p. 245, the pinch-off voltage is

$$V_P = V_{GB} - V_{fb} - \psi_{sT}. \quad \text{(Pinch-off voltage ignoring body effect)} \tag{7.92}$$

Equation (7.92) says that increasing the gate voltage increases $V_P$ and makes it harder to pinch off the channel (see Fig. 7.22 on p. 245). The threshold voltage $V_{TB}$ is, from (7.78) on p. 250,

$$V_{TB}(x) = V_{fb} + \psi_{sT} + V_{CB}(x). \tag{7.93}$$

The inversion charge can be written using (7.77) on p. 250 as

$$Q_{inv}(x) = \begin{cases} -C_{ox}[V_{GB} - V_{TB}(x)] & (V_{GB} \geq V_{TB}) \\ 0 & (V_{GB} < V_{TB}) \end{cases} \quad (V_{GB}\text{-controlled}) \tag{7.94}$$

$$= \begin{cases} -C_{ox}[V_{GB} - V_{fb} - \psi_{sT} - V_{CB}(x)] & (V_{GB} \geq V_{TB}) \\ 0 & (V_{GB} < V_{TB}) \end{cases}$$

$$= \begin{cases} -C_{ox}[V_P - V_{CB}(x)] & (V_{CB} \leq V_P) \\ 0 & (V_{CB} < V_P) \end{cases}. \quad (V_{CB}\text{-controlled}) \tag{7.95}$$

Equations (7.94) and (7.95) are similar in form, which corresponds to the similarity between Fig. 7.21 on p. 245 and Fig. 7.22 on p. 245.

The drain current equation also becomes simpler and easier to grasp. Assuming $\gamma = 0$ in (7.88) on p. 252 for the current in the nonsaturation region, we get

$$I_{DS} = \frac{\mu_n W C_{ox}}{L}\left[(V_{GB} - V_{fb} - \psi_{sT})(V_{DB} - V_{SB}) - \frac{1}{2}\left(V_{DB}^2 - V_{SB}^2\right)\right] \tag{7.96}$$

$$= \frac{\mu_n W C_{ox}}{L}\left[V_P(V_{DB} - V_{SB}) - \frac{1}{2}\left(V_{DB}^2 - V_{SB}^2\right)\right]$$
$$= \frac{\mu_n W C_{ox}}{2L}\left[(V_P - V_{SB})^2 - (V_P - V_{DB})^2\right]. \quad \text{(Nonsaturation)} \tag{7.97}$$

We used the pinch-off voltage equation (7.92) on p. 255 to arrive at the second line. If it is difficult to see how to go from the second line

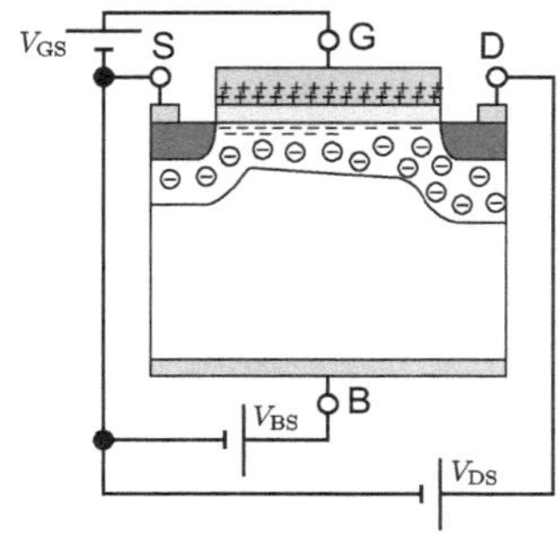


FIGURE 7.28 A source-referenced four-terminal nMOSFET.

to the third line (7.97), you can expand the latter to get the equation in the second line. In (7.97), the effect of gate voltage is incorporated into $V_P$ via (7.92). Equation (7.97) is highly symmetric with respect to the swapping of $V_{DB}$ and $V_{SB}$ (only the sign changes because the direction of the current reverses when they are swapped), and the relation to (7.95) for the inversion charge can be easily seen.

Setting $\gamma = 0$ in the saturation drain current equation (7.91) on p. 253 yields

$$I_{DSsat} = \frac{\mu_n W C_{ox}}{2L}(V_P - V_{SB})^2. \qquad \text{(Saturation drain current)} \quad (7.98)$$

From (7.90) on p. 253, the boundary between nonsaturation and saturation regions is $V_{DB} = V_P$, so the relation between (7.97) and (7.98) should also be clear.

### 7.4.2 Source-Referenced Analysis

Here, we will rewrite equations using source-referenced terminal voltages as shown in Fig. 7.28 on p. 256. This is the traditional treatment. This is more convenient when considering the correspondence with actual circuits, but the symmetry of the equations is broken, and the resulting equations are not as insightful as the back-gate-referenced case.

From Fig. 7.28, the equations needed for transforming the back-gate-referenced representation to the source-referenced representation are as follows:

$$\begin{cases} V_{GB} = V_{GS} + V_{SB}, \\ V_{DB} = V_{DS} + V_{SB}, \\ V_{SB} = -V_{BS}. \end{cases} \tag{7.99}$$

This situation is similar[2] to the three-terminal MOS structure we considered in Fig. 7.24 on p. 247, where the terminal C was the datum node. With this in mind, set $\gamma = 0$ in (7.70) on p. 247 for $V_{TC}$, and let the result be the threshold voltage $V_T$ (see Fig. 7.6 on p. 219).

$$V_T \equiv V_{fb} + \psi_{sT}, \quad \text{(Threshold voltage ignoring body effect)} \tag{7.100}$$

where the flat-band voltage $V_{fb}$ is given by (7.4) on p. 224 and $\psi_{sT}$ by (7.43) on p. 236. In (7.100), the $x$-dependence of the threshold voltage due to the fact that $V_{DB} > V_{SB}$ is obscured. Thus, the MOSFET threshold voltage is a convenient but rather ambiguous quantity.

Assuming $V_{SB} = 0$ in (7.96) for the drain current in the nonsaturation region (Fig. 7.5 on p. 219), since $V_{DB} = V_{DS}$ (Fig. 7.28), we can then use (7.100) to obtain (7.1) on p. 220 for the drain current.

## 7.5 SCALING AND SHORT-CHANNEL MOSFETS

### 7.5.1 MOSFET Scaling

In §1.4, it was mentioned that the smaller the MOSFET, the better the performance. Here, we will investigate how the characteristics of MOSFETs depend on their dimensions and the supply voltage. Changing device dimensions or supply voltage is called *scaling*.

Since the basic usage of MOSFETs is to operate them in the saturation region, we will use the saturation drain current equation (7.2) on p. 220. Inserting $V_{GS} = V_{dd}$ (supply voltage) into (7.2) leads to

$$I_{DSsat} = \frac{\mu_n W C_{ox}}{2L}(V_{dd} - V_T)^2. \quad \text{(Saturation drain current)} \tag{7.101}$$

Based on this equation, let us consider what happens if we change the parameters that can be manipulated (i.e., $L$, $W$, $t_{ox}$, $V_{dd}$, etc. in Fig. 7.29).

The scaling scenarios we consider are as follows:

[2] Not exactly the same unless $V_{DB} = V_{SB}$.

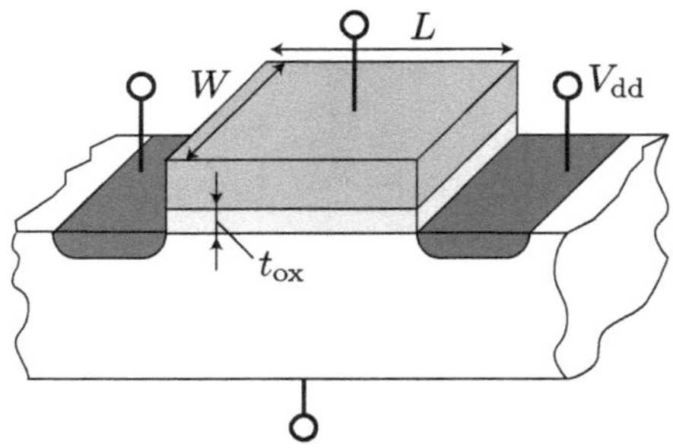


FIGURE 7.29 Scaling parameters of MOSFET.

TABLE 7.2 MOSFET Scaling Law

| Parameters | Constant-field | Constant-voltage | Dimensions $1/\kappa$ Voltage $1/U$ |
|---|---|---|---|
| | $(U = \kappa)$ | $(U = 1)$ | |
| Dimensions $(L, W, t_{ox})$ | $1/\kappa$ | $1/\kappa$ | $1/\kappa$ |
| Supply voltage $V_{dd}$ | $1/\kappa$ | 1 | $1/U$ |
| Doping density | $\kappa$ | $\kappa^2$ | $\kappa^2/U$ |
| Gate area $LW$ | $1/\kappa^2$ | $1/\kappa^2$ | $1/\kappa^2$ |
| Gate oxide capacitance per area $C_{ox}$ | $\kappa$ | $\kappa$ | $\kappa$ |
| Gate capacitance $C_{gate} = LWC_{ox}$ | $1/\kappa$ | $1/\kappa$ | $1/\kappa$ |
| Saturation drain current $I_{DSsat}$ | $1/\kappa$ | $\kappa$ | $\kappa/U^2$ |
| On resistance $R_{on} = V_{dd}/I_{DSsat}$ | 1 | $1/\kappa$ | $U/\kappa$ |
| Delay $\tau = R_{on}C_{gate}$ | $1/\kappa$ | $1/\kappa^2$ | $U/\kappa^2$ |
| Clock frequency $f_{clk} \propto 1/\tau$ | $\kappa$ | $\kappa^2$ | $\kappa^2/U$ |
| Switching energy $C_{gate}V_{dd}^2$ | $1/\kappa^3$ | $1/\kappa^2$ | $1/\kappa U^2$ |
| Switching power $f_{clk}C_{gate}V_{dd}^2$ | $1/\kappa^2$ | 1 | $\kappa/U^3$ |
| Devices per area $(\propto 1/(LW))$ | $\kappa^2$ | $\kappa^2$ | $\kappa^2$ |
| Power consumption per area | 1 | $\kappa^2$ | $\kappa^3/U^3$ |

1. *Constant-electric-field scaling*. Parameters are scaled such that the electric field intensity in the device stays constant. If device dimensions are multiplied by a factor of $1/\kappa$ $(\kappa > 1)$, the supply voltage must also be multiplied by $1/\kappa$.
2. *Constant-voltage scaling*. The supply voltage is kept constant and dimensions are multiplied by $1/\kappa$.
3. *Generalized scaling*. Dimensions are multiplied by $1/\kappa$, and the supply voltage is multiplied by $1/U$ $(U > 1)$.

Table 7.2 summarizes the results of the calculations based on (7.101).

Let us examine the contents of the table. Here we need to consider how to scale the doping density. When the device dimensions are multiplied by a factor of $1/\kappa$, we want the depletion layer thickness to also be multiplied by a factor of $1/\kappa$. Looking at expression (7.37) on p. 235 for the depletion layer thickness $d_b$, we see that for constant-electric-field scaling, $\psi_s$ is a quantity that should become $1/\kappa$ times. To make $d_b \to d_b/\kappa$, the acceptor ion density $N_A^-$ should be multiplied by $\kappa$. Noting that $C_{ox} \propto 1/t_{ox}$, with constant-electric-field scaling, the gate capacitance $C_{gate}$ and current become smaller ($\times 1/\kappa$), but the on resistance remains constant. Therefore, the delay, which is determined by the RC time constant, becomes $1/\kappa$ times. So the clock frequency $f_{clk}$, which determines the speed of digital circuits, can be multiplied by $\kappa$. Alternatively, digital circuits can be made lower-power by keeping $f_{clk}$ constant. Since the number of MOSFETs per unit area can be made $\kappa^2$ times, more complex circuits can be made in the same area. When $f_{clk}$ is multiplied by $\kappa$, the power consumption per unit area is the same as before scaling. This means that we can expect to create circuits with higher performance at the same power consumption by applying constant-electric-field scaling.

In the case of constant-voltage scaling, the power density increases, and it makes it difficult for the circuit to dissipate all the heat generated. The performance improvement brought about by constant-electric-field scaling motivated the miniaturization of MOSFETs and the higher levels of integration, and it became the guiding principle for the development of integrated circuits (§1.4). In practice, generalized scaling with $\kappa > U \geq 1$ is more realistic because it is difficult to lower the supply voltage proportionally to the reduction of device dimensions.

One thing that Table 7.2 does not properly take into account is the scaling of the threshold voltage $V_T$. In Table 7.2, it is implicitly assumed that $V_T$ is scaled in the same way as the supply voltage. However, (7.100) on p. 257 and (7.4) on p. 224 indicate that $V_T$ is not directly related to the supply voltage, but is determined by things like the work functions of materials. Therefore, the threshold voltage does not obey the scaling law. There are various nonideal factors of this kind, and various innovations are required for MOSFET miniaturization.

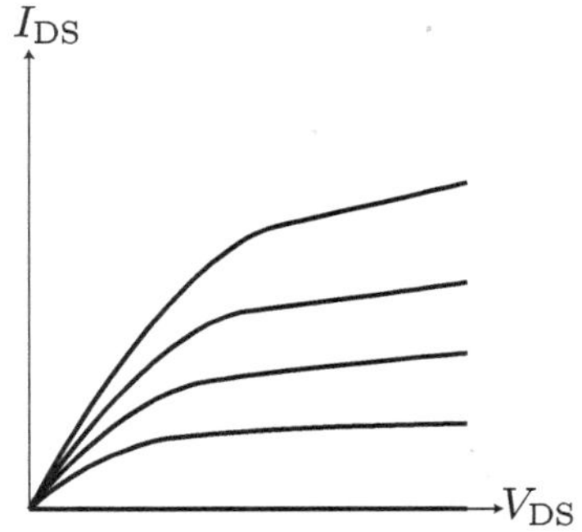


FIGURE 7.30 Channel-length modulation seen in $I_{DS}$-$V_{DS}$ characteristics.

### 7.5.2 Short-Channel Effects

As the MOSFET dimensions are reduced more and more, the channel length $L$ becomes too small to apply the gradual-channel approximation (p. 249). Such MOSFETs are called short-channel MOSFETs. Various characteristic degradations occur in short-channel MOSFETs. This is called the *short-channel effect* in a broad sense. In a narrow sense, the "short-channel effect" refers to the phenomenon in which the threshold voltage decreases as the channel length becomes shorter. Minimizing the degradation of characteristics of short-channel MOSFETs is an important issue in MOSFET research and development. There are a myriad of specific topics related to the degradation of short-channel MOSFET characteristics, but it is beyond the scope of this book to discuss them in detail. In the following, only two topics related to those already mentioned earlier will be discussed.

#### *7.5.2.1 Channel-Length Modulation*

As the channel length $L$ becomes shorter, the effective channel length $L'$ $(< L)$ (p. 253) cannot be regarded as sufficiently close to $L$ (i.e., $L'/L \simeq 1$), and *channel-length modulation* becomes significant. As a result, the drain current in the saturation region becomes dependent on the drain voltage $V_{DS}$ as shown in Fig. 7.30. Channel-length modulation leads to a reduction in circuit performance (e.g., reduction in the gain of amplifier circuits).

Gradual-channel approximation considers only the one-dimensional Poisson equation in the $y$-direction. When channel-length modulation is significant, the $x$-direction electric field cannot be

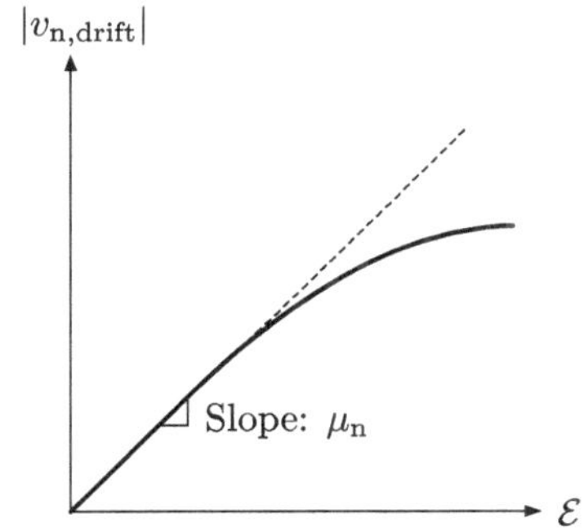


FIGURE 7.31 Carrier drift velocity versus electric field exhibiting velocity saturation.

ignored in the region $L' \leq x \leq L$. Therefore, a two-dimensional Poisson equation must be considered. This makes analytical treatment difficult.

### *7.5.2.2 Carrier Velocity Saturation*

When a MOSFET is miniaturized, the electric field in it becomes very large because constant-electric-field scaling is not possible. Carrier scattering then becomes so intense that the proportional relationship between the carrier drift velocity and the electric field in (5.47) and (5.49) on p. 136 no longer holds. As the electric field $\mathcal{E}$ increases, the drift velocity saturates $v_{\text{drift}}$ at a constant value as shown in Fig. 7.31. This phenomenon is the carrier *velocity saturation*. If $v_{\text{drift}} \propto \mathcal{E}$ does not hold, then all results derived using linear response mobility are not correct, as described in the Box on p. 139. Instead, equations must be derived using (5.27) through (5.30) on p. 130. The scaling law in Table 7.2 was also derived using mobility, so the results will be somewhat different when velocity saturation occurs.

It is safe to assume that velocity saturation always occurs in today's small MOSFETs. Moreover, around p. 251, we developed a rather dubious argument related to the channel potential, and on p. 248, we made various (not necessarily correct) assumptions. Therefore, there is not much point in trying to memorize the equations that appeared earlier and the results in Table 7.2.

Rather, what is more important is to acquire the ability to derive various results based on basic principles and make physical sense of

TCAD simulation data. To this end, we focused on the fundamentals and explained the meaning of the quasi-Fermi levels and how to read energy band diagrams in great detail (Chapter 5). As an application example of the basic principles, we also revisited the conventional assumptions of p-n junction analysis (Chapter 6), which can be said to be the origin of the somewhat "questionable" development above. As a result, we had to say some things that are different from the explanations in existing books. We hope that this book will help readers recognize the importance of fundamentals.

**MOSFET SCALING LAW**

As for the scaling law in Table 7.2 (p. 258), it might seem that the parameters were determined arbitrarily. However, the Poisson equation was used by Dennard, Gaensslen, and others to derive constant-electric-field scaling [10]. The dimensions and doping levels were determined so that the Poisson equation is conserved as much as possible before and after scaling.

Let us consider once again what the scaling law is. It gives guidelines on how to determine the channel length, oxide thickness, and doping density of the substrate to obtain the saturation characteristics when the channel length is shortened, assuming that the long-channel MOSFET to start with has ideal current-voltage characteristics (see Fig. 7.6 on p. 219). The current-voltage characteristics are maintained before and after scaling. This means that if the characteristics of the device before scaling are poor, the characteristics of the device after scaling (i.e., miniaturization) will remain poor. In scaling, only bad children are born from bad parent devices. However, in practice, good parent devices do not automatically produce good children by simple scaling—this is the short-channel effect. So various measures are needed to maintain good device characteristics.

If we try to generalize the scaling law, we need to consider three factors.

1. What we want to scale. (In ordinary device scaling, it is primarily the device dimensions that we want to scale.)
2. On what basis do we do it. (The basis is the Poisson equation.)

3. What is maintained before and after scaling? (Electric field distribution is preserved.)

Thinking along these lines, parameters other than dimensions and supply voltage, such as device operating temperature, can also be considered scalable. To reduce the operating temperature from 300 K (room temperature) to 77 K (liquid nitrogen temperature), we can derive a scaling law for the operating temperature based on the Fermi–Dirac distribution function while keeping the carrier distribution constant [39]. The author (Masu) first considered this *temperature scaling* theory in the late 1980s, but the need for MOSFETs operating at cryogenic temperatures is arising much more at the time of writing than it was then, especially due to the rapid development of quantum computers (see Problem 7.12 on p. 267). Table 7.2 may seem like just a list of parameters, but we hope you will see that it is based on physical phenomena and that there are other possibilities beyond those shown in this table.

## 7.6 SUMMARY

In this chapter, the structure and characteristics of MOSFETs are reviewed, and the DC current-voltage characteristics of planar long-channel MOSFETs are derived by considering the device physics of MOS capacitors, three-terminal MOS structures (i.e., gated diodes), and four-terminal MOSFETs.

- A standard planar MOSFET has four terminals: gate, back gate, drain, and source.
- The surface conditions on the semiconductor side of a MOS capacitor are classified as accumulation, flat band, depletion, weak inversion, and strong inversion.
- The channel of a MOSFET is an inversion layer formed on the surface of the semiconductor substrate by applying a gate voltage.
- The degree of inversion at the substrate surface increases as the gate voltage is increased, whereas the degree of inversion decreases as the channel potential is increased.

- In the nonsaturation region, the entire channel between the source and drain is strongly inverted, but in the saturation region, the channel near the drain is pinched off and weakly inverted.
- The MOSFET scaling law served as a powerful guiding principle for the development of MOSFETs and integrated circuits.

## 7.7 PROBLEMS

7.1 The "saturation region" in the transistor current-voltage characteristics has very different meanings for MOSFETs (Fig. 7.6 on p. 219) and bipolar transistors (p. 46). Search the Internet and find out the difference.

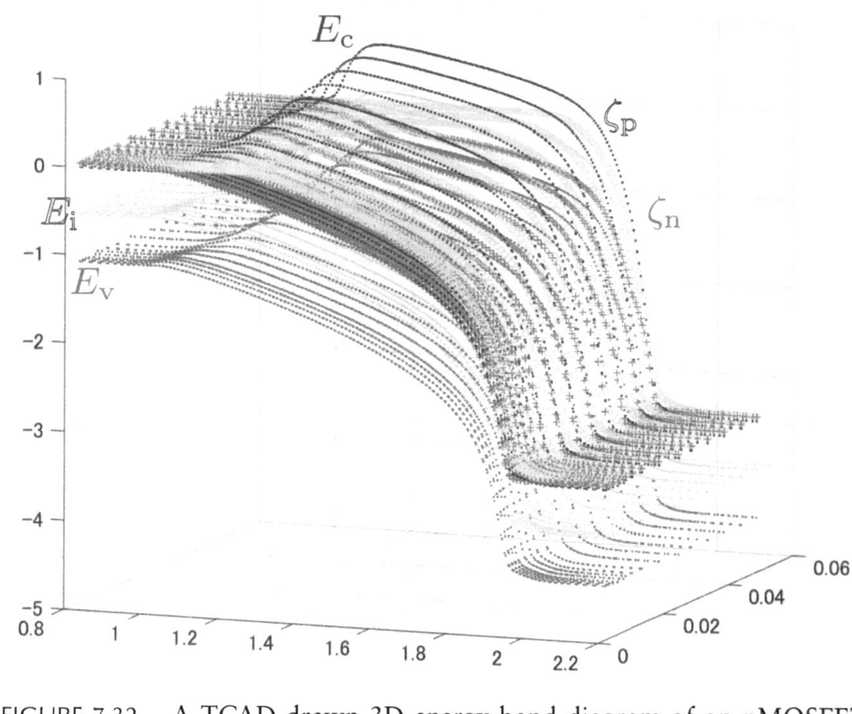


FIGURE 7.32 A TCAD-drawn 3D energy band diagram of an nMOSFET biased into the saturation region. Channel length is 1 µm. The channel is formed on the near side. The source is on the left-hand side and the drain is on the right-hand side.

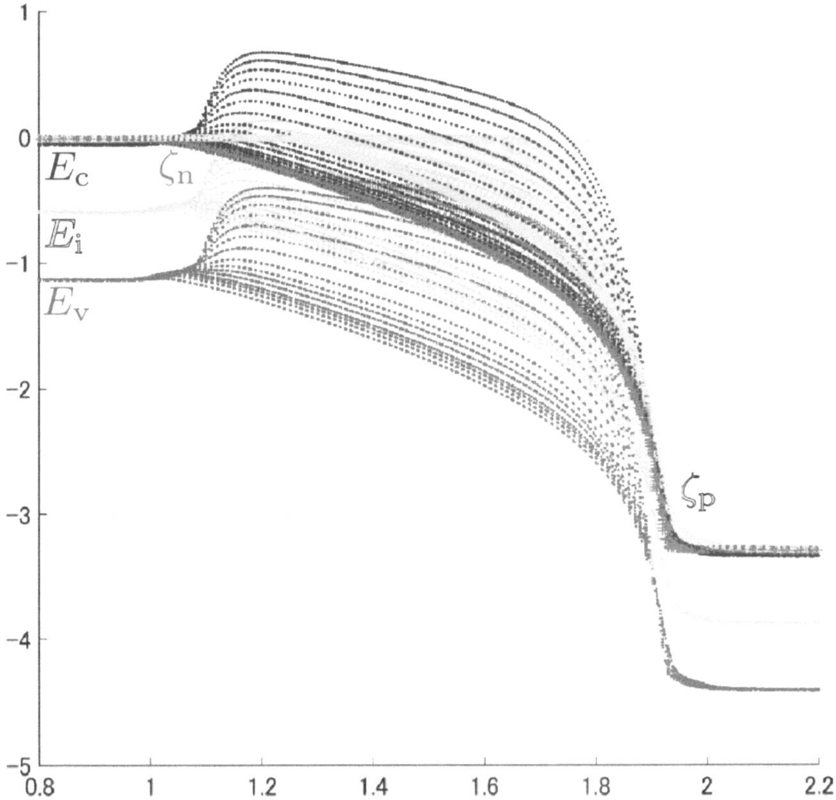


FIGURE 7.33 A TCAD-drawn 3D energy band diagram of an nMOSFET biased into the saturation region viewed from a different angle.

7.2 The drain voltage at the boundary between the nonsaturation and saturation regions in Fig. 7.6 on p. 219 is given by $V_{DS} = V_{GS} - V_T$. Derive this equation from (7.1) on p. 220 for the drain current in the nonsaturation region.

7.3 The energy band diagrams for a MOS capacitor in Fig. 7.12 (p. 226) to Fig. 7.16 (p. 232) do not include electrons, holes, and dopant atoms. Draw energy band diagrams including electrons, holes, and dopant atoms (both ionized and nonionized). Also, draw an energy band diagram for the case where the gate voltage is higher than in Fig. 7.16.

7.4 The energy band diagrams for a MOS capacitor in Fig. 7.12 (p. 226) to Fig. 7.16 (p. 232) are drawn as if there were no work

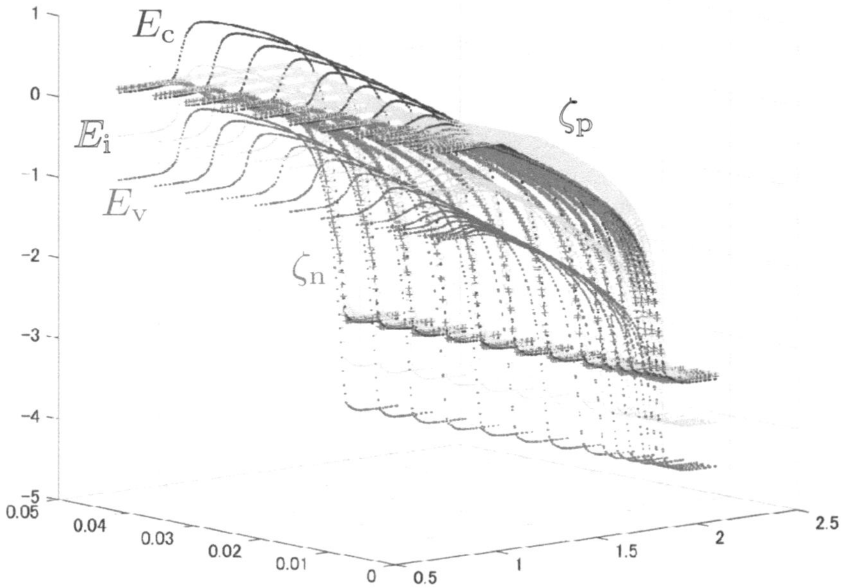


FIGURE 7.34 A TCAD-drawn 3D energy band diagram of an nMOSFET biased into the saturation region viewed from yet another angle.

function difference between the gate metal and the silicon substrate (p. 225). Redraw band diagrams for the cases where $\varphi_{SM} > 0$ and $\varphi_{SM} < 0$.

7.5 Use equations (7.56) on p. 239 and (7.35) on p. 235 to numerically find the surface potential $\psi_s$ as a function of the gate voltage $V_{GB}$. Plot $\psi_s$ and the induced charge $Q_{gi}$ as functions of $V_{GB}$.

7.6 Derive (7.67) on p. 245 for the pinch-off voltage.

7.7 Derive (7.88) on p. 252 for the drain current in the nonsaturation region.

7.8 The back gate is sometimes used as a second control terminal of the MOSFET. If the source-referenced back gate voltage $V_{BS}$ is raised or lowered (assuming $|V_{BS}| \ll V_{DS}$), how will the drain current change? Consider it using (7.96) on p. 255 and (7.88) on p. 252 in turn.

7.9 Figs. 7.32 through 7.34 show TCAD-drawn 3D energy band diagrams of an nMOSFET biased into the saturation region. If you have already gone through other books on semiconductor devices, you might have come across a somewhat similar-looking band diagram, which probably is Fig. 1(d) of [21]. It is reproduced in [30] as Fig. 8 of Chapter 6. Honestly, we had difficulty making good sense of this famous energy band diagram (see also Problem 6.7 on p. 213). Try taking a look at this famous band diagram and see if you can make sense of it. If you are like us, and if you have access to a device simulator, try drawing energy band diagrams like Figs. 7.32 through 7.34, and see if they make any better sense. Do the quasi-Fermi levels behave as assumed in §7.4.1.3? We are interested to hear how it turns out.

7.10 Use (7.57) through (7.59) on p. 243 and modify equation (7.56) on p. 239 for the induced charge $Q_{gi}$ so that the latter can be used for the four-terminal MOSFET. The result can be used to find the $I_{DS}$-$V_{GS}$ characteristics in all regions of operation, including the subthreshold region (see Fig. 7.7 (p. 219)).

7.11 The subthreshold current of an nMOSFET can be approximated as

$$I_{DS} = I_{DS0}e^{q(V_{GS}-V_T)/nkT} \qquad (V_{GS} < V_T), \tag{7.102}$$

where $I_{DS0}$ and $n \simeq \mathrm{d}V_{GS}/\mathrm{d}\psi_s\,(\geq 1)$ are constants.

$$S = \frac{\partial V_{GS}}{\partial \log I_{DS}} \qquad (V_{GS} < V_T) \tag{7.103}$$

is called the *subthreshold swing* (in millivolts per decade (mV/dec)). Ideally, $S$ is about 60 mV/dec at room temperature (with $n = 1$), but in practice, $S$ is about 80 mV/dec or larger. Try plotting a $\log I_{DS}$-$V_{GS}$ curve using (7.102) for $V_{GS} < V_T$ and (7.1) on p. 220 for $V_{GS} > V_T$. Does it go smoothly?

7.12 Consider the current-voltage characteristics of an nMOSFET at cryogenic temperatures. How does the drain current change compared to higher temperatures? How about the threshold voltage and the subthreshold current?

7.13 The contact potential $\varphi_{SM}$ between the gate metal and the silicon substrate (Fig. 7.9(b) on p. 222) is encoded in the flat-band voltage (7.4) on p. 224. The contact potential cannot be measured directly using a voltmeter (p. 6.7). Is there a way to somehow find $\varphi_{SM}$ via measurements of appropriately prepared MOS capacitor samples?

# Appendix

## A.1 MATRIX REPRESENTATIONS OF A TWO-PORT

### A.1.1 ABCD-Matrix

Let us define the *ABCD-matrix* of the *two-port network*[1] shown in Fig. A.1. A *port* is a terminal pair satisfying certain conditions.[2] An ABCD-matrix is also known as an F-matrix, chain matrix, or transmission matrix. Normally, the port current is defined as the current flowing into the positive terminal (and flowing out of the negative terminal), but the current $I_2$ at port 2 in Fig. A.1 is defined as the current flowing out from the positive terminal.

The ABCD-matrix **F** is a matrix that relates the voltage and current at port 1 ($V_1$ and $I_1$) to the voltage and current at port 2 ($V_2$ and $I_2$), as given in (A.1).

$$\begin{bmatrix} V_1 \\ I_1 \end{bmatrix} = \mathbf{F} \begin{bmatrix} V_2 \\ I_2 \end{bmatrix} = \begin{bmatrix} A & B \\ C & D \end{bmatrix} \begin{bmatrix} V_2 \\ I_2 \end{bmatrix}. \tag{A.1}$$

The elements of an ABCD-matrix are called *ABCD-parameters*. Note that the elements of **F** have different dimensions.

$$A = \left. \frac{V_1}{V_2} \right|_{I_2=0} \quad \text{(Dimensionless)} \tag{A.2}$$

$$B = \left. \frac{V_1}{I_2} \right|_{V_2=0} \quad \text{(Dimensions of resistance)} \tag{A.3}$$

[1] A "two-port network" is also abbreviated as a "two-port."

[2] A pair of terminals can be regarded as a *port* if (i) the current flowing into one of the terminals is equal to the current flowing out of the other terminal, and (ii) the two terminals are located close to each other compared with the wavelength.

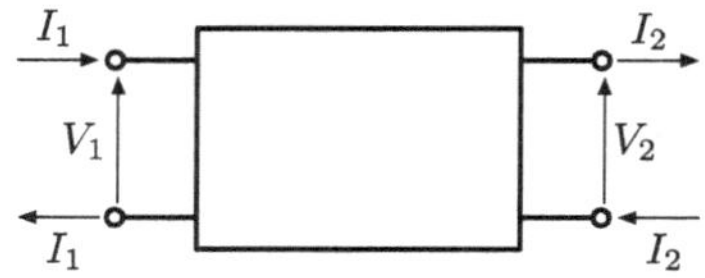


FIGURE A.1 Definitions of port voltages and port currents for an ABCD matrix.

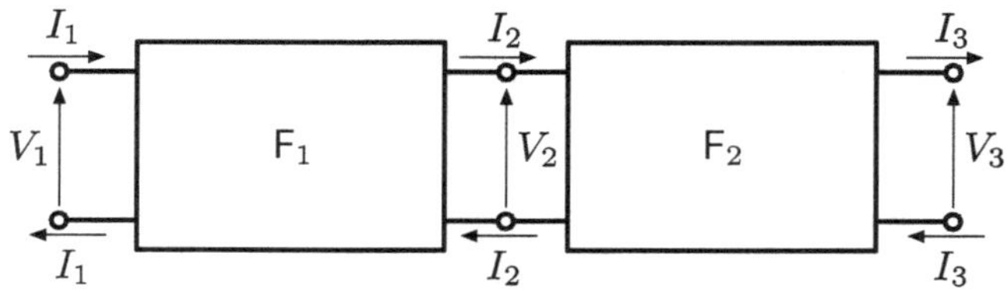


FIGURE A.2 Definitions of port voltages and port currents for cascaded two-ports.

$$C = \left.\frac{I_1}{V_2}\right|_{I_2=0} \qquad \text{(Dimensions of conductance)} \tag{A.4}$$

$$D = \left.\frac{I_1}{I_2}\right|_{V_2=0} \qquad \text{(Dimensionless)} \tag{A.5}$$

The ABCD-matrix is useful for calculating the characteristics of two-ports in a cascade connection as shown in Fig. A.2. Suppose that the ABCD-matrix of the left two-port is $\mathbf{F}_1$ and that of the right two-port is $\mathbf{F}_2$. Then the ABCD-matrix of the cascaded two-ports is given by the product of the two ABCD-matrices as follows:

$$\begin{bmatrix} V_1 \\ I_1 \end{bmatrix} = \mathbf{F}_1 \begin{bmatrix} V_2 \\ I_2 \end{bmatrix} = \mathbf{F}_1\mathbf{F}_2 \begin{bmatrix} V_3 \\ I_3 \end{bmatrix}. \tag{A.6}$$

Fig. A.3 shows some examples of ABCD-matrices.

### A.1.2 S-Matrix

Quantum mechanics, which is the basis of the band theory of solids, considers quantities such as the reflection coefficient and the transmission coefficient of electrons behaving as waves. To obtain the corresponding quantity in circuit theory, let us consider a matrix called the *scattering matrix* or *S-matrix*.

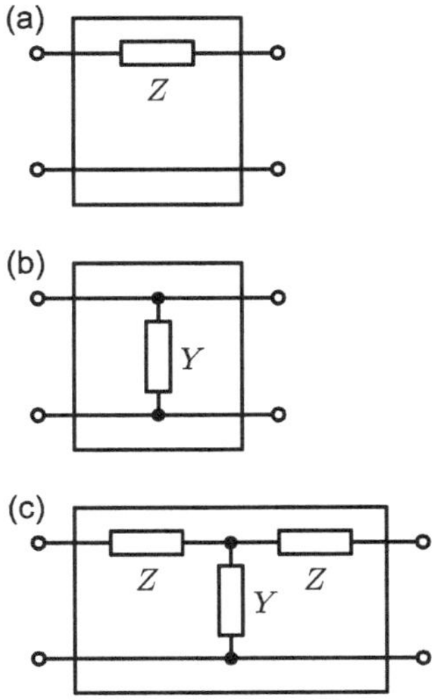


(a) $\mathbf{F} = \begin{bmatrix} 1 & Z \\ 0 & 1 \end{bmatrix}$

(b) $\mathbf{F} = \begin{bmatrix} 1 & 0 \\ Y & 1 \end{bmatrix}$

(c)
$$\mathbf{F} = \begin{bmatrix} 1 & Z \\ 0 & 1 \end{bmatrix} \begin{bmatrix} 1 & 0 \\ Y & 1 \end{bmatrix} \begin{bmatrix} 1 & Z \\ 0 & 1 \end{bmatrix} = \begin{bmatrix} 1+YZ & Z(2+YZ) \\ Y & 1+YZ \end{bmatrix}$$

FIGURE A.3 Examples of ABCD matrices. $Z$ is impedance. $Y$ is admittance. (a) Series impedance. (b) Parallel admittance. (c) T-network.

FIGURE A.4 S-matrix relates incoming traveling wave phasors, $a_i$, to outgoing traveling wave phasors, $b_i$.

Unlike most matrix representations of a two-port, including the ABCD-matrix, the input and output variables of the S-matrix are not the voltage and current at each port. Instead, the S-matrix uses as variables the *voltage traveling waves* flowing into and out of each port. The traveling wave entering port $i$ of a two-port is customarily denoted by $a_i$, and the traveling wave flowing out of port $i$ is denoted by $b_i$, as shown in Fig. A.4.

$$\begin{bmatrix} b_1 \\ b_2 \end{bmatrix} = \mathbf{S} \begin{bmatrix} a_1 \\ a_2 \end{bmatrix} = \begin{bmatrix} S_{11} & S_{12} \\ S_{21} & S_{22} \end{bmatrix} \begin{bmatrix} a_1 \\ a_2 \end{bmatrix}. \qquad \text{(A.7)}$$

The elements of an S-matrix are called *S-parameters*.

Since we are considering the frequency domain, both $a_i$ and $b_i$ are phasors. Note, however, that $a_i$ and $b_i$ are not exactly the voltage traveling wave phasors as in $V^{\pm}$ in (3.28) on p. 63, but are customarily defined as the voltage traveling wave phasors divided by $\sqrt{R_{\text{ref}}}$, a quantity that has the dimensions of the square root of resistance. $R_{\text{ref}}$ is called the *reference resistance* [1], and it is customary to set $R_{\text{ref}} = 50\,\Omega$. The value of $R_{\text{ref}}$ is related to the characteristic impedance of the transmission line connected to the device whose S-parameters are to be measured. The diagonal elements $S_{11}$ and $S_{22}$ of (A.7) are *reflection coefficients*, and the off-diagonal elements $S_{21}$ and $S_{12}$ are *transmission coefficients*. Since these are ratios between phasors, S-parameters are complex-valued. The reflection and transmission coefficients of electrons as waves, mentioned earlier, are real numbers and correspond to $|S_{11}|^2$ and $|S_{21}|^2$.

An ABCD-matrix can be converted to an S-matrix by the following formula.

$$\mathbf{S} = \frac{1}{AR_{\text{ref}} + B + CR_{\text{ref}}^2 + DR_{\text{ref}}} \times \begin{bmatrix} AR_{\text{ref}} + B - CR_{\text{ref}}^2 - DR_{\text{ref}} & 2\,(AD - BC)\,R_{\text{ref}} \\ 2R_{\text{ref}} & -AR_{\text{ref}} + B - CR_{\text{ref}}^2 + DR_{\text{ref}} \end{bmatrix}. \tag{A.8}$$

## A.2 *N*TH POWER OF UNIMODULAR MATRIX

For a $2 \times 2$ matrix

$$\mathbf{F} = \begin{bmatrix} A & B \\ C & D \end{bmatrix} \tag{A.9}$$

let $\xi$ be

$$\xi \equiv \frac{A + D}{2}. \tag{A.10}$$

By the Cayley–Hamilton theorem, the following equality holds:

$$\mathbf{F}^2 - 2\xi\mathbf{F} + \det\mathbf{F} \cdot \mathbf{1}_2 = \mathbf{F}^2 - (A + D)\,\mathbf{F} + (AD - BC)\,\mathbf{1}_2 = \mathbf{0}_2, \tag{A.11}$$

where

$$\mathbf{1}_2 \equiv \begin{bmatrix} 1 & 0 \\ 0 & 1 \end{bmatrix}, \qquad \text{(2 × 2 identity matrix)} \qquad (A.12)$$

$$\mathbf{0}_2 \equiv \begin{bmatrix} 0 & 0 \\ 0 & 0 \end{bmatrix}. \qquad \text{(2 × 2 zero matrix)} \qquad (A.13)$$

Let us assume now that the determinant of **F** equals unity as follows.

$$\det\mathbf{F} = AD - BC = 1 \qquad (A.14)$$

The F-matrix of a reciprocal two-port (see the Box on p. 48) is known to satisfy (A.14). A matrix with a unity determinant is called a *unimodular matrix*. Putting (A.14) in (A.11), we get

$$\mathbf{F}^2 = 2\xi\mathbf{F} - \mathbf{1}_2. \qquad (A.15)$$

Since the right-hand side of (A.15) contains only the first power of **F**, we can use (A.15) repeatedly to lower the order of higher powers of **F**. $\mathbf{F}^N$ can, therefore, be expressed as

$$\mathbf{F}^N = \mathbf{F}U_{N-1}(\xi) - \mathbf{1}_2 U_{N-2}(\xi), \qquad (A.16)$$

where $U_N(\xi)$ is a certain function. Equation (A.15) corresponds to the case where $N = 2$ in (A.16), with $U_1(\xi) = 2\xi$ and $U_0(\xi) = 1$. If we can find the function $U_N(\xi)$, then we have, in effect, computed $\mathbf{F}^N$. So we will look further into $U_N(\xi)$ in the following.

First, multiplying (A.16) by **F** yields

$$\mathbf{F}^{N+1} = \mathbf{F}^2 U_{N-1}(\xi) - \mathbf{F}U_{N-2}(\xi). \qquad (A.17)$$

Putting (A.15) into the right-hand side of (A.17) and lowering the order of the first term yields

$$\mathbf{F}^{N+1} = (2\xi\mathbf{F} - \mathbf{1}_2)\, U_{N-1}(\xi) - \mathbf{F}U_{N-2}(\xi). \qquad (A.18)$$

Next, with the substitution $N \rightarrow N+1$ in (A.16), we have

$$\mathbf{F}^{N+1} = \mathbf{F}U_N(\xi) - \mathbf{1}_2 U_{N-1}(\xi). \qquad (A.19)$$

Since the left-hand sides of (A.18) and (A.19) are equal, so are the right-hand sides. Thus,

$$(2\xi\mathbf{F} - \mathbf{1}_2)\, U_{N-1}(\xi) - \mathbf{F}U_{N-2}(\xi) = \mathbf{F}U_N(\xi) - \mathbf{1}_2 U_{N-1}(\xi) \qquad (A.20)$$

$$2\xi \mathbf{F} U_{N-1}(\xi) - \mathbf{F} U_{N-2}(\xi) = \mathbf{F} U_N(\xi). \qquad \text{(A.21)}$$

Focusing only on the coefficients of **F** and replacing $N$ with $N + 2$, we obtain the following recurrence formula:

$$U_{N+2}(\xi) - 2\xi U_{N+1}(\xi) + U_N(\xi) = 0.(N = 1, 2, \cdots) \qquad \text{(A.22)}$$

At this point, we need to invoke a known mathematical fact that, unfortunately, does not logically follow from the above derivation: the function $U_N(\xi)$ that satisfies (A.22) is a special function known as the *Chebyshev polynomial of the second kind*.

$$U_N(\xi) = \frac{\sin\left[(N+1)\cos^{-1}\xi\right]}{\sin\left(\cos^{-1}\xi\right)}. \qquad \text{(A.23)}$$

We can now use (A.23) in (A.16) to compute $\mathbf{F}^N$.

# Solutions to Selected Problems

## A.3.1 CHAPTER 1

1.1 (p. 26) Left to the reader.

1.2 (p. 26) Left to the reader.

1.3 (p. 26) 1 eV is the kinetic energy obtained by accelerating an elementary charge $q$ by a potential difference of 1 V and is equal to $1.6\times10^{-19}$ J. Therefore, an energy value in joules can be converted to that in electron volts by dividing the former by $q$. Thus,

$$\frac{kT}{q} \simeq 0.026\,\text{eV} = 26\,\text{meV}. \tag{A.24}$$

See also (7.44) on p. 236.

1.4 (p. 26) According to Table 1.3 (p. 5), 1 cm$^3$ of crystalline silicon contains $5 \times 10^{22}$ silicon atoms. So the boron atomic density is found by dividing this number by $10^5$ to be $5 \times 10^{17}\,\text{cm}^{-3}$. This is a typical value for the dopant density of nondegenerate silicon. Since the hole density is $p = 5 \times 10^{17}\,\text{cm}^{-3}$ and the intrinsic carrier density is $n_\text{i} = 1 \times 10^{10}\,\text{cm}^{-3}$, the hole density has become $p/n_\text{i} = 5\times10^7$ times after acceptor doping. According to (5.54) on p. 139, the conductivity due to hole conduction is proportional to the hole density $p$, so such a small amount of doping dramatically changes the conductivity. However, in reality, as doping density increases, Coulomb scattering by dopant ions increases (see Problem 5.7 on p. 164), and the mean free time (p. 135) becomes shorter. As a result, the mobility decreases. Also, as the doping density increases, the ionization rate of dopants decreases. So in practice, the conductivity does not increase proportionally to the doping density.

1.5 (p. 26) The equation for the balance between the Coulomb force and the centrifugal force acting on an electron orbiting a donor ion nucleus is given by

$$\frac{q^2}{4\pi\epsilon_{\mathrm{Si}} r^2} = m_{\mathrm{e}} r\omega^2, \tag{A.25}$$

where $r$ is the radius of the orbit, and $\omega$ is the angular frequency. From Table 1.3 (p. 5), the dielectric constant of silicon is $\epsilon_{\mathrm{Si}} = 12\epsilon_0$. Bohr's quantization condition can be written as

$$m_{\mathrm{e}} r^2 \omega = \frac{nh}{2\pi} \tag{A.26}$$

where $n$ is a positive integer. Eliminating $\omega$ from (A.25) and (A.26), we obtain

$$r = \frac{\epsilon_{\mathrm{Si}} n^2 h^2}{q^2 m_{\mathrm{e}} \pi} \tag{A.27}$$

This is $(\epsilon_{\mathrm{Si}}/\epsilon_0)(m_0/m_{\mathrm{e}})\,n^2$ times the Bohr radius, $r_{\mathrm{B}}$, of the hydrogen atom.

$$r_{\mathrm{B}} = \frac{\epsilon_0 h^2}{q^2 m_0 \pi} \simeq 0.53\text{Å}. \tag{A.28}$$

When $n = 1$, $(\epsilon_{\mathrm{Si}}/\epsilon_0)(m_0/m_{\mathrm{e}}) \simeq 12$ and $r \simeq 6.4$ Å.

The lattice constant of the silicon crystal is about 5.43 Å, and the nearest-neighbor atomic spacing is about 2.3 Å. Therefore, in Fig. 1.10 (p. 18, the conduction electron, which should be free from donor binding, is drawn too close to the donor ion.

Now, the potential energy of the electron in the ground state is given by

$$-\frac{q^2}{4\pi\epsilon_{\mathrm{Si}} r}. \tag{A.29}$$

The kinetic energy of the electron in the ground state is given by

$$\frac{m_{\mathrm{e}} r^2 \omega^2}{2} = m_{\mathrm{e}} r\omega^2 \frac{r}{2} = \frac{q^2}{4\pi\epsilon_{\mathrm{Si}} r^2}\frac{r}{2} = \frac{q^2}{8\pi\epsilon_{\mathrm{Si}} r}, \tag{A.30}$$

where we used (A.25). The sum of (A.29) and (A.30) is $-q^2/(8\pi\epsilon_{\mathrm{Si}} r)$, and therefore, the ionization energy is

$$\frac{q^2}{8\pi\epsilon_{\mathrm{Si}} r}. \tag{A.31}$$

Since the ionization energy of the ground-state hydrogen atom is

$$\frac{q^2}{8\pi\epsilon_0 r_\mathrm{B}} \simeq 13.6\,\mathrm{eV}, \tag{A.32}$$

the ionization energy of the donor is

$$\frac{q^2}{8\pi\epsilon_\mathrm{Si} r} = \frac{q^2}{8\pi\epsilon_0 r_\mathrm{B}} \frac{\epsilon_0 r_\mathrm{B}}{\epsilon_\mathrm{Si} r} \simeq 13.6 \times \frac{1}{12^2} \simeq 0.09\,\mathrm{eV}. \tag{A.33}$$

This result is roughly consistent with the statement on p. 19 that the donor level is several tens of millielectron volts (meV) below $E_\mathrm{c}$. Since $r \simeq 6.4\,\mathrm{Å}$ is not very much larger than the atomic spacing, it is questionable whether it is appropriate to use the dielectric constant of silicon in the calculation, but the result is quite reasonable. In other semiconductors (or even in silicon, depending on the crystal orientation), the effective mass of the electron is often much smaller, so the radius of the orbital is much larger.

1.6 (p. 27) The hole and electron densities of an intrinsic semiconductor, before adding any donors, are equal to the intrinsic carrier density (p. 8). When the intrinsic semiconductor is doped with donors, electrons are supplied by them, so the electron density becomes $n > n_\mathrm{i}$. However, since there is no supply of holes, one might think that $p = n_\mathrm{i}$ would persist. However, it turns out in §4.3.1 that that is not the case and $p < n_\mathrm{i}$.

1.7 (p. 27) The authors do not know the correct answer, and it should also depend on when the reader is trying to solve this problem. You should be able to find the yearly shipments of semiconductor integrated circuits and memory by searching.

As an example of a rough estimate, how about estimating how many MOSFETs there are by estimating the number of MOSFETs you have (infer from the product specifications of your electronic equipment), multiplying it by Earth's population, and then multiplying it by some "correction factor," considering penetration rates of such equipment? Has the number of MOSFETs you have had increased exponentially over the years?

This type of quick and rough estimation problem is sometimes called "Fermi estimation."

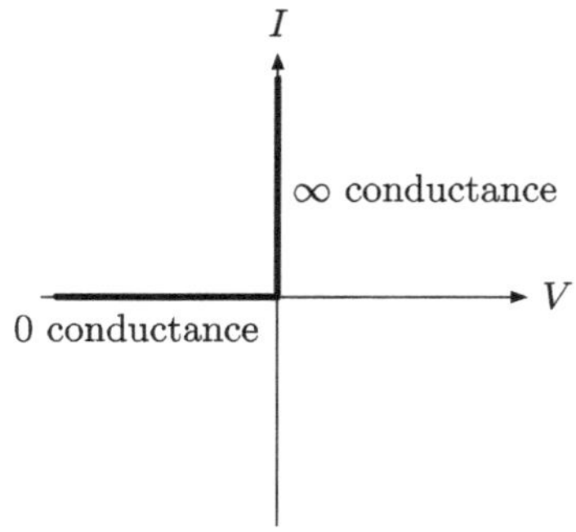


FIGURE A.5 Current-voltage characteristics of the ideal rectifier.

## A.3.2 CHAPTER 2

2.1 (p. 49) The graph is shown in Fig. A.5. The incremental conductance is given by

$$G_{\text{inc}}(V) = \begin{cases} 0 & (V < 0) \\ \infty & (V = 0) \end{cases}. \quad \text{(A.34)}$$

2.2 (p. 49)

$$V(t) = \frac{\mathrm{d}\Phi(t)}{\mathrm{d}t} = \frac{\mathrm{d}\Phi(I)}{\mathrm{d}I}\frac{\mathrm{d}I(t)}{\mathrm{d}t} = L_{\text{inc}}(I)\frac{\mathrm{d}I(t)}{\mathrm{d}t}. \quad \text{(A.35)}$$

2.3 (p. 49) Since the superposition principle does not hold for nonlinear circuits, all theorems derived from it cannot be applied to nonlinear circuits. Only Kirchhoff's voltage and current laws are applicable to nonlinear lumped circuits. Therefore, nonlinear circuit simulators rely on Kirchhoff's laws. However, it is not that theorems for linear circuits are completely useless for nonlinear circuits. In analog circuits, small-signal responses of nonlinear circuits are often considered. In such a case, a nonlinear circuit is linearized around a bias point and the response of the resulting linear circuit is used.

## A.3.3 CHAPTER 3

3.1 (p. 85) Notice that the infinite CL ladder shown in Fig. 3.25 (p. 85) obviously blocks DC voltage and current. Putting $Z = (\mathrm{j}\omega C)^{-1}$ and $Y = (\mathrm{j}\omega L)^{-1}$ in (3.7) on p. 57, we obtain

$$Z'_{\mathrm{in}} = \pm\frac{\sqrt{1-4\omega^2 LC}}{2\mathrm{j}\omega C}. \tag{A.36}$$

Since $1 - 4\omega_c^2 LC = 0$ holds at the cutoff angular frequency $\omega_c$, it is given by

$$\omega_c = \frac{1}{2\sqrt{LC}}. \text{ (Cutoff angular frequency of CL ladder)} \tag{A.37}$$

The input impedance $Z'_{\mathrm{in}}$ of an infinitely long CL ladder can be rewritten using $\omega_c$ as follows.

$$Z'_{\mathrm{in}} = \pm\frac{1}{2\mathrm{j}\omega C}\sqrt{1-\left(\frac{\omega}{\omega_c}\right)^2}. \tag{A.38}$$

The solution with $\Re\left(Z'_{\mathrm{in}}\right) \geq 0$ must be chosen. $Z'_{\mathrm{in}}$ is purely imaginary when $|\omega| < \omega_c$ and real when $|\omega| \geq \omega_c$. In other words, the CL ladder has high-pass characteristics.

3.2 (p. 85) Differentiate the telegrapher's equation (3.18) on p. 62 by $x$ and insert (3.19).

$$\begin{aligned}\frac{\partial^2 v(x,t)}{\partial x^2} &= -L\frac{\partial}{\partial x}\frac{\partial i(x,t)}{\partial t} = -L\frac{\partial}{\partial t}\frac{\partial i(x,t)}{\partial x}\\ &= LC\frac{\partial}{\partial t}\frac{\partial v(x,t)}{\partial t} = LC\frac{\partial^2 v(x,t)}{\partial t^2}.\end{aligned} \tag{A.39}$$

This is the wave equation (3.20).

Next, differentiate (3.19) by $x$ and insert (3.18).

$$\begin{aligned}\frac{\partial^2 i(x,t)}{\partial x^2} &= -C\frac{\partial}{\partial x}\frac{\partial v(x,t)}{\partial t} = -C\frac{\partial}{\partial t}\frac{\partial v(x,t)}{\partial x}\\ &= LC\frac{\partial}{\partial t}\frac{\partial i(x,t)}{\partial t} = LC\frac{\partial^2 i(x,t)}{\partial t^2}.\end{aligned} \tag{A.40}$$

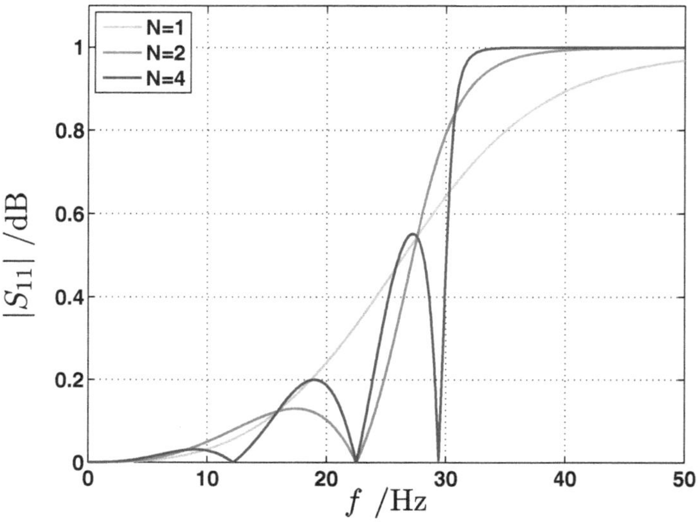


FIGURE A.6 Reflection coefficient $S_{11}$ for $N = 1, 2, 4$.

This is the wave equation (3.21).

3.3 (p. 85) From the expression (3.9) on p. 58 for the cutoff angular frequency $\omega_c$, the cutoff frequency for $N \to \infty$ is $f_c = \omega_c/(2\pi) \simeq$ 31.8 GHz. Therefore, the frequency should be calculated up to a frequency sufficiently higher than this $f_c$. In (3.36) on p. 66 for **F**, let $Z = j\omega L$ and $Y = j\omega C$, and use (3.37) on p. 66 to find $\mathbf{F}^N$ numerically using a computer. Then, put the result into (A.16) on p. 273 to obtain the reflection coefficient $S_{11}$ and the transmission coefficient $S_{21}$. Some results are shown in Figs. A.6 through A.9.

Looking at these graphs, it seems that no matter how large $N$ becomes, the "bumpy" pattern seen at frequencies below $f_c$ does not disappear. In fact, the "bumpiness" does not disappear as long as $N$ is finite. Mathematically, if $N \to \infty$, the bumpiness goes away. To understand this properly, it is necessary to consider the nature of the convergence of something called a *distribution* or *hyperfunction*, which is far beyond the scope of this book.

3.4 (p. 86) Left to the reader.

3.5 (p. 86) Investigate each of them by yourself. Photonic crystals and EBGs are often used under the condition (3.1) on p. 53. In

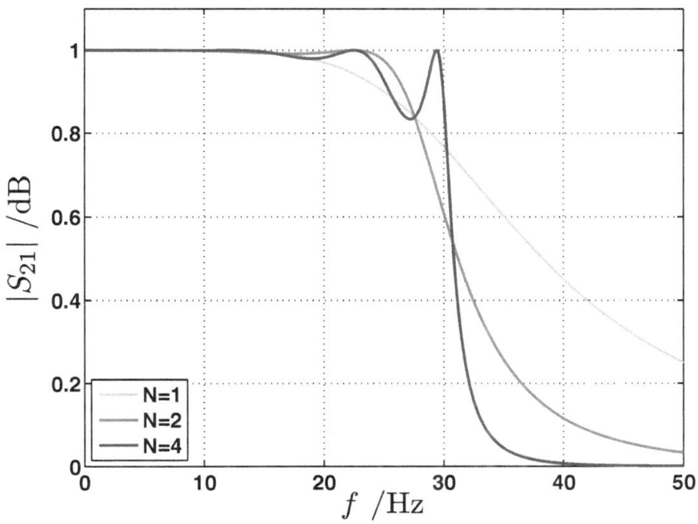


FIGURE A.7 Transmission coefficient $S_{21}$ for $N = 1, 2, 4$.

contrast, metamaterials are usually made so that (3.2) is satisfied. As a result, the "effective dielectric constant" and "effective permeability" can be manipulated.

## A.3.4 CHAPTER 4

4.1 (p. 112) At finite temperatures ($T > 0\,\mathrm{K}$), there are vacant states below the Fermi level $\zeta$ (see Fig. 4.2 on p. 90). This means that electrons can flow into those vacant states at $E < \zeta$ due to (vi) on p. 107. In contrast, at absolute zero, only the states at $E > E_F$ are vacant, so electrons can only flow into those states. From the above, $\zeta < E_F$.

4.2 (p. 113) Substituting the Maxwell–Boltzmann distribution function (4.8) into (4.7) on p. 92, which gives the electron density, we obtain

$$n = \frac{1}{\pi^2}\frac{m_c\sqrt{2m_c}}{\hbar^3}\int_{E_c}^{\infty}(E - E_c)^{1/2}\exp\left(-\frac{E-\zeta}{kT}\right)\mathrm{d}E. \tag{A.41}$$

Let $x \equiv (E - E_c)/kT$, and then $\mathrm{d}x = \mathrm{d}E/kT$. Using these, we get

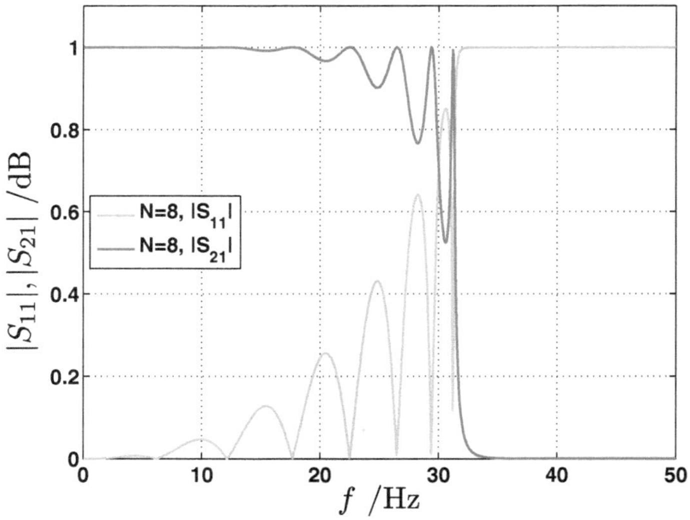


FIGURE A.8 Reflection coefficient $S_{11}$ and transmission coefficient $S_{21}$ for $N = 8$.

$$\begin{aligned} n &= \frac{1}{\pi^2}\frac{m_c\sqrt{2m_c}}{\hbar^3}(kT)^{3/2}\exp\left(\frac{\zeta - E_c}{kT}\right)\int_0^\infty x^{1/2}e^{-x}dx \\ &= \frac{1}{\pi^2}\frac{m_c\sqrt{2m_c}}{\hbar^3}(kT)^{3/2}\exp\left(\frac{\zeta - E_c}{kT}\right)\frac{\sqrt{\pi}}{2} \\ &= \frac{1}{\sqrt{2}}\left(\frac{m_c kT}{\pi\hbar^2}\right)^{3/2}\exp\left(-\frac{E_c - \zeta}{kT}\right). \end{aligned} \tag{A.42}$$

Introducing

$$N_c \equiv \frac{1}{\sqrt{2}}\left(\frac{m_c kT}{\pi\hbar^2}\right)^{3/2}, \tag{A.43}$$

we obtain (4.12) on p. 94.

4.3 (p. 113) Left to the reader.

4.4 (p. 113) From (4.21) on p. 97,

$$\frac{N_c}{N_v} = \exp\left(\frac{E_c - \zeta}{kT}\right)\exp\left(-\frac{\zeta - E_v}{kT}\right) = \exp\left(\frac{E_c + E_v - 2\zeta}{kT}\right). \tag{A.44}$$

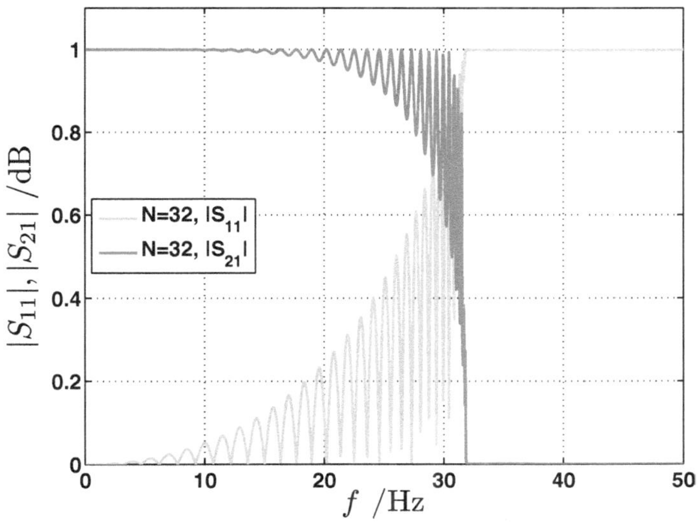


FIGURE A.9 Reflection coefficient $S_{11}$ and transmission coefficient $S_{21}$ for $N = 32$.

Taking the natural logarithm of both sides,

$$\ln\left(\frac{N_c}{N_v}\right) = \frac{E_c + E_v - 2\zeta}{kT}, \tag{A.45}$$

$$2\zeta = E_c + E_v - kT \ln\left(\frac{N_c}{N_v}\right). \tag{A.46}$$

From the above, we obtain equation (4.22) for the intrinsic Fermi level.

4.5 (p. 113) According to Table 1.3 (p. 5), the effective densities of states of gallium arsenide are $N_c \simeq 4.7 \times 10^{17}$ cm$^{-3}$ and $N_v \simeq 7.0 \times 10^{18}$ cm$^{-3}$. From these, the second term of (4.22) on p. 98 is

$$-\frac{kT}{2} \ln\left(\frac{N_c}{N_v}\right) \simeq 35\,\text{meV}. \tag{A.47}$$

Therefore, $E_i$ is about 35 meV above the midgap. This deviation from the midgap is much smaller than $E_g \simeq 1.4$ eV (Table 1.3).

4.6 (p. 113) The hole density of an intrinsic semiconductor is $p = n_\mathrm{i}$ and the Fermi level is $\zeta = E_\mathrm{i}$. Putting these in the hole density expression (4.17) on p. 95, we get

$$n_\mathrm{i} = N_\mathrm{v} \exp\left(-\frac{E_\mathrm{i} - E_\mathrm{v}}{kT}\right). \tag{A.48}$$

Solving this equation for $N_\mathrm{v}$, the effective density of states in terms of $n_\mathrm{i}$ is given by

$$N_\mathrm{v} = n_\mathrm{i} \exp\left(\frac{E_\mathrm{i} - E_\mathrm{v}}{kT}\right). \tag{A.49}$$

Substituting the right-hand side for $N_\mathrm{v}$ in (4.17) yields (4.28) on p. 99.

4.7 (p. 113) Left to the reader.

4.8 (p. 113) Fig. A.10(a) and A.10(b) on p. 285 show n-type and p-type semiconductors, respectively. The charge neutrality condition for Fig. A.10(a) is given by (p. 98) N

$$\mathrm{D}_+ + p = n. \tag{A.50}$$

Likewise, the charge neutrality condition for Fig. A.10(b) is given by

$$p = N_\mathrm{A}^- + n. \tag{A.51}$$

As a supplementary note, as shown on p. 20 and in Problem 1.5, the donor level ($E_\mathrm{c} - E_\mathrm{D}$ in Fig. A.10(a)) and the acceptor level ($E_\mathrm{A} - E_\mathrm{v}$ in Fig. A.10(b)) are several tens of millielectron volts (meV) in silicon. Given that $E_\mathrm{g} = 1.1\,\mathrm{eV}$ (see Table 1.3), $E_\mathrm{D}$ and $E_\mathrm{A}$ are drawn too far from the band edges $E_\mathrm{c}$ and $E_\mathrm{v}$, respectively.

4.9 (p. 114) First, comparing Figs. 4.10 and 4.11 on p. 110, the electron densities of the left and right metal pieces are unchanged. Therefore, what we want to do is not to move any electrons even if the left and right metals are connected via a voltage source. In other words, we want to match the left and right electrochemical

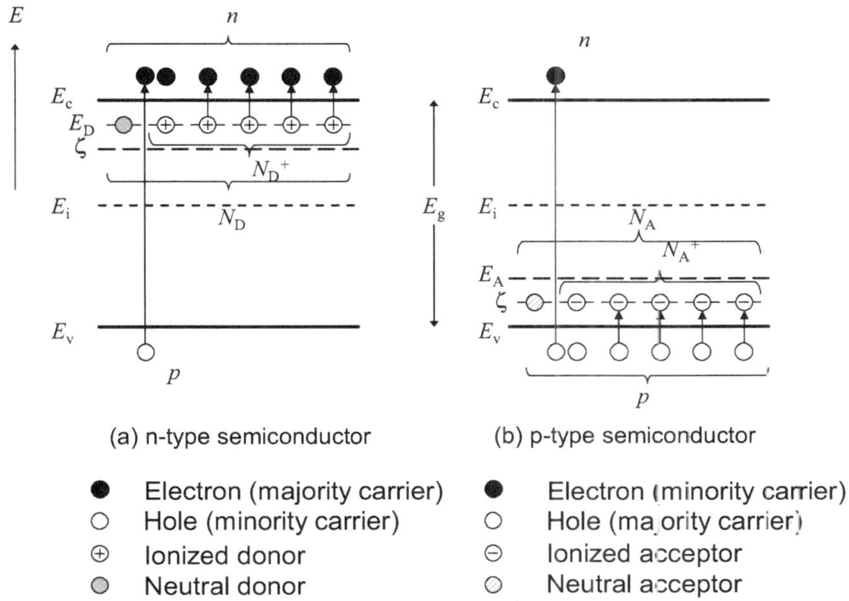


FIGURE A.10 Energy band diagrams of semiconductors. (a) n-type. (b) p-type.

potentials (i.e., Fermi levels) without changing the left and right internal chemical potentials. To do so, it is necessary to manipulate the external chemical potential (i.e., the potential energy of electrons).

As a preparation, let us consider what would happen if the left and right metals were brought into direct contact with each other. When in contact, electrons diffuse from right to left, and when equilibrium is reached, the metal on the left is negatively charged and the metal on the right is positively charged. That is, the voltage (or electrostatic potential) of the metal on the right becomes higher relative to the metal on the left. And even if more electrons try to leave the metal on the right, electrostatic force prevents them from doing so. The situation is like Fig. 6.3(b) on p. 169 flipped horizontally.

Now, in order to connect the left and right metals via a voltage source and still prevent electrons from moving to the other metal, we need a potential energy difference that prevents electrons from moving, just as we considered above. Since it can become confusing if we think about what is going on inside the voltage source, we only consider the positive and negative electrodes on its surface. The role of the voltage source is as shown in Fig. 2.8 on p. 36, where the potential difference between the positive and negative electrodes is maintained at a constant value. In the equilibrium state reached after the direct connection between the left and right metals considered above as preparation, the voltage of the right metal was higher, which prevented further electron transfer. Therefore, to prevent the diffusion of electrons out of the right metal in Fig. 4.10, the external chemical potential of the right metal must be lowered, or equivalently, the voltage of the right metal must be raised. To that end, the positive electrode of the voltage source must be connected to the right metal and the negative electrode to the left metal, as correctly shown in Fig. 4.11.

The setup for this thought experiment is not defined in precise detail. For more in-depth consideration, it is necessary to refine the setup (what the material of the electrodes of the voltage source is, how they are connected to the metals, etc.). As a result, some modifications to the discussion might become necessary.

## A.3.5 CHAPTER 5

5.1 (p. 164) According to the discussion on p. 122, we must define separate quasi-Fermi levels, $\zeta_n$ and $\zeta_p$, for electrons and holes in nonequilibrium. The reason for $\zeta_n \neq \zeta_p$ in nonequilibrium is the inequality

$$\text{(Minority carrier lifetime)} \gg \text{(Dielectric relaxation time)}. \tag{A.52}$$

Therefore, if the minority carrier lifetime becomes shorter and approaches the dielectric relaxation time, the value of $|\zeta_n - \zeta_p|$ is expected to become smaller. Numerical examples demonstrating this can be found in §6.9.2.

5.2 (p. 164) This is a difficult question. Dopants and traps can exchange carriers with both conduction and valence bands (see Fig. 5.7 on p. 145). It does not seem possible to describe the occupancy of these states by some distribution function written in terms of $\zeta_n$ or $\zeta_p$, as in (5.15) or (5.16) on p. 125, because it is not clear which of the quasi-Fermi levels to use, $\zeta_n$ or $\zeta_p$. On second thought, (125) is the "distribution function for conduction band states" and (5.16) is the "distribution function for valence band states." One possibility would be to introduce a new "quasi-Fermi level" for each type of dopant or trap and use it to introduce a corresponding distribution function [17], which may be of a different form from the Fermi-Dirac distribution function. In view of this, maybe we should call $\zeta_n$ the "quasi-Fermi level for the conduction band" and $\zeta_p$ the "quasi-Fermi level for the valence band."

5.3 (p. 164) In (4.26) on p. 98, $\zeta_n = E_c$ and the "reference energy" is $E_i$. See also the discussions in §4.3.2.

5.4 (p. 164) Comparing (5.33) on p. 131 with the first line of (5.37) on p. 132, we see that $\frac{dE_c}{dx}$ corresponds to $\frac{d\zeta_{n,ext}}{dx}$ and $\frac{kT}{n}\frac{dn}{dx}$ corresponds to $\frac{d\zeta_{n,int}}{dx}$.

5.5 (p. 164) From (5.43) on p. 134,

$$\frac{d\langle v\rangle(t)}{dt} = -\frac{\langle v\rangle(t)}{\tau_e} - \frac{q\mathcal{E}}{m_e}. \tag{A.53}$$

If $\langle v\rangle(t)$ is differentiated by $t$, $\langle v\rangle(t)$ is multiplied by a factor $-1/\tau_e$, so the solution can be written as

$$\langle v\rangle(t) = A\exp(-t/\tau_e) + B, \tag{A.54}$$

where $A$ and $B$ are constants. The given initial condition is $\langle v\rangle(0) = 0 = A + B$. From (A.53),

$$\frac{d\langle v\rangle(0)}{dt} = -\frac{q\mathcal{E}}{m_e} = -\frac{A}{\tau_e}. \tag{A.55}$$

Thus, $A = q\mathcal{E}\tau_e/m_e$ and $B = -A = -q\mathcal{E}\tau_e/m_e$. From these, we obtain (5.45).

5.6 (p. 164) When the current density is calculated based on the motion of electrons in the "punch hole model" shown in Fig. 5.5(b) (p. 137), the direction of current flow is to the right, opposite to the electric field $\mathcal{E}$, and the magnitude would be given by

$$|J_n| = qn_v\mu_n\mathcal{E}, \tag{A.56}$$

where $n_v$ is the electron density of the valence band.

The current should actually flow rightward, with only the hole density $p$ contributing to it (one hole in Fig. 5.5). Therefore, the "bubble model" seems to be better as far as current density is concerned. However, the "bubble model" also has the problem described in the Box on p. 136. In short, both of these models are just poor man's models and cannot be said to be correct.

The correct theoretical description of hole conduction is free from the above problems. The pictures offered by the "bubble model" and the "punch hole model" do not faithfully represent what the theory and associated mathematical equations say.

5.7 (p. 164) Coulomb interaction is involved in the processes (b), (c), (e), and (h). Coulomb interaction is not involved in (a), (d), (f), and (g). In (c) and (h), a carrier with the opposite charge is attracted to the charged trap. In (b) and (e), the trap gets charged as a result of a carrier leaving the trap, so again Coulomb attraction acts on the carrier. In contrast, in (a), (d), (f), and (g), there is no Coulomb interaction with the carrier because the traps are neutral in these cases. The presence of Coulomb interactions significantly affects the generation-recombination rate in the presence of an external electric field. This phenomenon is known as *field-enhanced barrier lowering*.

5.8 (p. 164) Left to the reader.

5.9 (p. 165) If only the excess hole density $\Delta p$ violates the charge neutrality condition, the situation is equivalent to a shortage of electrons by $\Delta n = -\Delta p$. The rest can be considered the same as on p. 155.

5.10 (p. 165) Substituting the numerical values into (5.107) on p. 158, the dielectric relaxation time corresponding to a resistivity of $10^0 \Omega\cdot$cm ($10^{-2} \Omega\cdot$m) is about $10^{-12}$ s, and that corresponding to a resistivity of $10^{-2} \Omega\cdot$cm ($10^{-4} \Omega\cdot$m) is about $10^{-14}$ s. Therefore, the representative value of $10^{-12}$ s given in Table 5.2 (p. 122) seems OK.

5.11 (p. 165) Left to the reader.

5.12 (p. 165) Left to the reader.

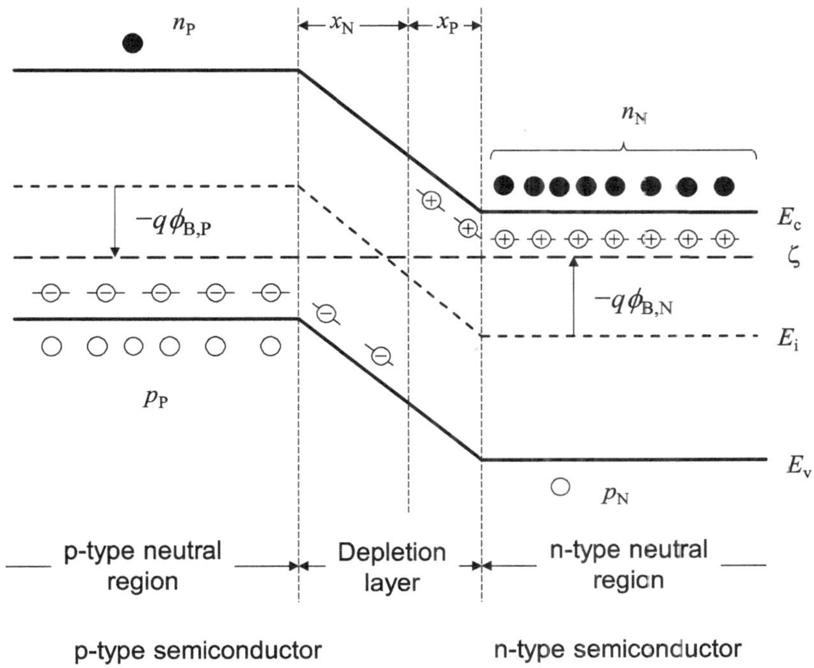


FIGURE A.11 Energy band diagram of zero-biased p-n junction.

## A.3.6 CHAPTER 6

6.1 (p. 212) In Fig. 6.4 (p. 170), since $\zeta_A > \zeta_B$, electrons flow from A to B when they make contact, and A becomes positively charged. In other words, this example is the case where $\varphi_{AB} > 0$ as in Fig. 6.3(b) (p. 169). Since $\varphi_{W,A} < \varphi_{W,B}$ in Fig. 6.4, $\varphi_{AB} = -(\varphi_{W,A} - \varphi_{W,B}) (> 0)$.

6.2 (p. 212) See Fig. A.11. For simplicity, the band bends are drawn using straight line segments in this figure, rather than as quadratic curves.

6.3 (p. 213) The built-in potential of the p-n junction is the contact potential between the p-type and n-type regions (p. 176), and is determined by the work function difference between the two regions according to (6.3) on p. 170. The values of the Fermi level $\zeta$ of nondegenerate semiconductors are in the range of (4.47) on p. 103 for n-type semiconductors and (4.52) on p. 104 for p-type semiconductors. According to Table 1.3 (p. 5), the energy gap of silicon is $E_g = 1.1\,\text{eV}$, so the built-in potential of the silicon p-n junction is somewhat below 1 V.

6.4 (p. 213) Put (6.29) and (6.30) in (6.32).

$$d_{\text{dep}} = x_{\text{P}} + x_{\text{N}} = \sqrt{\frac{2\epsilon_{\text{Si}}\varphi_{\text{bi}}}{q\left(N_{\text{A}}^{-} + N_{\text{D}}^{+}\right)}}\left(\sqrt{\frac{N_{\text{D}}^{+}}{N_{\text{A}}^{-}}} + \sqrt{\frac{N_{\text{A}}^{-}}{N_{\text{D}}^{+}}}\right)$$

$$= \sqrt{\frac{2\epsilon_{\text{Si}}\varphi_{\text{bi}}}{q\left(N_{\text{A}}^{-} + N_{\text{D}}^{+}\right)}}\frac{N_{\text{A}}^{-} + N_{\text{D}}^{+}}{\sqrt{N_{\text{A}}^{-}N_{\text{D}}^{+}}} = \sqrt{\frac{2\epsilon_{\text{Si}}\left(N_{\text{A}}^{-} + N_{\text{D}}^{+}\right)\varphi_{\text{bi}}}{qN_{\text{A}}^{-}N_{\text{D}}^{+}}}. \quad \text{(A.57)}$$

6.5 (p. 213) First, suppose that the bias voltage is $V = \varphi_{\text{bi}}$, and there is no electrostatic potential difference between the p-type and n-type regions. Next, lower $V$ so that the electrostatic potential difference between the two regions equals the thermal voltage $\varphi_{\text{th}} = kT/q \simeq 26\,\text{mV}$ ((7.44) on p. 236). Equations (6.44) on p.189 and (5.108) on p. 158 imply that the depletion layer thickness on the lowly doped (p-type) side in this condition is the Debye length.

6.6 (p. 213) First, let $n_{\text{N0}}$ be the "reference density" and express $n_{\text{P}}(-x_{\text{P}})$ in the form of (5.23) on p. 128. In Fig. 6.19 (p. 195), the "reference energy" corresponding to $n_{\text{N0}}$ is the "relative energy" of $\zeta_{\text{nN}}$ in the n-type neutral region. This "relative energy" is shifted up, parallel to $E_{\text{c}}$, by $q(\varphi_{\text{bi}} - V)$ at $x = -x_{\text{P}}$. Thus, we obtain the middle formula of (6.66) on p. 196. Next, let us express $n_{\text{P}}(-x_{\text{P}})$ in terms of $n_{\text{P0}}$. Fig. 6.19 shows that the "reference energy" corresponding to $n_{\text{P0}}$ is the "relative energy" of $\zeta_{\text{nP}}$ at the far left of the p-type neutral region. Since the value of $\zeta_{\text{nP}}$ at $x = -x_{\text{P}}$ that makes the exponent of (5.23) 0 is higher by $qV$, the right-hand side of (6.66) is obtained.

6.7 (p. 213) In a zero-biased p-n junction, diffusion and drift are balanced and no net current flows (p. 175). Under a forward bias, diffusion becomes dominant and current flows against the electric field in the depletion layer (p. 179). Then, drift should be dominant in the depletion layer under a reverse bias. Note that diffusion cannot be dominant in both cases (positive and negative biases) because it contradicts the fact that drift and diffusion are in balance under zero bias. For the drift current to be dominant, the quasi-Fermi levels, $\zeta_{\text{n}}$ and $\zeta_{\text{p}}$, and the band edge energies,

$E_c$ and $E_v$, must slope down in the same direction (see p. 119, p. 132, and Problem 5.4 on p. 164). In Fig. 6.33, however, $\zeta_n$ and $\zeta_p$ are flat where $E_c$ and $E_v$ are sloping. This produces carrier density gradients, but there is no place where the drift current is dominant.

In contrast, in Fig. 6.15 (p. 181) and Fig. 6.24 (p. 206), $\zeta_n$ and $\zeta_p$ have the same direction of slopes where $E_c$ and $E_v$ have slopes. Drift current is dominant in that region. In short, Assumption 7 on p. 192 (the assumption of quasi-equilibrium [12]) was saying that drift current cannot become dominant in the depletion layer, which is physically impossible.

Moreover, when the magnitude of the reverse bias is greater than a few hundred millivolts, the slopes of $\zeta_n$ and $\zeta_p$ are almost the same as the slopes of $E_c$ and $E_v$ (see Figs. 6.24 (p. 206), 6.25 (p. 206), 6.28 (p. 210), and 6.29 (p. 211)). To see why, note that, from (5.33) on p. 131 and (5.37) on p. 132, the electron current density, for example, is given by

$$\mu_n n \frac{d\zeta_n}{dx} = \mu_n n \frac{dE_c}{dx} + \mu_n kT \frac{dn}{dx}. \tag{A.58}$$

Under the reverse-bias condition, the absolute value of the second term on the right-hand side (the diffusion term) is much smaller than the absolute value of the first term (the drift term). Therefore, the following approximate equality holds.

$$\frac{d\zeta_n}{dx} \simeq \frac{dE_c}{dx}. \tag{A.59}$$

And likewise for holes. Energy band diagrams in which drift is dominant in the depletion layer but do not satisfy (A.59) are also found in the literature. Note that the fact that the quasi-Fermi levels penetrate into allowed bands is not in itself a problem (see Figs. 7.32–7.34).

The fact that TCAD-drawn energy band diagrams of a reverse-biased p-n junction do not look like those found in authoritative books was pointed out by Yang and Schroder [37], too, without discussion of physics.

6.8 (p. 214) The lengths of these slopes are related to diffusion lengths. The diffusion lengths are given by (6.58) and (6.59) on p. 194 and are proportional to the square root of the diffusion coefficient. The diffusion coefficient is proportional to the mobility according to Einstein's relation (5.62) and (5.63) on p. 141. Since electron mobility is greater than hole mobility (p. 136), $L_n > L_p$ if the lifetimes of electrons and holes are the same (Table 6.1 on p. 199).

6.9 (p. 214) A comparison of Fig. 6.28 ($\tau_n = \tau_p = 10^{-7}$ s) and Fig. 6.29 ($\tau_n = \tau_p = 10^{-11}$ s) shows that in the region around the depletion layer where $\zeta_p > \zeta_n$ and hence carrier generation is dominant (p. 151), Fig. 6.29 shows a smaller value of $\zeta_p - \zeta_n$. This means that the minority carrier density contributing to the current is higher in Fig. 6.29 (read the minority carrier densities from Figs. 6.28 and 6.29). This is a reason for the larger reverse current density for shorter lifetimes.

## A.3.7 CHAPTER 7

7.1 (p. 264) The "saturation region" of the $I_{CE}$-$V_{CE}$ characteristic of a bipolar transistor corresponds to the "nonsaturation region" of the $I_{DS}$-$V_{DS}$ characteristic of a MOSFET (Fig. 7.6 on p. 219). Be careful not to mix these up.

7.2 (p. 265) In Fig. 7.6, the boundary between the nonsaturation and saturation regions is the vertex of the parabola, given by (7.1), so we "complete the square" so that the coordinates of the vertex can be found.

$$\begin{aligned}\frac{I_{DS}}{\mu_n C_{ox} W/L} &= (V_{GS} - V_T)\, V_{DS} - \frac{1}{2} V_{DS}^2 \\ &= -\frac{1}{2}\left[V_{DS}^2 - 2(V_{GS} - V_T)\, V_{DS}\right] \\ &= -\frac{1}{2}\left\{[V_{DS} - (V_{GS} - V_T)]^2 - (V_{GS} - V_T)^2\right\}. \qquad \text{(A.60)}\end{aligned}$$

Thus, the abscissa of the vertex of the parabola is $V_{DS} = V_{GS} - V_T$.

7.3 (p. 265) The energy band diagram for the flat-band condition, corresponding to Fig. 7.12 (p. 226), is shown in Fig. A.12. The numbers of ionized acceptors (negatively charged) and holes (positively charged) are the same, and the system as a whole satisfies the

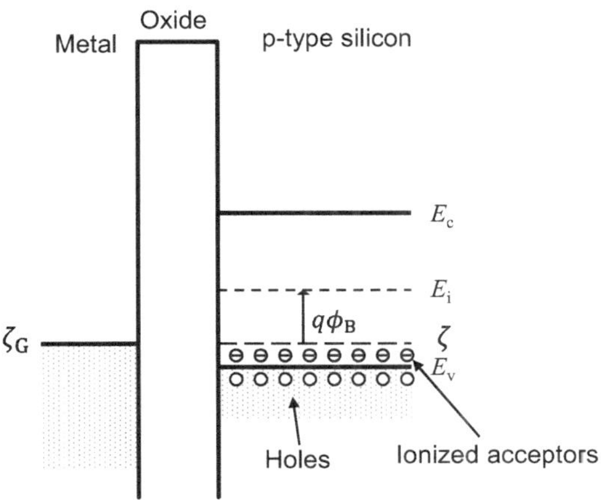


FIGURE A.12 Flat band.

charge neutrality condition. In Fig. A.12, electrons in the conduction band, which are minority carriers of the p-type substrate, are neglected because their density is very low. In the following figures, too, minority carriers (electrons) in the conduction band of the neutral region of the p-type substrate are not drawn. In the accumulation condition shown in Fig. A.13 (p. 294), holes are induced at the substrate surface, and electrons are induced on the metal surface, as was also shown on the left-hand side of Fig. 7.13 (p. 228). If the gate material is an ideal monovalent metal (a metal with one conduction electron per atom), conduction electrons of the metal are induced on the gate metal surface (more precisely, on the metal side of the metal-oxide interface).

Figs. A.14 (p. 294) through A.17 (p. 296) show the band bending $\varphi_S^{(j)}$ $(j = 1, 2, 3, 4)$ and the potential difference $\varphi_{OX}^{(j)}$ $(j = 1, 2, 3, 4)$ across the oxide. $\varphi_S^{(j)}$ in the figures has the same meaning as the surface potential $\psi_S$, but the datum point for measuring the potential is not the same as for $\psi_S$.

In the depletion condition shown in Fig. A.14 (p. 294), there are almost no electrons at the substrate surface (they are pushed to the right due to the electric field at the surface), and the negative charges of the ionized acceptor atoms appear as space charges, as shown on the left-hand side of Fig. 7.14 (p. 229). Correspondingly, the gate metal surface becomes positively charged. Incidentally, how do the positive charges appear on the metal surface?

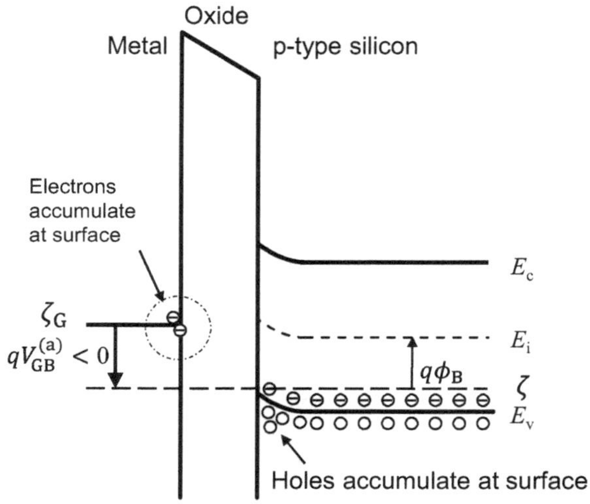


FIGURE A.13 Accumulation.

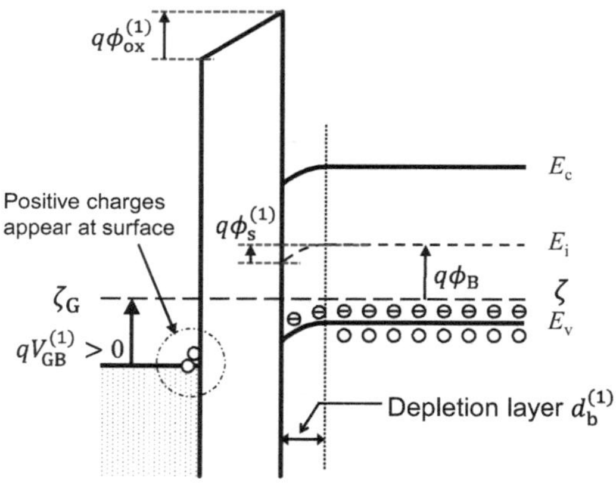


FIGURE A.14 Depletion.

These are the metal cations that appear as a result of conduction electrons being displaced. Note that $\varphi_B > \varphi_s^{(1)} > 0$.

In the weak inversion condition shown in Fig. A.15 on p. 295 (corresponding to Fig. 7.15 on p. 230), inversion electrons are induced in the conduction band at the substrate surface. Note that the amount of negative charges due to acceptor ions in the depletion layer and electrons in the inversion layer is the same as the amount of positive charges induced on the gate metal surface.

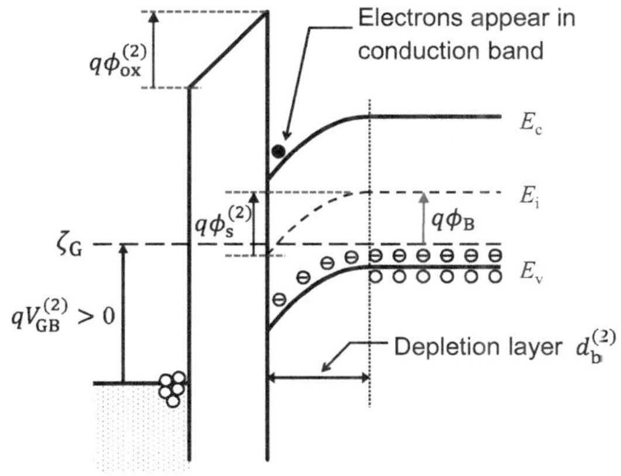


FIGURE A.15 Weak inversion.

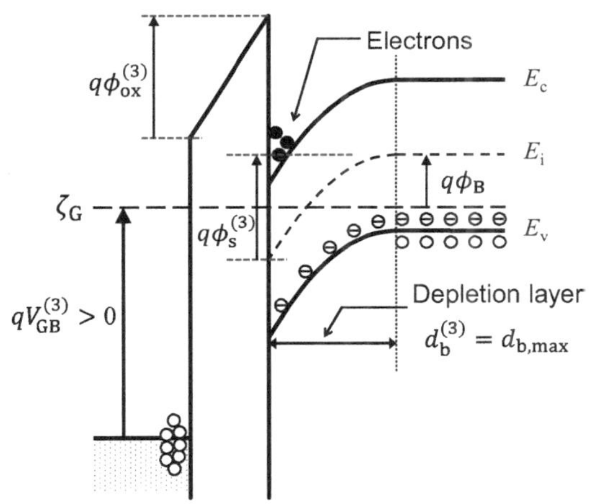


FIGURE A.16 Strong inversion.

In the strong inversion condition shown in Fig. A.16 on p. 295 (corresponding to Fig. 7.16 on p. 232), $\varphi_s^{(3)} = 2\varphi_B$ (the second term of (7.43) on p. 236 is ignored). Just as in Fig. A.15, the amount of negative charges due to acceptor ions and inversion layer electrons is the same as the amount of positive charges induced on the gate metal surface. Note that, as shown in Fig. 7.17 (p. 233), the number of electrons induced in the inversion layer in the strong inversion condition is orders of magnitude larger than in the weak inversion condition (Fig. A.15).

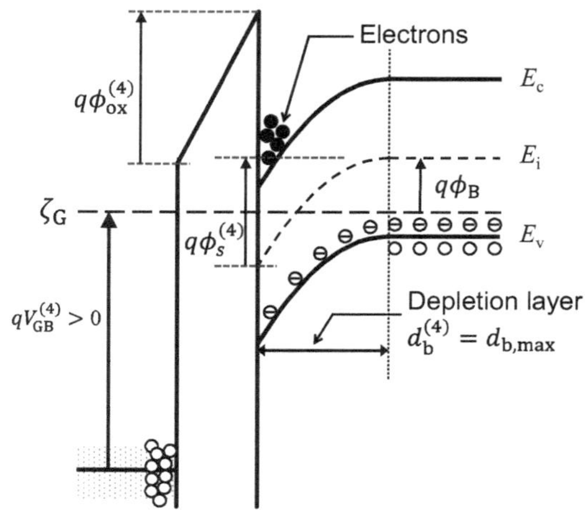


FIGURE A.17 Stronger inversion.

Fig. A.17 (p. 296) shows the energy band diagram when a larger gate voltage ($V_{GB}^{(4)} > V_{GB}^{(3)}$) is applied than in Fig. A.16. The band bending in the semiconductor is almost the same as in the previous case of Fig. A.16 ($\varphi_s^{(4)} = \varphi_s^{(3)} = 2\varphi_B$). So the extra gate voltage applied relative to the previous case is applied to the gate oxide: $V_{GB}^{(4)} - V_{GB}^{(3)} = \varphi_{ox}^{(4)} - \varphi_{ox}^{(3)}$. $V_{GB}^{(j)} > \varphi_s^{(j)} + \varphi_{ox}^{(j)}$ holds in Figs. A.14 through A.17. Note that all these figures are drawn as such. Strictly speaking, $\varphi_s^{(4)} > \varphi_s^{(3)}$, rather than $\varphi_s^{(4)} = \varphi_s^{(3)}$, but when the gate voltage changes, the amount of positive charge induced on the gate metal surface changes linearly. On the other hand, since the inversion electron density changes exponentially ($\propto \exp(\varphi_s/kT)$) as $\varphi_s$ changes, it is safe to approximate that $\varphi_s$ does not change.

7.4 (p. 265) Left to the reader.

7.5 (p. 266) Left to the reader.

7.6 (p. 266) From (7.70) on p. 247,

$$\psi_{sT} + V_P + \gamma\sqrt{\psi_{sT} + V_P} - V_{GB} + V_{fb} = 0,$$
$$\sqrt{V_P + \psi_{sT}}^{\,2} + \gamma\sqrt{V_P + \psi_{sT}} - (V_{GB} - V_{fb}) = 0. \tag{A.61}$$

Since this is a quadratic equation for $\sqrt{V_P + \psi_{sT}}$, the roots are given by

$$\sqrt{V_{\mathrm{P}} + \psi_{\mathrm{sT}}} = -\frac{\gamma}{2} \pm \sqrt{\frac{\gamma^2}{4} + V_{\mathrm{GB}} - V_{\mathrm{fb}}}. \tag{A.62}$$

Since $\gamma > 0$ from the body-effect coefficient equation (7.41) on p. 235, the plus sign should be taken. Thus (7.71) is obtained.

7.7 (p. 266) Put (7.83) on p. 251 in (7.93) and integrate.

$$\begin{aligned}
I_{\mathrm{DS}} &= -\frac{\mu_{\mathrm{n}} W}{L} \int_{V_{\mathrm{SB}}}^{V_{\mathrm{DB}}} Q_{\mathrm{inv}}(V_{\mathrm{CB}})\, \mathrm{d}V_{\mathrm{CB}} \\
&= \frac{\mu_{\mathrm{n}} W C_{\mathrm{ox}}}{L} \int_{V_{\mathrm{SB}}}^{V_{\mathrm{DB}}} \left(V_{\mathrm{GB}} - V_{\mathrm{fb}} - \psi_{\mathrm{sT}} - V_{\mathrm{CB}} - \gamma\sqrt{\psi_{\mathrm{sT}} + V_{\mathrm{CB}}}\right) \mathrm{d}V_{\mathrm{CB}} \\
&= \frac{\mu_{\mathrm{n}} W C_{\mathrm{ox}}}{L} \left[ (V_{\mathrm{GB}} - V_{\mathrm{fb}} - \psi_{\mathrm{sT}})\, V_{\mathrm{CB}} - \frac{1}{2} V_{\mathrm{CB}}^2 - \frac{2}{3}\gamma(\psi_{\mathrm{sT}} + V_{\mathrm{CB}})^{3/2} \right] \Bigg|_{V_{\mathrm{SB}}}^{V_{\mathrm{DB}}} \\
&= \frac{\mu_{\mathrm{n}} W C_{\mathrm{ox}}}{L} \Bigg\{ (V_{\mathrm{GB}} - V_{\mathrm{fb}} - \psi_{\mathrm{sT}})(V_{\mathrm{DB}} - V_{\mathrm{SB}}) - \frac{1}{2}(V_{\mathrm{DB}}^2 - V_{\mathrm{SB}}^2) \\
&\qquad - \frac{2}{3}\gamma \left[ (\psi_{\mathrm{sT}} + V_{\mathrm{DB}})^{3/2} - (\psi_{\mathrm{sT}} + V_{\mathrm{SB}})^{3/2} \right] \Bigg\}.
\end{aligned} \tag{A.63}$$

7.8 (p. 266) First, substitute (7.99) on p. 257 into (7.96) and use (7.100).

$$\begin{aligned}
I_{\mathrm{DS}} &= \frac{\mu_{\mathrm{n}} W C_{\mathrm{ox}}}{L} \left[ (V_{\mathrm{GS}} - V_{\mathrm{BS}} - V_{\mathrm{T}})\, V_{\mathrm{DS}} - \frac{1}{2} \left[ (V_{\mathrm{DS}} - V_{\mathrm{BS}})^2 - V_{\mathrm{BS}}^2 \right] \right] \\
&= \frac{\mu_{\mathrm{n}} W C_{\mathrm{ox}}}{L} \left[ (V_{\mathrm{GS}} - V_{\mathrm{BS}} - V_{\mathrm{T}})\, V_{\mathrm{DS}} - \frac{1}{2} \left( V_{\mathrm{DS}}^2 - 2 V_{\mathrm{DS}} V_{\mathrm{BS}} \right) \right] \\
&= \frac{\mu_{\mathrm{n}} W C_{\mathrm{ox}}}{L} \left[ (V_{\mathrm{GS}} - V_{\mathrm{T}})\, V_{\mathrm{DS}} - \frac{1}{2} V_{\mathrm{DS}}^2 \right].
\end{aligned} \tag{A.64}$$

The dependence on the back gate voltage $V_{\mathrm{BS}}$ has disappeared. This is the same formula as (7.1) on p. 220. If we want to know the effect of the back gate voltage, it seems inappropriate to set $\gamma = 0$.

Next, let us try to do the same using (7.94) on p. 255. Since only a term containing $\gamma$ needs to be added to (A.64),

$$\begin{aligned}
I_{\mathrm{DS}} &= \frac{\mu_{\mathrm{n}} W C_{\mathrm{ox}}}{L} \Bigg\{ (V_{\mathrm{GS}} - V_{\mathrm{T}}) V_{\mathrm{DS}} - \frac{1}{2} V_{\mathrm{DS}}^2 \\
&\qquad - \frac{2}{3}\gamma \left[ (\psi_{\mathrm{sT}} + V_{\mathrm{DS}} - V_{\mathrm{BS}})^{3/2} - (\psi_{\mathrm{sT}} - V_{\mathrm{BS}})^{3/2} \right] \Bigg\}.
\end{aligned} \tag{A.65}$$

Although it is a little difficult to figure out what happens when $V_{BS}$ becomes nonzero from the form of the equation alone, $I_{DS}$ increases if $V_{BS} > 0$, and decreases if $V_{BS} < 0$. This is known as the *body-bias effect* or *substrate-bias effect*. This is analogous to the increase in $I_{DS}$ when $V_{GS}$ is increased and the decrease in $I_{DS}$ when $V_{GS}$ is decreased.

7.9 (p. 266) Some observations on Figs. 7.32 through 7.34:

- $E_c$ is sloped downward toward the drain not only in the channel (the near side of Fig. 7.32) but also on the far side.
- $E_c$ and $\zeta_n$ very nearly overlap with each other throughout the channel, and therefore the channel is strongly inverted.
- $\zeta_n < E_i$ on the far side of Fig. 7.32, and therefore that part of the substrate is not inverted.
- $\zeta_p$ at the substrate surface (the near side of Fig. 7.32) goes into the conduction band but runs parallel to $E_c$ in a significant part of the channel.
- On the far side of Fig. 7.32, $\zeta_p > E_i$ in some parts, so that part of the substrate is strongly depleted of holes.
- $\zeta_p$ and $\zeta_n$ are sloped downward from the far side toward the near side.

Are these reasonable?

7.10 (p. 267) Left to the reader.

7.11 (p. 267) Left to the reader.

7.12 (p. 267) Left to the reader.

7.13 (p. 267) Left to the reader.

# References


1. S. Amakawa, "Scattered reflections on scattering parameters—demystifying complex-referenced S parameters—," *IEICE Trans. Electron.*, vol. E99-C, no. 10, pp. 1100–1112, Oct. 2016, doi: 10.1587/transele.E99.C.1100.
2. B.-L. Anderson and R. Anderson, *Fundamentals of Semiconductor Devices*, 2nd edition, McGraw-Hill, 2017.
3. N. W. Ashcroft and N. D. Mermin, *Solid State Physics*, Thomson Learning, 1976.
4. L. Brillouin, *Wave Propagation and Group Velocity*, Academic Press, 1960.
5. L. T. Bruton, *RC-Active Circuits: Theory and Design*, Prentice Hall, 1980.
6. L. Chua, "Resistance switching memories are memristors," *Appl. Phys. A*, vol. 102, pp. 765–783, 2011, doi: 10.1007/s00339-011-6264-9.
7. L. O. Chua, "Memristor—the missing circuit element," *IEEE Trans. Circuit Theory*, vol. 18, no. 5, pp. 507–519, Sep. 1971, doi: 10.1109/TCT.1971.1083337.
8. L. O. Chua, C. A. Desoer, and E. S. Kuh, *Linear and Nonlinear Circuits*, McGraw-Hill, 1987.
9. R. E. Collin, *Foundations for Microwave Engineering*, 2nd edition, Wiley, 2001.
10. R. H. Dennard, F. H. Gaensslen, H.-N. Yu, V. L. Rideout, E. Bassous, and A. R. LeBlanc, "Design of ion-implanted MOSFET's with very small physical dimensions," *IEEE J. Solid-State Circuits*, vol. 9, no. 5, pp. 256–268, Oct. 1974, doi: 10.1109/JSSC.1974.1050511.
11. E. Episkopou, S. Papantonis, W. J. Otter, and S. Lucyszyn, "Defining material parameters in commercial EM solvers for arbitrary metal-based THz structures," *IEEE Trans. Terahertz*

*Sci. Technol.*, vol. 2, no. 5, pp. 513–524, Sep. 2012, doi: 10.1109/TTHZ.2012.2208456.

12. A. S. Grove, *Physics and Technology of Semiconductor Devices*, Wiley, 1967.
13. M. Kawakami, *Kaiban Kiso Denkikairo I (Basic Electric Circuits I)* (in Japanese), Corona Publishing Co., 1967.
14. C. Kittel, *Introduction to Solid State Physics*, 8th edition, Wiley, 2004.
15. C. Kittel and H. Kroemer, *Thermal Physics*, 2nd edition, Freeman, 1980.
16. H. Kroemer, "Quasi-electric fields and band offsets: Teaching electrons new tricks," Nobel Lecture, Dec. 2000, https://www.nobelprize.org/prizes/physics/2000/kroemer/lecture/.
17. P. T. Landsberg, *Recombination in Semiconductors*, Cambridge University Press, 1991.
18. T. H. Lee, "The (pre-) history of the integrated circuit: a random walk," *IEEE Solid-State Circuits Soc. Newsletter*, vol. 12, no. 2, pp. 16–22, 2007, doi: 10.1109/N-SSC.2007.4785573.
19. S. A. Maas, *Microwave Mixers*, 2nd edition, Artech House, 1993.
20. K. Masu and S. Amakawa, *Denshi bussei to debaisu (Elementary Solid-State Device Physics)*, in Japanese, Corona Publishing Company, 2020.
21. H. C. Pao and C. T. Sah, "Effects of diffusion current on characteristics of metal-oxide (insulator)-semiconductor transistors," *Solid-State Electron.*, vol. 9, no. 10, pp. 927–937, Oct. 1966, doi: 10.1016/0038-1101(66)90068-2.
22. C.-T. Sah, *Fundamentals of Solid-State Electronics*, World Scientific, 1991.
23. D. K. Schroder, *Semiconductor Material and Device Characterization*, 3rd edition, Wiley, 2006.
24. E. F. Schubert, *Doping in III-V Semiconductors*, Cambridge University Press, 1993.
25. K. Seeger, *Semiconductor Physics*, 9th edition, Springer, 2010.
26. S. Seely and M. K. Goldstein, "State function approach to system analysis," *J. Franklin Institute*, vol. 283, no. 3, pp. 187–202, Mar. 1967, doi: 10.1016/0016-0032(67)90023-3.
27. W. Shockley, *Electrons and Holes in Semiconductors*, D. Van Nostrand Company, 1950.

28. W. Shockley, "The path to the conception of the junction transistor," *IEEE Trans. Electron Devices*, vol. 23, no. 7, pp. 597–620, July 1976, doi: 10.1109/T-ED.1976.18463.
29. J. A. Stratton, *Electromagnetic Theory*, McGraw-Hill, 1941.
30. S. M. Sze, Y. Li, and K. K. Ng, *Physics of Semiconductor Devices*, 4th edition, Wiley, 2021.
31. Y. Taur and T. H. Ning, *Fundamentals of Modern VLSI Devices*, 2nd edition, Cambridge University Press, 2009.
32. Y. Taur and T. H. Ning, *Fundamentals of Modern VLSI Devices*, 3rd edition, Cambridge University Press, 2022.
33. Y. Tsividis, *Operation and Modeling of the MOS Transistor*, 2nd edition, McGraw-Hill, 1999.
34. Y. Tsividis and C. McAndrew, *Operation and Modeling of the MOS Transistor*, 3rd edition, Oxford University Press, 2011.
35. F. Z. Wang, "A triangular periodic table of elementary circuit elements," *IEEE Trans. Circuits Syst. I*, vol. 60, no. 3, pp. 616–623, Mar. 2013, doi: 10.1109/TCSI.2012.2209734.
36. D. A. Wells, "Application of the Lagrangian equations to electrical circuits," *J. Appl. Phys.*, vol. 9, pp. 312–320, May 1938, doi: 10.1063/1.1710422.
37. X. Yang and D. K. Schroder, "Some semiconductor device physics considerations and clarifications," *IEEE Trans. Electron Devices*, vol. 59, no. 7, pp. 1993–1996, July 2012, doi: 10.1109/TED.2012.2195011.
38. Y.-W. Yi, "Basic research for cold electronics semiconductor integrated circuits: Measurement of Hall coefficients of Al, Mo, and W thin films," Graduation thesis, Faculty of Engineering, Tohoku University, 1987 (in Japanese).
39. Y.-W. Yi, K. Masu, K. Tsubouchi, and N. Mikoshiba, "Temperature-scaling theory for low-temperature-operated MOSFET with deep-submicron channel," *Jpn. J. Appl. Phys.*, vol. 27, no. 10, pp. L1958–L1961, Oct. 1988, doi: 10.1143/JJAP.27.L1958.

# Index

FSC
www.fsc.org
MIX
Papier | Fördert
gute Waldnutzung
FSC® C083411

Kolibri 360 GmbH
Ferdinand-Jühlke-Straße 7
99095 Erfurt, Deutschland
produktsicherheit@kolibri360.de